T0227731

NANO- AND BIOCOMPOSITES

NANO- AND BIOCOMPOSITES

EDITED BY

ALAN KIN-TAK LAU
FARZANA HUSSAIN
KHALID LAFDI

CRC Press
Taylor & Francis Group
Boca Raton London New York

CRC Press is an imprint of the
Taylor & Francis Group, an **informa** business

CRC Press
Taylor & Francis Group
6000 Broken Sound Parkway NW, Suite 300
Boca Raton, FL 33487-2742

First issued in paperback 2017

ISBN 13: 978-1-138-11212-4 (pbk)
ISBN 13: 978-1-4200-8027-8 (hbk)

Library of Congress Cataloging-in-Publication Data

Nano- and biocomposites / [edited by] Alan Kin-tak Lau, Farzana Hussain, and Khalid Lafdi.
p. cm.
Includes bibliographical references and index.
ISBN 978-1-4200-8027-8 (hard back : alk. paper)
1. Nanotechnology. 2. Biotechnology. 3. Composite materials. I. Lau, Alan K. T. II. Hussain, Farzana. III. Lafdi, Khalid. IV. Title.

TP248.25.N35N265 2010
620.1'18--dc22   2009022129

Visit the Taylor & Francis Web site at
http://www.taylorandfrancis.com

and the CRC Press Web site at
http://www.crcpress.com

Contents

Section I Nanostructured Polymer Composites

Section II Nano-Bio Composites

Preface

Advanced polymer–based nanocomposite materials have gained in popularity for a wide range of engineering applications, with improvement of virtually all types of products and commercialization of products that exploit their unique mechanical, thermal, and electrical properties. However, these properties present new challenges in understanding, predicting, and managing potential adverse effects, such as toxicity and the impact of exposure on human lives and the environment. Thus, widespread applications of nanomaterials have enormous potential to both positively and negatively affect humans and the environment. The federal budget emphasizes these implications, and it is expected that the total annual budget for various sectors from the National Nanotechnology Initiatives will increase substantially in coming years.

In the past few years of research, biological applications of nanostructural resins have been conducted in *in vitro* and *in vivo* environments. The evaluation involved how the resins can bond for biocompatibility to bone for repair after breaking, to teeth for filling, to various types of tissues for wound healing, and so on. Natural and synthetic polymeric materials have been found to be suitable for tissue engineering applications. For an example, silk (like cocoon or spider) fiber, and biodegradable polymer biocomposites have been used for tissue engineering (scaffolding) for bone repair. Many researchers have also demonstrated the use of nanostructural materials as reinforcements, such as nano-apatite, nanoclay, and nanofibers (polymer-based or carbon nanotubes) to enhance the mechanical properties and thermal stability of biocompatible polymers for artificial joints and scaffolding. Tissue engineering is one such aspect that utilizes both engineering and life science disciplines to either maintain existing tissue structures or enable tissue growth. Furthermore, tissue-engineered organs can be used in testing procedures, reducing or eliminating the need for animal subjects. Nano-biotechnology is an interdisciplinary field resulting from the interfaces between biotechnology, materials science, and nanotechnology.

This book includes works from different aspects of nanomaterial and biomaterial technologies to contribute to the advanced materials and biomedical industries. In fact, nano- and biotechnology are the two foremost research areas that govern the majority of research in the science and engineering fields.

Included in this book are 12 chapters organized into two main sections: "Nanostructured Polymer Composites" and "Nano-Bio Composites." All contributing authors have been working in these fields for many years. The works addressed in this book will give important guidelines and new insights for readers and will stimulate investigation of anticipated research.

In the first section, a basic understanding of nanomaterial and nanocomposite research will be provided, to give fundamental knowledge on how these nanostructured fillers strengthen polymer-based materials. The second section will emphasize the use of nanostructured fillers and natural fiber to reinforce biodegradable and biocompatible polymers to form new types of biomedical and bioengineered composites for biomedical applications. The last chapter will focus on the toxicity impact of using nanostructured materials, which is an important topic that most researchers have ignored in their research in the past years.

Here I would like to give my sincere thanks to all contributors to this book, as the time and effort involved to make such a comprehensive work as this is enormous. The great help given by the publisher, CRC Press/Taylor & Francis Group, is also important to bring the book finally to the market. We believe that this book will give many researchers, scientists, and academics important information in the fields of nanomaterials, biomaterials, and the up-and-coming topic—nano-biomaterials research.

Alan K.T. Lau
The Hong Kong Polytechnic University
Hong Kong, SAR
Chonbuk National University
Korea

Farzana Hussain
Oregon State University
Corvallis, Oregon

Khalid Lafdi
University of Dayton Research Institute
Dayton, Ohio

Editors

Alan K.T. Lau, Ph.D., received his bachelor and master degrees of Engineering in Aerospace Engineering from the Royal Melbourne Institute of Technology (RMIT) University, Australia, in 1996 and 1997, respectively. Within this period, he also worked at General Aviation Maintenance Pty. Ltd. Australia and the Corporative Research Centre for Advanced Composite Structure (CRC-ACS) in Australia, designing a repair scheme and an advanced manufacturing process for composite structures. He then received his Ph.D. at The Hong Kong Polytechnic University in 2001. Thereafter, he was appointed Assistant Professor and then promoted to Associate Professor in 2002 and 2005, respectively. Currently, he is also an Adjunct Professor at the University of New Orleans (Louisiana), Lanzhou University (China), Ocean University of China (China), and the University of Southern Queensland (Centre of Excellence in Engineered Fiber Composites, Australia).

Based on his outstanding research performance in the fields of advanced composites, FRP for infrastructure applications, smart materials and structures, and nanomaterials, he has received numerous awards, including: Best Paper Awards on Materials (1998); Sir Edward Youde Memorial Fellowship Award (2000); Young Scientist Award (2002); Young Engineer of the Year Award (2004); Faculty Outstanding Award for Research and Scholarly Activities (2005); Award for Outstanding Research in Nanocomposites for Space Applications (2006); Chemical Physics Letters, Most Cited Paper 2003–2007 Award (2007), and the President Award in Teaching (2008). In 2007, due to his significant contribution to the field of science and engineering, he was elected as a member of the European Academy of Sciences, with the citation "for profound contributions to materials science and fundamental developments in the field of composite materials." He is also the winner of the Ernest L. Boyer International Award for Excellence in Teaching, Learning, and Technology in the 20th International Conference on College Teaching and Learning, United States (2009).

Dr. Lau has published more than 190 scientific and engineering articles, and his publications have been cited in over 1000 instances since 2002. Three of his articles have placed in the Thomson Reuters Top 1% Most Cited Paper in Its Field in 2007 and 2008 according to *Essential Science Indicators*. Dr. Lau has successfully converted his research findings into real-life practical tools, and as a result, has been granted eight patents. Currently, he is serving on more than 40 local and international professional bodies as chairman, committee member, editor, and key officer to promote the engineering profession to the public. In 2007, Dr. Lau was elected as Fellow of Engineers Australia (FIEAust) and Institution of Mechanical Engineers (FIMechE); Chair of the Institution of Engineering Designers (IED) (United Kingdom), Hong

Kong Chapter; President of Engineers Australia, Hong Kong Chapter; and Vice Chair of the American Society of Mechanical Engineers, Hong Kong Section. He is also the Chairman of the 1st International Conference on Multifunctional Materials and Structures.

Farzana Hussain has been involved with scientific endeavors of composite materials, focusing on man-made fibrous polymeric composites, nanocomposite and biocomposite materials since 2000. Her current focus is to process, model, and characterize biocomposite materials at Oregon State University. In 2006, Hussain performed biopolymer research designing microfabrication for biomedical application at ONAMI (Oregon Nanoscience and Micro-Nanotechnology Institute). Prior to her work at ONAMI, she was a technical project lead for aerospace polymer nanocomposite materials research at the Aerospace Manufacturing Technology Centre, Institute for Aerospace Research, National Research Council (NRC), Canada. At NRC, Hussain was also involved in product designing, evaluation of different processing and manufacturing techniques, and mechanical behaviors of advanced composite aircraft structures using liquid composite molding techniques.

Hussain has authored numerous journal articles, technical and conference papers, book chapters, and a comprehensive review on polymer composite and nanocomposite materials. She is the invited reviewer of the *Journal of Composite Materials, Composite Science and Technology* and the *Journal of Advanced Materials*. In addition, her comprehensive review based on polymer nanocomposites placed the most-read rankings article in 2006, 2007, 2008, and 2009 in the *Journal of Composite Materials*. She is an active member of the Society of Women Engineers (SWE) and American Institute for Aeronautics and Astronautics (AIAA). Her contribution to the book is dedicated to her parents and her husband.

Khalid Lafdi, Ph.D., is a professor at the University of Dayton (UD) and carbon group leader at University of Dayton Research Institute (UDRI), Ohio. From 1994 to 2000, Dr. Lafdi worked at the Center for Advanced Friction Studies (CAFS), Southern Illinois University at Carbondale (SIUC). He is a carbon specialist with expertise in the fields of carbon processing, physical properties, and structural characterizations at all scale levels. From 1987 to 1991, he completed his habilitation (physics) and Ph.D. (Chemical Engineering) under the supervision of Dr. Agnes Oberlin with a focus on carbon science and technology of carbons from macro- to nanometric nanoscales.

In January 1994, Dr. Lafdi was invited to stimulate carbon materials research at Southern Illinois University and to develop various applications including the development of new electrode supercapacitors, nanostructural materials, nanocomposites, carbon–carbon composites, and aircraft friction materials. He joined the University of Dayton as a UDRI scientist and professor (Mechanical Engineering and Aerospace Department) in July 2001, to help with the nano research activity. He was involved in building the nano research vision at the University of Dayton. In 2004, he was responsible for establishing the Nanoscale Engineering, Science, and Technology (NEST) facility at UD. Also in 2007, Dr. Lafdi established a new Carbon Research Laboratory (CRL) at UDRI. CRL is directly involved in the processing, characterization, and modeling of various aspects of carbon hybrid research. In the last four years, he had a state-of-the-art thermal and energy management laboratory built. Dr. Lafdi has established a manufacturing transition facility located at the National Composite Center (Kettering, Ohio) to facilitate the nano-artifacts scale-up processes and technology transfer.

At this time, Dr. Lafdi has more than 140 articles and chapters published in refereed journals and four patents. He has gained valuable experience in developing new ideas and collaborations to develop a world-class carbon research program on material hybrids from fundamental understanding to manufacturing.

Contributors

Luc Avérous Laboratoire d'Ingénierie des Polymères pour les Hautes Technologies (LIPHT), Ecole Européenne de Chimie, Polymères et Matériaux (ECPM), Université de Strasbourg (UdS), Strasbourg, France

Brigida Bochicchio Department of Chemistry, Universitá della Basilicata, Potenza, Italy

Perrine Bordes Laboratoire d'Ingénierie des Polymères pour les Hautes Technologies (LIPHT), Ecole Européenne de Chimie, Polymères et Matériaux (ECPM), Université de Strasbourg (UdS), Strasbourg, France

Guohua Chen Institute of Polymer and Nanomaterials, Huaqiao University, Fujian, China

Hoi-Yan Cheung The Hong Kong Polytechnic University, Kowloon, Hong Kong

Hans-Peter Fink Fraunhofer Institute for Applied Polymer Research, Potsdam, Germany

Johannes Ganster Fraunhofer Institute for Applied Polymer Research, Potsdam, Germany

Christin Grabinski University of Dayton Research Institute, Dayton, Ohio

Deanna Guerra University of Modena and Reggio Emilia, Modena, Italy

Yogita Krishnamachari University of Iowa, Iowa City, Iowa

Alexander B. Morgan University of Dayton Research Institute, Dayton, Ohio

Masami Okamoto Advanced Polymeric Nanostructured Materials Engineering, Toyota Technological Institute, Nagoya, Japan

Antonietta Pepe   Department of Chemistry, Universitá della Basilicata, Potenza, Italy

Eric Pollet   Laboratoire d'Ingénierie des Polymères pour les Hautes Technologies (LIPHT), Ecole Européenne de Chimie, Polymères et Matériaux (ECPM), Université de Strasbourg (UdS), Strasbourg, France

Daniela Quaglino   University of Modena and Reggio Emilia, Modena, Italy

Ivonne Pasquali Ronchetti   University of Modena and Reggio Emilia, Modena, Italy

Aliasger K. Salem   Department of Chemistry, University of Iowa, Iowa City, Iowa

Antonio Mario Tamburro   Department of Chemistry, Universitá della Basilicata, Potenza, Italy

Weifeng Zhao   College of Materials Science and Engineering, Huaqiao University, Fujian, China

Section I

Nanostructured Polymer Composites

1

Carbon Nanotube Polymer Composites

Alan Kin-Tak Lau

The Hong Kong Polytechnic University, Hong Kong

CONTENTS

1.1 Introduction

Carbon nanostructured materials like carbon blacks and nanotubes have been well accepted as the strongest nanoreinforcement for polymer-based composite materials. Since the discovery of nanotubes (almost 15 years ago) by a Japanese experimentalist Sumio Ijima, extensive research focusing on different scales and methodologies has emerged to study the feasibility, in terms of the enhancements of mechanical and thermal properties, and economic viability of using the nanotubes for polymer and polymer-based composite structures. Experimental studies at both nanoscopic and microscopic levels, computational analyses through molecular dynamics (MD) simulations and then simplified finite element analysis (FEA), and, last, theoretical analysis have been conducted to discover different extraordinary properties of the nanotubes. Through these approaches, it was found that their mechanical, electrical, and thermal properties are basically governed by their carbon–carbon structural arrangement. Such an arrangement is called a "chiral arrangement." Ideally, all carbon atoms are covalently bonded in the nanotubes and form repeated, close-packed hexagonal structures in each layer. Due to these naturally chemically formed atomic arrangements, the nanotubes possess superior mechanical properties that make them one of the strongest materials in the world. Many critical results have been reported recently on the use of nanotubes as an atomic force microscope (AFM) probe, conductive devices in artificial muscles, nanothermometers, and a storage compartment for hydrogen atoms to form fuel cells [1–4]. In the United States, investment in the development of fuel cells by storing hydrogen atoms inside the cavity of nanotubes to supply electricity to microelectromechanical (MEM) or even nanoelectromechanical (NEM) devices has been increasing. In the near future, the use of hydrogen fuel cells may be one of the alternatives to the use of existing diesel fuel because of the emission of carbon dioxide that worsens the global climate.

Since 10 years ago, the use of nanotubes as nanoreinforcements for polymer and polymer-based fiber reinforced composite materials has emerged. The ultimate goal is to alter or enhance their mechanical, thermal, electrical, and abrasion properties and electrostatic behavior for space and infrastructure applications. More recently, some works have also been focused on the use of these nanostructural materials to enhance the biodegradability and thermal stability of biodegradable polymers. In order to achieve these goals, certain aspects have to be studied in detail, such as (1) understanding the mechanical properties of both single-walled nanotubes (SWNTs) and multiwalled nanotubes (MWNTs); (2) investigating appropriate fabricating processes of nanotube/polymer composites; (3) clarifying the interfacial bonding properties between the nanotubes and surrounding matrix; and (4) justifying the benefit based on the strength improvement of composites after mixing with the nanotubes. Research related to the aforementioned issues still has a long way to go due to many uncertainties about the properties of nanotubes and their structural integrity in nanotube/polymer composites which are not yet well understood.

In this chapter, a critical review on the above aspects is given based on recent research findings obtained by the authors and other researchers. All these aspects cannot be considered individually in the development of nanotube/polymer composites; therefore, a detailed discussion of each of these aspects is given, and how they co-link is also addressed. As this chapter is also focused on the mechanical properties of nanotube/polymer composites, the fundamental physics that governs the bonding behavior between carbon–carbon atoms of nanotubes will also be briefly explained.

1.2 Properties of Carbon Nanotubes

1.2.1 Experimental Measurements

Ideally, a SWNT is similar to a flat graphene sheet rolled up to form a tube sealed by two semi-hemisphere caps. All carbon atoms are chemically

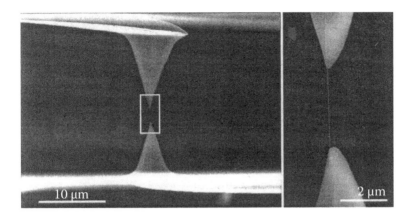

FIGURE 1.1
Scanning electron microscope (SEM) images of a nanotube linked between two opposing atomic force microscope (AFM) tips before tensile loading [5].

bonded to each other to form a close-packed hexagonal structure in the form of a circular tube. If more than one layer of graphene layers is rolled together, a coaxial tube is formed, and this type of nanotube is called a MWNT. The space between each layer of nanotube is dependent on the size and potential energies between carbon atoms. This varies slightly with different MWNTs. The mechanical and electrical properties of nanotubes are governed by their atomic structure. At the beginning of the surge in interest in nanotube research, many works attempted to differently measure the mechanical properties of nanotubes. Unfortunately, due to their size at the nanoscale level, it is impossible to conduct any direct mechanical property test to measure the strength of nanotubes by using traditional testing methods. The production of identical nanotubes with the same dimension and atomic arrangements for the tests is an issue. Yu et al. [5] attempted to grow MWNTs on an AFM tip and conducted a tensile test between two tips by moving one end upward. The extension of the nanotubes and the load applied were measured simultaneously. The whole stretching process was captured inside a scanning electron microscope (SEM) as shown in Figure 1.1. As mentioned previously, obtaining identical nanotubes was almost impossible for the tests. The result obtained from the tests shown in Figure 1.1 indicates that the Young's moduli measured have a great discrepancy, ranging from 0.32 TPa to 1.47 Pa. It is noteworthy that almost an entire outer layer of the nanotubes was first broken during the test followed by the pullout of the layer. This means that during the loading condition, the outer layer of the nanotubes took on a large portion of the load. Therefore, a MWNT with large diameter can theoretically bear more load than that of others with a smaller diameter under a tensile force. This conclusion may not be satisfied based on some estimates of strength from continuum models and MD simulations.

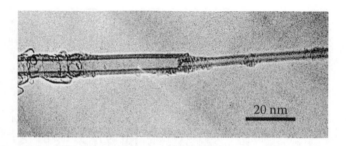

FIGURE 1.2
A telescoping action occurring in a multiwalled nanotube [7].

An indirect tensile test on MWNTs conducted by Demczyk et al. [6] demonstrated that the tensile strength of MWNTs was about 0.15 TPa, which is far below the theoretical estimation of the known properties of a graphene layer (~1 TPa). A telescoping action was seen in some fractured nanotubes. This finding agreed with the results addressed in Reference [7], wherein a weak van der Waals attractive force between individual shells of the MWNTs resulted in generating an extremely low frictional force among the shells, and all the inner shells slid freely in their longitudinal direction (Figure 1.2). The principle is that all inner shells cannot contribute any strength to the nanotubes while they are subject to tension. Qi et al. [8] also measured the Young's modulus of MWNTs using a nano-indentation method. An indentor was pressed onto vertically aligned nanotubes, and the bending stiffness of the nanotubes was measured and also calculated using classical bending theory. The effective bending and axial elastic moduli ranged from 0.91 ~ 1.24 TPa and 0.90 ~ 1.23 TPa, respectively. Because the ineffectiveness of the stress transfer among different shells of the nanotubes was not considered in the calculation, particularly in the bending case, the moduli estimated by this method do not represent the true tensile modulus of the nanotubes.

1.2.2 Theoretical Study and Molecular Dynamics Simulation

As previously mentioned, nanotubes are extremely small, and the measurement of their mechanical properties through traditional testing is almost impossible. Due to the substantial increase of storage and memory capacities, as well as the running speed of computers, the use of computer modeling to solve problems for different scale projects have become very popular. In the past few years, much effort has been spent on using MD simulations to predict the properties of nanotubes. Because the accuracy of simulated results is highly dependent on the size of the models, the capacity (memory size), and the running speed of computers, there are great discrepancies in the basic assumptions used for different scenarios, the results addressed in different papers, and the approaches. Tu and Ou-Yang [9] have used the local density approximation method associated with elastic shell theory to estimate the mechanical properties of SWNTs and MWNTs. They found that the Young's modulus of the MWNTs decreases with an increasing number of layers. The Young's modulus of the nanotubes can be determined by the following equation:

$$E_m = \frac{n}{n-1+{t}/{d}} \cdot \frac{t}{d} \cdot E_s \tag{1.1}$$

where E_m and E_s represent the Young's moduli of MWNTs and SWNTs, respectively, and n, t, and d denote the number of layers, average thickness of the layer (≈ 0.75 Å), and spacing between the layers (≈ 0.34 Å), respectively. They also reduced Equation 1.1 to the continuum limit form, and the classic shell theory can be used to describe the deformation of the nanotubes. Comparing their result with the previously mentioned experimental finding, we find that the increase in the number of layers does not represent that the effective cross-sectional area increase of the nanotubes is in proportion to their diameter. As in a macro-level, the spaces between the layers of the nanotubes do not contribute to any of the load taking substance. So, the overall effective cross-sectional area is less than that expected. Lau et al. [10] used the Tersoff–Brenner bond order potential to represent the interaction between carbon atoms to study the stretching motion of MWNTs. In their study, it was found that only an outer layer of the nanotubes takes all applied loads when nanotubes were subject to tensile and torsion motions. Because only a weak van der Waals interaction exists between the layers, external motions of the outer layer cannot be easily transferred to inner layers. Eventually, fracture in the outer layer is initiated, and this phenomenon is similar to that mentioned in Reference [6] where a pullout of nanotubes occurs. Lau [12] also pointed out that the size and Young's modulus of nanotubes are highly dependent on their chiral arrangement. The diameter of a zigzag nanotube is generally smaller than that of an armchair type. The radius of the first layer (an inner layer) of nanotubes can be determined by using the rolled graphene sheet model:

$$\rho_o = \frac{\sqrt{m^2+n^2+mn}}{\pi} \cdot \sqrt{3a_o} \tag{1.2}$$

where ρ_o, m, n, and a_o are the nonrelaxed radius and indices of the SWNTs and C–C bond distance (1.42 Å), respectively.

The MD simulation has been popularly adopted in recent years to predict the properties of nanostructural materials at the atomic scale with high accuracy. At the early stage, several studies used the empirical-force-potential molecular dynamic simulations to estimate the Young's modulus of nanotubes and found that its value was four times that of the diamond. However, those calculations were based on the SWNTs of several Å in radius. To determine the mechanical properties of nanotubes, the details of their atomic arrangements have to be clearly understood. In general, there are three types of nanotube structures: zigzag (n, 0), armchair (n, n), and chiral (n, m where $n \neq m$). Lau and Hui [12] provided a comprehensive review on the structures of nanotubes. To investigate the mechanical properties of materials at the atomic scale by using MD simulations, the interactions between neighboring atoms have to be accurately calculated.

Two common approaches based on quantum mechanics and molecular mechanics are used to simulate these interactions. Both approaches attempt to capture the variation of system energy associated with the change in atomic positions by following Newton's second law, $F = ma$. In carbon nanotubes, the mutual interactions are described by force potentials from both bonding and nonbonding interactions. The nonbonding interactions are due either to the van der Waals force (that can be attractive or repulsive depending on the distance between atoms) or to electrostatic interactions. The van der Waals force F_{VDW} is most often modeled using the Lennard-Jones potential function [11], originally derived for inert gases. The general form of this potential is

$$\Phi(r) = \frac{\lambda_n}{r^n} - \frac{\lambda_m}{r^m} \tag{1.3}$$

For van der Waals forces arising from dipole–dipole interactions, the attractive part corresponds to $m = 6$. The most common form of this potential is the (6-12) form:

$$\Phi(r) = 4\varepsilon \left[\left(\frac{\sigma}{r} \right)^{12} - \left(\frac{\sigma}{r} \right)^{6} \right] \tag{1.4}$$

The minimum of $\Phi(r)$ is determined by equating to zero the first-order derivative of $\Phi(r)$ versus r. The van der Waals force between two carbon atoms can be estimated from

$$F_{VDW} = -\frac{d\phi}{dr} = \frac{24\varepsilon}{r} \left[2\left(\frac{\sigma}{r} \right)^{12} - \left(\frac{\sigma}{r} \right)^{6} \right] \tag{1.5}$$

The two parameters σ and ε can be estimated from experimental data such as the equilibrium bond length (lattice parameters at equilibrium), equilibrium bond energy (cohesive energy), and bulk modulus at equilibrium. The bonding energy (E_{bond}) is the sum of four different interactions among

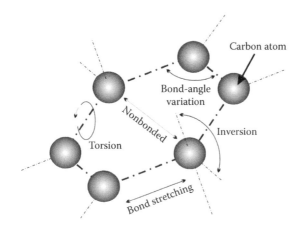

FIGURE 1.3
Bond structures and corresponding energy terms of a graphene cell.

atoms—namely, bond stretching (U_ρ), angle variation (U_θ), inversion (U_ω), and torsion (U_τ) [12], written as

$$E_{Bond} = U_\rho + U_\theta + U_\omega + U_\tau \qquad (1.6)$$

A schematic illustration of each energy term and corresponding bond structure for a graphene cell is shown in Figure 1.3. The most commonly used functional forms are

$$U_\rho = \frac{1}{2}\sum_i K_i (dR_i)^2$$

$$U_\theta = \frac{1}{2}\sum_i C_j (d\theta_j)^2$$

$$\qquad (1.7a\text{–}d)$$

$$U_\omega = \frac{1}{2}\sum_k B_k (d\omega_k)^2$$

$$U_\tau = \frac{1}{2}\sum_i A_i [1 + \cos(n_i \tau_i - \phi_i)]$$

where dR_i is the elongation of the bond identified by the label i; K_i is the force constant associated with the stretching of the "i" bond; and $d\theta_j$ and $d\omega_k$ are the variance of bond angle j and inversion angle k, respectively. C_j and B_k are force constants associated with angle variance and inversion, respectively. A_i is the "barrier" height to rotation of the bond i; n_i is the multiplicity that gives the number of minimums as the bond is rotated through 2π [13].

To determine the tensile modulus of a SWNT subject to a uniaxial loading, it is useful to observe that at small strains the torsion, the inversion, the van der Waals, and the electrostatic interaction energy terms are relatively small compared with the bond stretching and the angle variation terms. Thus, the total energy of the SWNT can be reduced to

$$E_{Total} = \frac{1}{2}\sum_i K_i(dR_i)^2 + \frac{1}{2}\sum_j C_j(d\theta_j)^2 \tag{1.8}$$

The force constants K_i and C_i can be obtained from quantum mechanics (*ab initio*). The average macroscopic elastic modulus and Poisson's ratio were estimated to be about 1.347 TPa and 0.261, respectively [14]. Such calculations can be performed by using either the force or the energy approach to measure the mechanical forces developed between carbon atoms in nanotubes with different chiral arrangements. Lu [15] has used the empirical-force potential molecular dynamic simulation to investigate the properties of nanotubes. The structure of the nanotubes was obtained by the conformational mapping of a graphene sheet onto a cylindrical surface. The nanotube radius was estimated by Equation 1.2. The average estimated tensile modulus of SWNTs and MWNTs is about 1 TPa. The elastic properties are the same for all nanotubes with a radius larger than 1 nm. Zhou et al. [16] used the first principles cluster model within the local density approximation to evaluate the mechanical properties of a SWNT. The estimated values for tensile modulus, tensile strength, and Poisson's ratio were 0.764 TPa, 6.248 GPa, and 0.32 [17], respectively. The binding energy of the nanotube is less than that of graphite due to the curvature effect. Lier et al. [17] calculated the tensile modulus of zigzag and armchair SWNTs using the *ab initio* multiplicative integral approach (MIA), which was based on the energy of elongation of nanotubes in a simple tension (that is not constrained laterally or in any other way). They found that the modulus of SWNTs or MWNTs was larger than that of a graphene sheet. The MD simulations showed that the fracture behavior of zigzag nanotubes is more brittle than the fracture behavior of armchair nanotubes [18]. The formation of a local Stone–Wales defect (5-7-7-5) in the deformed armchair nanotube induced ductile deformation.

1.2.3 Finite Element Modeling

In the past few years, the demand for the development of faster methods to compute the mechanical properties of nanostructures has been increasing. The classical shell theory has been judged as too simple and less accurate because it is limited by some unrealistic boundary conditions. The finite element modeling (FEM) method associated with MD or equivalent-continuum (EC) model was recently adopted to calculate the mechanical properties of

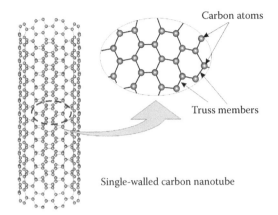

FIGURE 1.4
Truss model of a carbon nanotube.

nanotubes. Odegard et al. [19] developed an equivalent-continuum tube model to determine the effective geometry and effective bending rigidity of a graphene structure. Molecular mechanics considerations (see Equations 1.6 and 1.7) were first used to determine linking forces between individual carbon atoms. This molecular force field was simulated by using a pin-joint truss model (i.e., each truss member represents the force between two atoms as shown in Figure 1.4). Therefore, the truss model allows for the accurate simulation of the mechanical behavior of nanotubes in terms of atom displacements (similar to the nonlinear modeling of structures). As the nanotube was subject only to a uniaxial tensile load [20], the bond stretching and bond-angle variation energies (see Equation 1.8) were considered. The strain energy of the whole system was used in the FEM computation to estimate the effective thickness of the nanotube layer. It was found that the effective thickness of the nanotubes (0.69 Å and 0.57 Å) was significantly larger than the interlayer spacing of graphite, estimated to be ~0.34 Å.

Li and Chou [20,21] worked out the contributions of van der Waals interactions between individual carbon atoms within nanotubes in a FEM truss model. The relationships between the structural mechanics parameters EA, EI, and GJ and the molecular mechanics parameters $K\rho$, $C\theta$, and $A\tau$ as shown in Equations 1.7a, 1.7b, and 1.7d for each truss member:

$$\frac{EA}{L} = K_\rho$$

$$\frac{EI}{L} = C_\theta \qquad \qquad (1.9a\text{–}c)$$

$$\frac{GJ}{L} = A_\tau$$

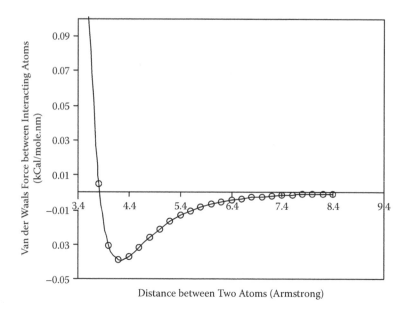

FIGURE 1.5
Van der Waals force versus the distance between two carbon atoms.

In Figure 1.5, the dependence of the van der Waals force versus the distance between two carbon atoms is plotted. Nonlinear truss elements were used in simulations, as the force between two carbon atoms is also nonlinear. A uniaxial load was applied uniformly at the end of the nanotubes; the effects due to the end caps were neglected. It was found that the Young's modulus of the nanotubes increases with increasing the diameter of the nanotubes. The Young's and shear moduli of MWNTs estimated were in the range of 1.05 ± 0.05 and 0.4 ± 0.05 TPa, respectively. For SWNTs, the Young's modulus was almost constant when the nanotubes' diameters were larger than 1 nm. The average Young's modulus of the zigzag nanotubes was slightly higher than the armchair type. Also, the Young's modulus of MWNTs was generally higher (~7%) than that of SWNTs.

1.3 Fabrication Process of Nanotube/Polymer Composites

To achieve desirable properties of nanotube polymer-based composites, control of the fabrication process is an essential factor that governs the production of well-dispersed composites as well as ensures a successful bonding between the nanotubes and their surrounding matrix, which results in providing good stress transfer from the matrix to the nanotubes. Several parameters such as the selection of a dispersion solvent (with the proper control of holding time for evaporating the solvent before hardener is added), sonication time, stirring speed, viscosity of the resin, and mixing and curing temperatures are important to the properties of resultant composites. On the other hand, the strength of the composites is highly dependent on the alignment of the nanotubes. In reality, the control of the alignment of the nanotubes to a desirable direction is still a challenge to date, as these tiny structured materials are not easily controlled by any means of our existing practices that are mainly for micro- and macroscale structures.

For general practice, sonication is one of the most popular ways to disperse nanotubes into polymer-based resin. Park et al. [22] first demonstrated the use of *in situ* polymerization to disperse SWNT bundles in the polymer matrix. A dilute SWNT solution, typically around 0.05 wt% of nanotubes in dimethylformamide (DMF), subject to sonication for one and half hours in an ultrasonic bath (40 kHz) was followed by mixing with the hardener. It was found that the long sonication time may cause the nanotubes to entangle and form bundles, which may result in losing the benefit of their high strength properties in composite materials. Entanglement of the high aspect ratio nanotubes may also cause poor impregnation of resin into the bundle unless the viscosity of the resin can be controlled to a relatively low limit. Mulhopadhyay et al. [23] reported that if the sonication time is greater than 4 hours, it may destroy most of the graphene layers of the nanotubes and, possibly, cause the formation of junctions in the nanotubes. Such distortion may be due to overviolent motions of the nanotubes, and thus, they crash to each other. Localized heat may be induced during the sonication process and subsequently cause the breaking of carbon–carbon bonds of the nanotubes.

Lu et al. [24] found that the use of different chemical solvents as dilute solutions for separating nanotubes in epoxy-based resin would greatly influence the integrity of resultant composites. Acetone, ethanol, and DMF were used for the test, and their boiling temperatures are 56°C, 78°C, and 130°C, respectively. Because DMF has a high boiling temperature compared with ethanol and acetone, this solvent would remain inside the resin due to the difficulty of evaporation under room temperature and thus alter the rate of chemical reaction between the resin and the hardener. Eventually, the overall mechanical and chemical properties of the composite decreased. In Table 1.1, the Vicker's hardness of different nanotube/epoxy composites mixed by using different types of solvent is shown. It is obviously seen that the mechanical properties of the composites are directly affected by the melting temperatures of the solvent.

In Figure 1.6, Fourier transform infrared (FTIR) spectra of different types of composites are shown. Spectra **b**, **c**, and **d** represent the solvents acetone, ethanol, and DMF, respectively, used to disperse the nanotubes into an epoxy matrix. Spectrum **a** is for pure epoxy. The most prominent feature in the spectra **b**, **c**, and **d** are the appearance of a new absorption band located at ~1650 cm^{-1}. Considering that epoxide and hydroxyl (OH) groups are the

TABLE 1.1

Vicker's Hardness Values of Different Samples Determined
from a Load Force of 100 g and a Dwell Time of 15 s

Composites	Dispersion Solvent	Vicker's Hardness Reading
CNTs/Epoxy	Acetone	18.0 ± 0.11
CNTs/Epoxy	Ethanol	14.4 ± 0.08
CNTs/Epoxy	DMF	7.7 ± 0.10
Pure Epoxy	—	17.8 ± 0.06

only two reaction groups in the epoxy molecule, the ~1650 cm^{-1} band can be assigned to an amino group formed by the intermolecular nucleophilic substitution of hydroxyl at the amide functionality, which can therefore be used to estimate the relative amount of the product of the cure reaction. It is also noticed that the band locations are different in the spectra b, c, and d, with values of 1645 cm^{-1}, 1649 cm^{-1}, and 1664 cm^{-1}, respectively. This indicates that there are functional and curing differences among epoxy resins with different solvent treatments, which could be related to the variation of the mechanical properties discussed above. Blanchet et al. [25] directly dispersed nanotubes into polyaniline (PANI) using a sonication method. The nanotubes were first sonicated in xylene, and that dispersion afterward was sonicated in the DNNSA-PANI solution. It was found that all the nanotubes were

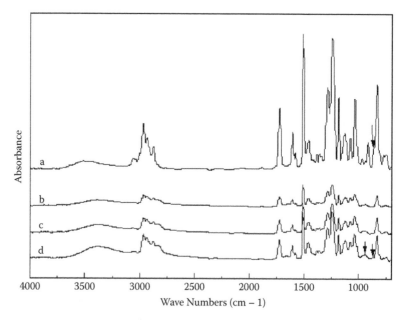

FIGURE 1.6

The representative Fourier transform infrared (FTIR) spectra measured from samples (a) through (d).

well dispersed into the PANI, and enhancement of the electrical conductivity was achieved. Tang et al. [26] also introduced the use of a melt processing technique for making nanotube/high-density polyethylene (HDPE) composites. The nanotubes and HDPE were premelted together to form pellets, followed by feeding them into a twin-screw extruder to make composites. In their work, it was difficult to control the uniformity of the nanotubes in the HDPE, because once the pellets were melted, the nanotubes may be trapped inside the injection head of the extruder or may get tangled in the extruder, forming bundles. Although research related to nanotube/polymer composites has been conducted for more than a decade, detailed study on how to fabricate well-dispersed and well-aligned nanotube/polymer composites is still in progress. Sonication using a water bath and mechanically mixing by using stirrers of a nanotube/polymer mixture to ensure its homogeneity, and vacuum and humidity control of the mixture to avoid the formation of voids inside resultant composites are the common ways to produce uniformly dispersed nanotube/polymer composites. However, within the process, due to the fact that the control of viscosity and curing time of resin are typically ignored by many researchers, sinking and entanglement of the nanotubes in a low-viscosity environment may result due to the gravitation force.

1.4 Interfacial Bonding Properties of Nanotube/Polymer Composites

1.4.1 Experimental Investigation

It has been recognized that the high mechanical strength of nanotube/polymer composites could be achieved only when all nanotubes are aligned parallel to the load direction, which is similar to glass or carbon fiber reinforced composites (GFRP or CFRP). All external stresses applied to the composites should be effectively transferred to the nanotubes via matrix through their interfacial bonding shear.

Jin et al. [27] and Wood et al. [28] studied the control of the alignment of nanotubes in a polymer matrix to produce ultra-high-strength nanocomposites. However, such products can only be produced inside a laboratory environment, and it is difficult to approach the level of mass production. The control of alignment of all nanotubes in a desirable direction is still a challenge to the composite community. Such alignment affects not only the basic properties of the nanocomposites but also their electrical conductivity, thermal stability, and somehow, sensorability.

The stress transfer mechanisms of different types of nanotubes in nanotube/polymer composites were investigated in the past few years through experimental, theoretical, and numerical approaches. Wagner et al. [29] and

Qian et al. [30] reported that the shear strength at the interface between the nanotubes and surrounding matrix reached 43.3 MPa. However, in several experimental investigations, it was obvious that a poor adhesion property in nanotube/polymer composites was found [31]. The pullout of the nanotubes was observed on the fracture surface of broken specimens. In addition, a weak bonding force between nanotube layers of a multiwalled nanotube causes the failure initiated in its outermost layer. The latest experimental study conducted by Cooper et al. [32] demonstrated that the shear strength between a nanotube and epoxy matrix varied with its diameter, length, and number of layers. Two sections of an epoxy matrix were formed with a nanotube bridging across the sections. The two sections were pulled apart using the tip of a scanning probe microscope. The maximum shear strengths ranging from 38 MPa to 376 MPa were recorded, and a high shear strength was reported for SWNT ropes. A pullout of the nanotubes from a solid polyethylene matrix was demonstrated using an AFM [33], and the average interfacial stress required to remove a SWNT from the matrix measured was about 47 MPa. This experiment was closer to a realistic situation because all the nanotubes were physically bonded to the matrix, and a typical pullout test was actually conducted. However, as described in the previous section, the mechanical properties of different kinds of nanotubes could influence their bonding strength to the matrix. To better understand the interfacial bonding behavior between the nanotube and matrix, an improved experimental setup to perform pullout tests for different types and geometries of nanotubes is needed.

1.4.2 Theoretical Study and Molecular Dynamics Simulation

The MD simulation is a powerful technique to predict the bonding properties among different layers in nanotubes, as previously mentioned, and also between the nanotubes and surrounding atoms and molecules. Liao and Li [34] simulated a pullout action of nanotubes in a polystyrene (PS) matrix using the commercial software Hyerchem®. The adhesion strength between PS molecules and a graphene sheet was studied using a molecular mechanics model employing an empirical MM+ force field. A random coil of PS $[(-CH_2CHC_6H_6-)_n]$ molecules with n = 2, 4, 10, 20, 40, and 80 was constructed and located near the surface layer of the nanotubes (Figure 1.7). The nanotubes were then subject to a pullout force to measure their interfacial shear strength. In their study, the shear stress between the nanotubes and polymer estimated from the simulation was about 160 MPa. Lordi and Yao [35] calculated the binding energies and sliding frictional stresses between nanotubes and polymer matrix by using a force-field-based molecular mechanics. They found that the frictional stress between the polymer substrate and the surface of nanotubes was much higher than that of the stress between the nanotube layers of MWNTs. The key factor in forming a strong bond between the polymer and nanotube lies in the polymer's morphology, specifically its ability to form large-diameter helices around individual nanotubes. Frankland

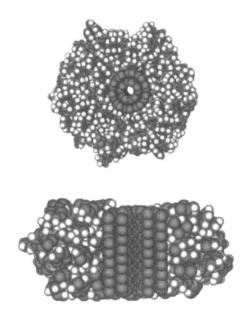

FIGURE 1.7
A molecular model of a double-walled carbon nanotube in a polystyrene (PS) matrix.

et al. [36] studied the effect of chemical cross-links on the interfacial bonding strength between a SWNT and polymer matrix using MD simulations. Their models were composed of single-walled (10,10) armchair nanotubes embedded into either a crystalline or amorphous matrix. The nonbond interactions within the polyethylene matrix and between the matrix and nanotubes were modeled with Lennard-Jones 6-12 potentials. They found that even a relatively low density of cross-links can have a large influence on the properties of the nanotube–polymer interface. Ren et al. [37] also found that pullout energies are affected by the thermal condition of composites. However, in those simulated studies, due to the reduction of complexity, chemical interactions between the nanotubes and matrix were generally ignored. Only nonbond interactions and electrostatic and van der Waals forces were assumed.

In the early stages of nanotube-related research, many diverse results on the interfacial-bonding characteristic between nanotubes and polymer-based matrices have been found. Xu et al. [38] addressed that high interfacial shear stress between a MWNT and an epoxy matrix was observed from a fractured sample. Wagner [39] first used the Kelly–Tyson model, which has been widely used to study the matrix-fiber stress transfer mechanism in micron-size fiber composites, to study the interfacial shear strength between the nanotube and polymer matrix. Because it was found that the binding force between inner layers is very low, and sliding failure always occurs, only a single-walled system is of interest in his work. In his study, it was assumed that an externally applied stress to a nanotube/polymer composite

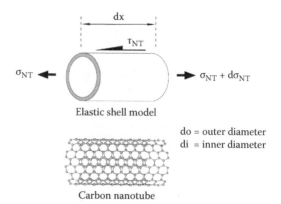

FIGURE 1.8
Stress transfer study using the Kelly–Tyson model.

is fully transferred to the nanotubes via a nanotube-matrix interfacial shear mechanism at the molecular level—that is, the length of the nanotubes (l) is larger than the critical length (lc). A single nanotube cylindrical model was used to study the stress transfer properties of the composite shown in Figure 1.8. To consider the force balance in the composite system, the following equation is formed:

$$\tau_{NT} d_i dx = (\sigma_{NT} + d\sigma_{NT})\left(\frac{d_o^2 - d_i^2}{4}\right) - \sigma_{NT}\left(\frac{d_o^2 - d_i^2}{4}\right) \tag{1.10}$$

where τ_{NT} is the interfacial shear strength between the nanotube and matrix, σ_{NT} is the tensile strength of a nanotube segment of length dx, and d_o and d_i are outer and inner diameters of the nanotube, respectively. After integrating Equation 1.10 and considering the critical length of a typical short fiber system in composite structures, the interfacial shear strength can be written as follows:

$$\tau_{NT} = \sigma_{NT}\left(\frac{1}{2}\left(\frac{l_c}{d_o}\right)^{-1}\left(1 - \frac{d_i^2}{d_o^2}\right)\right) \tag{1.11}$$

where l_c/d_o is the critical aspect ratio of the nanotube, and d_i/d_o is the diameter ratio. As an externally applied stress of 50 GPa was used in the literature, the interfacial strength was calculated for critical length values of 100, 200, and 500 nm. It is concluded that the interfacial shear stress is affected by several factors: the critical length and the outer diameter of the nanotubes. Increasing the diameter of the nanotube also increases the interfacial shear strength at the bond interface.

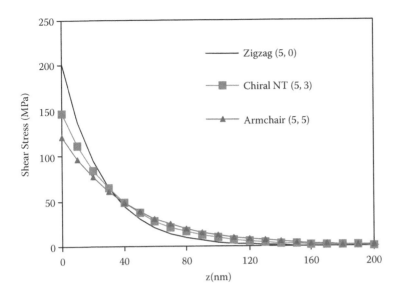

FIGURE 1.9
Plot of interfacial shear stress versus different types of nanotubes.

Lau [11] conducted an analytical study of the interfacial bonding properties of nanotube/polymer composites by using the well-developed local density approximation model, as described previously, and classical elastic shell theory and the conventional fiber-pullout model. In his study, several important parameters such as the nanotube wall thickness, Young's modulus, volume fraction, and chiral vectors of the nanotubes were considered. It was found that the decrease of the maximum shear stress occurs with increasing the size of the nanotubes. Increasing the number of the walls of the nanotubes will cause a decrease of Young's modulus of the nanotubes, an increase of the effective cross-sectional area, and an increase in the total contact surface area at the bond interface and the allowable pullout force of the nanotube/polymer system. In Figure 1.9, a plot of the interfacial shear stress of SWNTs with different chiralities is shown. Shown in Figure 1.9 is that the maximum shear stress of a zigzag nanotube (5,0) is comparatively higher than those of chiral (5,3) and armchair (5,5) nanotubes.

1.5 Summary

In this chapter, a review on the strength of nanotubes, fabrication processes, and interfacial bonding properties of nanotube/polymer composites is given. Although many works have been done recently, it was found that there are still many uncertainties in measuring the mechanical properties of nanotubes. The difficulty in getting good equipment to conduct nanoscale property tests remains. MD and theoretical analyses are mainly based on certain assumptions that may not be practically applied to a real situation in nanoscale composites. Dispersion properties and control of the alignment of nanotubes are also important issues that govern the global properties of the composites. To date, no comprehensive work has been conducted to report an exact solution for producing large-scale uniformly dispersed nanotube/polymer composites.

However, due to increased demand for light and high strength composites, including the massive usage of glass and carbon fiber reinforced composites for AIRBUS 380 and Streamliner Boeing 787, the investigation on how to reinforce polymer-based materials and their composites would be of great interest to many industries. It is necessary for certain questions to be further investigated before actually applying these nanoscale composites to real-life applications. These questions include the following: (1) Does the chemical bonding between nanotubes and the matrix exist? (2) Do the nanotubes maintain their extraordinary mechanical, electrical, and thermal properties if chemical bonding exists between the nanotube and the matrix?

References

1. Snow E.S., Campbell P.M., and Novak J.P. 2002. Single-Wall Carbon Nanotube Atomic Force Microscope Probes *Appl. Phys. Lett.* 80(11) 2002–2004.
2. Kiernan G., Barron V., Blond D., Drury A., Coleman J., Murphy R., Cadek M., and Blau W. 2003. Characterization of Nanotube-Based Artificial Muscles Materials *Proc. SPIE* 4876 775–781.
3. Gao Y. and Bando Y. 2002. Carbon Nanothermometer Containing Gallium *Nature* 415–599.
4. Liu C., Yang L., Tong H.T., Cong H.T., and Cheng H.M. 2002. Volumetric Hydrogen Storage in Single-Walled Carbon Nanotubes *Appl. Phys. Lett.* 80(13) 2389–2391.
5. Yu M.F., Files B.S., Arepalli S., and Ruoff R.S. 2000. Tensile Loading of Ropes of Single-Wall Carbon Nanotubes and Their Mechanical Properties *Phys. Rev. Lett.* 84(24) 5552–5555.
6. Demczyk B.G., Wang Y.M., Cumings J., Hetman M., Han W., Zettl A., and Ritchie R.O. 2002. Direct Mechanical Measurement of the Tensile Strength and Elastic Modulus of Multiwalled Carbon Nanotubes *Mater. Sci. Eng. A* 334 173–178.
7. Cumings J. and Zettl A. 2000. Low-Friction Nanoscale Linear Bearing Realized from Multiwall Carbon Nanotubes *Science* 289 602–604.
8. Qi H.J., Teo K.B.K., Lau K.K.S., Boyce M.C., Milne W.I., Robertson J., and Gleason K.K. 2003. Determination of Mechanical Properties of Carbon Nanotubes and Vertically Aligned Carbon Nanotube Forests Using Nanoindentation *J. Mech. Phys. Solids* 51 2213–2237.

9. Tu Z.C. and Ou-Yang Z.C. 2002. Single-Walled and Multiwalled Carbon Nanotubes Viewed as Elastic Tubes with the Effective Young's Moduli Dependent on Layer Number *Phys. Rev. B* 65 233–407.

10. Lau K.T., Gu C., Gao G.H., Ling H.Y., and Reid S.R. 2004. Stretching Process of Single- and Multiwalled Carbon Nanotubes for Nanocomposite Applications *Carbon* 42 423–460.

11. Lau K.T. 2003. Interfacial Bonding Characteristics of Nanotube/Polymer Composites *Chem. Phys. Lett.* 370 399–405.

12. Lau K.T. and Hui D. 2002. The Revolutionary Creation of New Advanced Materials: Carbon Nanotube Composites *Compos. Pt. B: Eng.* 33 263–277.

13. Lennard-Jones J.E. 1924. The Determination of Molecular Fields: From the Variation of the Viscosity of a Gas with Temperature *Proc. R. Soc.* A106 441.

14. Chang T.C. and Gao H.J. 2003. Size-dependent Elastic Properties of a Single-Walled Carbon Nanotube via a Molecular Mechanics Model *J. Mech. Phys. Solid* 51 1059–1074.

15. Lu J.P. 1997 Elastic Properties of Single and Multi-Layered Nanotubes *J. Phys. Chem. Solids* 58(11) 1649–1652.

16. Zhou G., Duan W.H., and Gu B.L. 2000. First-Principles Study on Morphology and Mechanical Properties of Single-Walled Carbon Nanotubes and Graphene *Chem. Phys. Lett.* 326 181–185.

17. Lier G.V., Alsenoy C.V., Doren VV., and Geerlings P. 1998. *Ab Initio* Study of the Elastic Properties of Single-Walled Nanotubes *Phys. Rev. B* 58(20) 14013–14019.

18. Nardelli M.B., Yakobson B.I., and Bernholc J. 1998. Brittle and Ductile Behaviour in Carbon Nanotubes. *Phys. Rev. Lett.* 81(21) 4656–4659.

19. Odegard G.M., Gates T.S., Nicholson L.M., and Wise K.E. 2002. Equivalent-continuum Modelling of Nano-Structured Materials *Comp. Sci. Tech.* 62 1869–1880.

20. Li C.Y. and Chou T.W. 2003. A Structural Mechanics Approach for the Analysis of Carbon Nanotubes *Inter. J. Solid & Struct.* 40 2487–2499.

21. Li C.Y. and Chou T.W. 2003. Elastic Moduli of Multi-Walled Carbon Nanotubes and the Effect of Van der Waals Forces *Comp. Sci. Tech.* 63 1517–1524.

22. Park C., Ounaies Z., Watson K.A., Crooks R.E., Smith Jr. J., Lowther S.E., Connell J.W., Siochi E.J., Harrison J.S., and Clair T.L. 2002. Dispersion of Single-Wall Carbon Nanotubes by *In Situ* Polymerization under Sonication *Chem. Phys. Lett.* 364 303–308.

23. Mukhopadhyay K., Dwivedi C.D., and Mathur G.N. 2002. Conversion of Carbon Nanotubes to Carbon Nanofibers by Sonication *Carbon* 40 1369–1383.

24. Lu M., Lau K.T., Ling H.Y., Zhou L.M., and Li H.L. 2004. Effect of Solvents Selection for Carbon Nanotubes Dispersion on the Mechanical Properties of Epoxy-Based Nanocomposites *Comp. Sci Tech.* Accepted.

25. Blanchet G.B., Fincher C.R., and Gao F. 2003. Polyaniline Nanotube Composites: A High-Resolution Printable Conductor *Appl. Phys. Lett.* 82(8) 1290–1292.

26. Tang W.Z., Santare M.H., and Advani S.G. 2003. Melt Processing and Mechanical Property Characterization of Multiwalled Carbon Nanotube/High Density Polyethylene (MWNT/HDPE) Composite Films *Carbon* 41 2779–2785.

27. Jin L., Bower C., and Zhou O. 1998. Alignment of Carbon Nanotubes in a Polymer Matrix by Mechanical Stretching *Appl. Phys. Lett.* 73(9) 1197–1199.

28. Wood J.R., Zhao Q., and Wagner H.D. 2001. Orientation of Carbon Nanotubes in Polymers and Its Detection by Raman Spectroscopy *Compos. Pt. A.* 32 391–399.

29. Wagner H.D., Lourie O., Feldman Y., and Tenne R. 1998. Stress-Induced Fragmentation of Multiwall Carbon Nanotubes in a Polymer Matrix *Appl. Phys. Lett.* 72(2) 188–190.
30. Qian D., Dickey E.C., Andrews R., and Rantell T. 2000. Load Transfer and Deformation Mechanisms in Carbon Nanotube-Polystyrene Composites *Appl. Phys. Lett.* 76(20) 2868–2870.
31. Lau K.T. and Hui D. 2002. Effectiveness of Using Carbon Nanotubes as Nano-Reinforcement for Advanced Composite Structures *Carbon* 40 1597–1617.
32. Cooper C.A., Cohen S.R., Barber A.H., and Wagner H.D. 2002. Detachment of Nanotubes from a Polymer Matrix *Appl. Phys. Lett.* 81(20) 3873–3875.
33. Barber A.H., Cohen S.R., and Wagner H.D. 2003. Measurement of Carbon Nanotube-Polymer Interfacial Strength *Appl. Phys. Lett.* 82(23) 4140–4142.
34. Liao K. and Li S. 2001. Interfacial Characteristics of a Carbon Nanotube-polystyrene Composite System *Appl. Phys. Lett.* 79(25) 4225–4227.
35. Lordi V. and Yao N. 2000. Molecular Mechanics of Binding in Carbon-Nanotube-Polymer Composites *J. Mater. Res.* 15(12) 2770–2779.
36. Frankland S.J.V., Caglar A., Brenner D.W., and Griebel M. 2002. Molecular Simulation of the Influence of Chemical Cross-Links on the Shear Strength of Carbon Nanotube-Polymer Interfaces *J. Phys. Chem. B.* 106 3046–3048.
37. Ren Y., Fu Y.Q., Li F., Cheng H.M., and Liao K. 2004. Fatigue Failure Mechanisms of Single-Walled Carbon Nanotube Ropes Embedded in Epoxy *Appl. Phys. Lett.* 84(15) 2811–2813.
38. Xu X.J., Thwe M.M., Shearwood C., and Liao K. 2002. Mechanical Properties and Interfacial Characteristics of Carbon-Nanotube-Reinforced Epoxy Thin Film *Appl. Phys. Lett.* 81 2833.
39. Wagner H.D. 2003. Nanotubes-Polymer Adhesion: A Mechanic's Approach *Chem. Phys. Lett.* 361 57.

2

Processing, Properties, and Flow Behavior of Carbon Nanofiber–Based Polymeric Nanocomposites

Khalid Lafdi

University of Dayton Research Institute, Ohio

CONTENTS

2.1 Introduction

The impetus behind a polymeric composite is its ability to be tailored to any system, its high strength to weight ratio, and its improved fatigue and corrosion resistances. As a load is applied to the continuous phase, it is then transferred to the discontinuous, reinforcement phase. The choice and degree of reinforcing filler can then be assigned according to the requirements of the particular application. Superior performance and the resulting cost savings have allowed composites to be incorporated into nearly every aspect of our lives.

Carbon composite research has made great strides in the last 40 years due to international collaboration. Modern carbon fiber history began in the 1960s with the work of A. Shindo of the Industrial Research Institute of Osaka, Japan. Western scientists began using carbon fibers as a reinforcement later that decade, and the first commercial-grade fiber was produced in England in 1967 [1]. Traditional carbon composites used today consist of tows of micrometer carbon fibers incorporated into various resin matrices.

Aerospace composites consist of prepreg, or fibers impregnated with resin. Layers of prepreg are then aligned and arranged according to the desired form and then heated to complete the cure of the resin. The resin facilitates load transfer to the reinforcement and protects the filler from detrimental environmental effects. Polymer resins constitute the largest group of resins and are further divided into thermosets and thermoplastics [1]. Polymers can be distinguished by their intermolecular structure, as thermosets are chemically cross-linked in their cured state, and thermoplastics are melted and then cooled into their final form. Thermosets are generally more rigid and perform better at elevated temperatures. These advantages have lent thermosets to historically serve as the principal matrix material for aerospace composites [2]. Epoxy resins are thermosets and are commonly used experimentally because of their excellent adhesion, strength, adaptability to manufacturing methods, and processing versatility [2].

Scientific advancement has spurned many improvements in the composite industry, and the latest to impact the field is the incorporation of carbon nanotubes and nanofibers with their impressive electrical, mechanical, and thermal properties into composites. The discovery of carbon nanotubes in 1991 by Iijima [3] has brought about this latest evolution in carbon composites from the microscale to the nanometric-scale. The mechanical strength of carbon nanotubes/fibers is due to the carbon–carbon double bond of graphite, the strongest possible bond occurring in nature [4]. As the physical scale of carbon fibers is reduced, the theoretical behavior of carbon at the molecular level is approached [5]. This improvement in the physical properties is due to the reduced possibility of vacancies and imperfections.

When a mechanical load is applied to the fiber, it is felt across the entire body of the fiber. As an imperfection is reached, there can be no further transference of the load. As a result, there is a localized buildup at the site of the imperfection. Therefore, these imperfections and vacancies serve as nucleation sites for mechanical failure [6]. This adverse behavior at imperfections and vacancies can also be seen when thermal and electrical loads are applied to the carbon fiber. A single-walled carbon nanotube (SWNT), which lacks imperfections, serves as the ultimate reinforcing fiber with regard to its strength in the direction of the nanotube axis. Unfortunately, SWNTs are difficult and expensive to manufacture [4]. The difficulties in synthesizing SWNTs have caused researchers to pursue multiwalled carbon nanotubes (MWNTs) and carbon nanofibers as a viable substitute for the SWNTs for many applications [4].

Multiwalled nanotubes consist of an inner single-walled nanotube surrounded by coaxial tubes of increasing diameter. The different layers of the MWNT are held together by pi-bonding. Due to the dominant intralayer sigma bonding, the weaker interlayer pi bonds serve as possible sights of mechanical failure. This type of failure is called sword-in-sheath failure, because the outer layers are "sheathed" from the inner layers of the nanotube [7]. A carbon nanofiber is similar to a MWNT in that both tubes consist of layers of carbon surrounding an inner hollow core. The surface of carbon nanofibers consists of exposed graphene edges that serve as locations for both chemical and physical bonding. The edges are more chemically reactive due to their unfulfilled bonding requirements and allow the resin matrix to essentially infiltrate the fiber, increasing the interfacial surface area between the two phases. The inferior physical properties of carbon nanofibers are offset by their cost savings in applications and their ability to be "wetted" by resins, thereby improving the matrix/reinforcement adhesion.

The objective of this chapter is to analyze the mechanical, thermal, and electrical effects of carbon nanocomposites created by the addition of a series of carbon nano-artifacts, such as nanofibers, to a polymeric matrix system. The primary focus is to determine the level of nanofiber addition at which the physical improvement of the resultant nanocomposite is thereby optimized. Various experiments were carried out to study the effect of heat treatment of the nanofibers; surface functionalization with specific functional groups as a function of the host polymer matrix; and flow behavior of this nanocomposite during conventional composite processing, such as resin transfer molding (RTM).

2.2 Purification and Heat Treatment of Nanofibers

Heat treatment (carbonization and graphitization) is an effective method of removing defects from carbon nanofibers, which diminish their electrical and mechanical properties. It is important to understand that graphitization causes structural changes that enhance lattice perfection through heat treatment, and this may improve specific applications of the nanofibers.

Endo et al. [8] heat treated stacked cup–type carbon nanofibers from 1800 to 3000°C to examine structural changes. The truncated cones cause a high chemical reactivity in the outer surface and the inner hollow core, because these end planes of graphene layers are active edge sites. Heat treatment to 3000°C resulted in transformation to a rugged surface and the formation of energetically stable loops between adjacent graphene layers from the unstable edge planes in both the outer surface and the inner hollow cores. Examination of the fibers by X-ray diffraction and Raman spectroscopy identified that the interlayer spacing of the graphitized samples increased, possibly due to the large number of loop formations between adjacent graphene layers. The absence of separation of (101) and (100) lines and low intensity of (004) lines indicated that the nanofibers achieved relatively low graphitization following heat treatment to 3000°C. The formation of loops began below 2100°C and was followed by a few changes up to 3000°C on the outer surface of the carbon nanofibers. With increasing heat treatment temperature, there was a progressive decrease in the electrical resistivity in the bulk state for carbon nanofibers due to the amount of loops, especially on the outer surface of the carbon nanofibers. There was a significant decrease in the electrical resistivity from the as-grown nanofiber to the nanofiber heat treated to 1800°C due to the evolution of volatile material.

Lim et al. studied stacked coin–type nanofibers and the impact of mechanical and chemical treatments on the morphology of the fibers [9]. Heat treatment at 2800°C induced closed-loop ends on the surface of the nanofibers formed by the folding of some planar hexagons at their edges. The heat treatment removed C–H bonds and densely stacked hexagonal layers of graphene, forming chemically active sites on the edges. The edges were stabilized by bonding to each other, even though the bonding caused high tension through the formation of a sharp curvature. Acidic oxidation of the nanofibers cut off the closed-loop ends, resulting in improved overall alignment of the graphene layers. Both treatments generated many free edges and a high graphitization extent, indicating the possibility of improved interfacial bonding with a polymer matrix.

Katayama et al. examined the effect of heat treatment on MWNTs with bamboo-like structures [10]. Heat treatment at 2800°C reduced the interlayer spacing to that of graphite. In addition, encapsulates in the bamboo structure with a small diameter were opened, and metal impurities were entirely removed from the nanotube. The decrease in interlayer spacing lies in contrast to the observance reported by Endo et al. for carbon nanofibers heat treated to 3000°C.

Kiselev et al. examined by high resolution electron microscopy the structural changes to MWNTs following heat treatment [11]. Links formed between the neighboring open edges by loops along the tube sides on both the insides and outsides of the nanotubes. The radii of curvature of the loops were noted at 1.05 to 1.40 nm, which is close to the diameter of SWNTs. The degree of carbon layer linking was dependent on the temperature of the heat treatment. After treatment at 1200°C, only a small amount of linked layers were observed along the external sides. At treatment above 2000°C, the number of loops per unit volume increased, and loops were observed on both the external and internal surfaces.

Results were reported on the effect of heat treatment of carbon nanofibers on nanocomposite properties. Xu et al. included nanofiber/vinyl ester composite with nanofibers heat treated to 3000°C in an examination of electrical properties of nanocomposites [12]. The percolation threshold was found to be between 2 and 3 wt% fiber loading. The heat-treated fibers were better electrical conductors than pyrolytically stripped fibers. The higher surface activity of the heat-treated fibers should lead to thicker resin coating and more complete resin infusion into the fibers during mixing. However, the measured flexural modulus of the nanocomposite was equal to that of the pure resin. This may have been due to poor fiber dispersion in the matrix. In addition, the measured glass transition temperature of a nanocomposite with heat-treated fibers was nearly 20°C higher than that of a nanocomposite with pyrolytically stripped fibers.

Finegan et al. [13] noted that more graphitic fibers of higher graphitization index tend to make composites having both lower strength and modulus. Graphitized fibers projected much farther from the fracture surface of a nanofiber/polypropylene composite than air-etched fibers, implying that the interfacial shear strength of the graphitized fibers was significantly lower. The polymer did not appear to wet the fiber surface in either case, indicating that low interfacial strength is not singularly dependent on surface wetting. The results from this study may have been impacted by ball milling of all nanofiber samples, resulting in lower fiber aspect ratios.

Kuriger et al. examined the thermal and electrical properties of heat-treated nanofiber/polypropylene composites [14]. The electrical resistivity decreased with fiber volume fraction. This is characteristic of the electrical conductivity of composites which is governed by percolation and quantum mechanical tunneling between the fibers. The electrical resistivity was significantly lower than glass fiber reinforced polymers. The thermal conductivity increased with fiber loading and reached a maximum with 23% fiber volume fraction at 5.38 W/mK in the longitudinal direction.

Ma et al. included heat-treated nanofiber/polyester composites in their study of nanocomposite properties [15]. Thermogravimetric analysis showed that the onset of degradation for nanofibers heat treated at 3000°C was nearly 700°C compared to approximately 300°C for as-grown nanofibers. The tensile modulus of the nanocomposite was slightly higher than that of the neat resin, and the tensile strength was comparable to the neat resin and significantly higher than its pyrolytically stripped-fiber nanocomposite counterpart. The inability of the heat-treated fibers to improve the tensile properties of the neat resin may be due to the graphitic planes not being aligned parallel to the nanofiber axis. This counters the alignment of the nanofibers in the neat resin. In addition, the heat-treated nanofibers were ball milled, leading to attrition of the nanofiber aspect ratio, which is detrimental to the tensile properties. The compressive strength and torsional modulus of the heat-treated nanofiber composite was significantly higher than that of the neat resin. The nanofibers may have acted as a barrier to halt propagation, thereby improving compressive properties.

In this study, carbon nanofibers were produced by Applied Sciences, Inc. (Cedarville, Ohio). This group of nanofibers (labeled PR-24) has diameters between 60 and 100 nanometers and lengths ranging from 30 to 100 microns. Nanofibers were subjected to heat-treatment temperatures ranging from 1500°C to 3000°C to alter the properties of the reinforcement and the physical and chemical interaction between the nanofiber and the epoxy polymer. The strength of adhesion between the fiber and an epoxy (thermoset) matrix was characterized by the flexural strength and modulus (ASTM D790-00), and the electrical (ASTM B193-87) and thermal properties (ASTM E1269-89, ASTM C693-74) of the composites were investigated from the viewpoint of heat-treatment temperature of the carbon nanofibers. The carbon nanofibers were dispersed within the epoxy resin at loading rates of 4%, 8%, and 12%

FIGURE 2.1
The carbon nanofiber. It consists of stacked conic sections of graphite lattice. (Courtesy M. Endo.)

by weight. Previous work [16] indicated that loading beyond 12% by weight actually led to a decrease in nanocomposite mechanical properties. In addition, processing nanocomposites with loads beyond 12% by weight becomes difficult in establishing a homogenous mixture. The as-received carbon nanofibers, designated as PS, and the neat resin were used as a baseline for the study.

In order to fully understand the properties that make carbon nanofibers unique, both the microstructure and surface of nanofibers samples were characterized. The pristine and surface-treated nanofibers consisted of nested and straight carbon nanofibers (Figure 2.1).

Nested carbon nanofibers have an orientation similar to that of a set of stacked Dixie® cups with a hollow core and are also referred to as fish-bone-type carbon nanofibers [17]. At low temperatures (below 1200°C), carbon exhibits only local molecular ordering. As carbon nanofibers are heat treated, an increase in temperature results in the aromatic molecules becoming stacked in a column structure. Further heat treatment causes these columns to coalesce, forming a distorted, wavy structure. After surpassing a temperature of 2500°C, the distorted graphene layers of carbon become flattened, forming an aligned structure. If the material is graphitic, it will attain the minimum interlayer spacing in the graphite order between graphene layers. By analogy, after heat treating the pristine nanofibers to a temperature of 3000°C, graphene layers became straight, and minimum interlayer spacing was reached. As shown by transmission electron microscopy (TEM) (Figure 2.2 and Figure 2.3), the layers within the "Dixie cup" carbon nanofiber have coalesced following heat treatment.

At this magnification, the inclination angle of each cup is apparent. Within each cup, it can be seen that the localized ordering of the graphene planes has been changed due to coalescence, resulting in continuous planes. The stacking effect is shown through the use of a gray scale. The walls of the nanofibers are dark due to their high electronic density. The surrounding regions are starkly lighter with low electronic densities.

At high magnification, the graphene layers appear very straight without any disclination defects. However, there is no change in the inclination angle

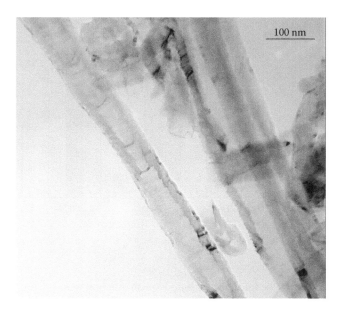

FIGURE 2.2
Bright field micrograph of a "Dixie cup" carbon nanofiber structure.

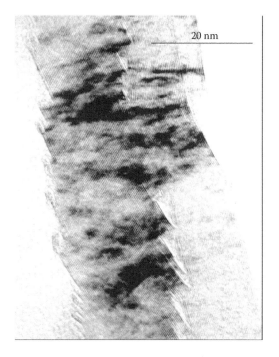

FIGURE 2.3
High-resolution imaging of localized area of a "Dixie cup" structure.

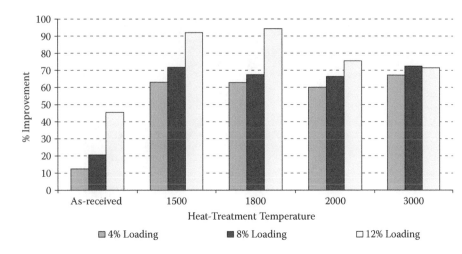

FIGURE 2.4
Nanocomposite flexural modulus percent improvement over neat resin as a function of nanofiber heat-treatment temperature.

to the central core axis. The edges of any pair of graphene layers have been rounded, encapsulating the carbon planes' exposed edge. This allows the exposed graphene planes to attain a level of maximum structural stability.

The improvements in mechanical properties are shown in Figure 2.3 and Figure 2.5. The flexural modulus increased with increasing nanofiber load. The maximum modulus of 3.5 GPa was achieved for the nanocomposite sample with 12 wt% loading of nanofibers heat treated to 1800°C. With increasing heat treatment temperature, the modulus for 12 wt% loading of

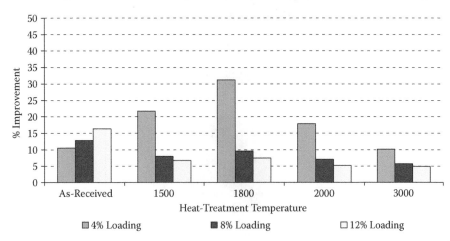

FIGURE 2.5
Nanocomposite flexural strength percent improvement over neat resin as a function of nanofiber heat-treatment temperature.

nanofibers heat treated to 2000°C and 3000°C decreased to 3.1 and 3.0 GPa, respectively. This indicated that by increasing the amount of graphitization, the adhesion between the carbon nanofiber and the epoxy resin decreased. The increased graphitization of the nanofibers may cause the previously truncated graphene layers to loop together, thereby eliminating free edges for bonding with the polymer matrix. However, the heat treatment of the carbon nanofibers showed an improvement over the pristine nanofibers. For the as-received PS nanofibers, the modulus at 12 wt% loading was 2.6 GPa. This increase in modulus may be due to the removal of "dirt"—polyaromatic hydrocarbons, sulfur, and oxygen and nitrogen-based functional groups— allowing for free ends of the graphene sheets to bind with the epoxy.

The ultimate strength of the nanocomposites decreased with increasing nanofiber load for the heat-treated nanofiber composites. Again, the nanofibers heat treated to 1800°C exhibited the highest strength values for the heat-treated samples at all nanofiber loading levels. Heat treatment had a positive effect on strength for low (4 wt%) nanofiber loading; however, at higher nanofiber loadings (8 and 12 wt%), heat treatment of nanofibers actually led to a lower strength than the pristine PS nanofiber composites. The minimal improvements in strength compared to that of modulus may be an indication that the chemical adhesion between the nanofiber and the epoxy matrix is not optimized. This can be improved through chemical functionalization of the nanofiber. In addition, the strength of the nanocomposite could be improved in one direction by alignment of the nanofibers in the matrix.

The improvement in electrical and thermal properties of the nanocomposites composed of heat-treated nanofibers was significant. This drastic improvement of thermal and electrical properties of the high heat-treated nanofiber-based composite over that of the nanofiber-based composite is due to the structural changes that occur during heat treatment. The alignment of the graphene layers within the nanofibers allows for a more efficient transfer of phonons and electrons.

The improvement in electrical resistivity is shown in Figure 2.6. Even though the addition of pristine PS nanofibers at 4 wt% loading decreased the resistivity of the nanocomposite by five orders of magnitude over the neat epoxy resin, heat treatment of the nanofibers to 1500°C reduced the resistivity by eight orders of magnitude, and heat treatment of the nanofibers to 3000°C decreased the resistivity by nine orders of magnitude. In addition, with increasing heat-treatment temperature, the electrical resistivity decreased from 440 Ω-cm at 4 wt% loading of nanofibers heat treated to 1500°C, to 94 Ω-cm at 4 wt% loading of nanofibers heat treated to 3000°C. Increasing the nanofiber loading further decreased the electrical resistivity. The electrical resistivity of the nanocomposite with 12 wt% loading of nanofibers heat treated to 1500°C was 6.5 Ω-cm, and the electrical resistivity of the nanocomposite with 12 wt% loading of nanofibers heat treated to 3000°C was 1.2 Ω-cm.

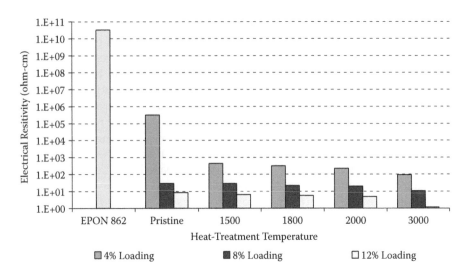

FIGURE 2.6
Nanocomposite resistivity as a function of nanofiber heat-treatment temperature.

In a similar manner, the thermal conductivity of neat epoxy resin increased significantly with the addition of nanofibers (Figure 2.7). The addition of as-received PS nanofibers at 4 and 12 wt% loading increased the thermal conductivity of the neat resin by 21% and 105%, respectively. The heat treatment of the nanofibers to 1500°C increased the thermal conductivity by 64% at 4 wt% loading and 194% at 12 wt% loading. Heat treatment of the nanofibers

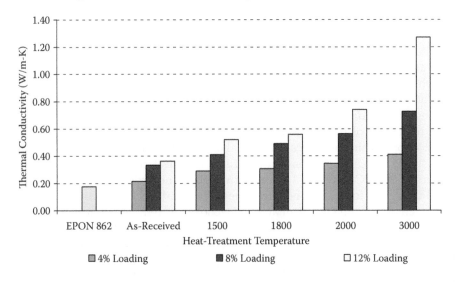

FIGURE 2.7
Nanocomposite thermal conductivity as a function of nanofiber heat-treatment temperature.

to 3000°C increased the thermal conductivity by 133% at 4 wt% loading and 6191% at 12 wt% loading. The degree of alignment and reduction in the inter-layer spacing between graphene layers led to a great increase in thermal conductivity and eventually to low electrical resistivity.

The heat treatment of carbon nanofibers led to the removal of impurities from the nanofiber and resulted in altered physical properties of a nanofiber reinforced epoxy composite. During heat treatment, the structure within the carbon nanofibers is altered from local molecular ordering to that of coalesced, flattened, graphene layers. Heat treatment up to 1800°C resulted in improved flexural modulus and strength of the nanocomposite. Additional heat treatment to higher temperatures led to increased conversion to graphene layers and resulted in lower mechanical properties due to poor adhesion to the epoxy matrix caused by the elimination of the free truncated edges of the graphene layers available to bond to the polymer matrix. The alignment of graphene layers within the nanofibers allowed for a more efficient transfer of phonons and electrons, resulting in significant decreases in electrical resistivity and increases in thermal conductivity of the neat epoxy resin. Heat treatment of the nanofibers allowed for the attainment of superior electrical and thermal properties at low fiber loadings that are not possible with pristine pyrolytically stripped carbon nanofibers.

The testing of heat-treated nanofiber-based nanocomposites demonstrated that the resulting nanocomposite properties must be compromised. If the end result of the nanocomposite is a desired high mechanical property, the nanofibers should not be heat treated above 1800°C. However, if the end result is a desired high electrical or thermal property, the nanofiber must be heat treated to 3000°C. At this higher temperature, while the graphene layers are aligned with minimal interlayer spacing to allow efficient electron and phonon transfer, the ends of the graphene planes tend to loop together and reduce the number of available sites for bonding with the resin. Therefore, a better method for improving interfacial adherence may be to chemically functionalize the nanofiber surface. This would impart groups on the surface that would bind to the resin without changing the structure of the nanofiber, resulting in reduced chemical bonding sites.

2.3 Nanofibers–Matrix Adhesion on Polymeric Nanocomposite Properties

Fiber functionalization is considered necessary to improve mechanical properties in nanofiber-reinforced composites by increasing the stress transfer between the nanofiber and the matrix of a nanocomposite structure. Fiber–matrix adhesion is governed by the chemical and physical interactions at the interface. Extensive literature exists on surface treatment of conventional carbon fibers by methods such as oxidation in gas and liquid phases and anodic etching. Poor fiber-matrix adhesion may result in composite failure at the interface, resulting in decreased longitudinal and transverse mechanical properties of the composite.

Surface modification of carbon nanofibers changes the graphitization extent of the fiber and increases the surface area of the fiber. Lim et al. studied "stacked coin"–type nanofibers and the impact of mechanical and chemical treatments on the morphology of the fibers [9]. Heat treatment at 2800°C induced closed-loop ends on the surface of the nanofibers formed by the folding of some planar hexagons at their edges. The heat treatment removed C–H bonds and densely stacked hexagonal layers of graphene, forming chemically active sites on the edges. The edges were stabilized by bonding to each other, even though the bonding caused high tension through the formation of a sharp curvature. Acidic oxidation of the nanofibers cut off the closed-loop ends, resulting in improved overall alignment of graphene layers. Both treatments generated many free edges and a high graphitization extent, indicating the possibility of improved interfacial bonding with a polymer matrix.

Toebes et al. [17] examined the effect of liquid phase oxidation of carbon nanofibers in nitric acid and mixtures of nitric and sulfuric acid for times up to 2 hours. The graphitic structure of the nanofibers was not altered by the treatments, but the texture of the fibers was significantly changed through an increase in the specific surface area and pore volume due to opening of the fiber inner tubes. The total oxygen content and surface oxygen functional groups were affected by the treatment time and acid type. Oxygen groups were also formed in the first 2 to 3 nm of the subsurface of the fibers.

Bubert et al. [18] investigated the influence of plasma treatment on the surface properties of carbon nanofibers by X-ray photoelectron spectroscopy (XPS) in combination with ion sputtering, acid–base titration, derivatization of carbonyl groups, pyrolysis, and CH analysis. The results indicated that the fiber surface is covered by a monomolecular oxygen-containing layer and that plasma treatment allows complete oxygen functionalization of the uppermost surface layer. XPS provides an average value for the content of functional groups of the first ten to fifteen molecular layers.

A number of results have been reported on the effect of chemical modification of carbon nanotubes on nanocomposite properties. Xu et al. included a nitric acid-oxidized nanofiber/vinyl ester composite in an examination of electrical properties of nanocomposites [19]. The resistivities of the oxidized fiber nanocomposites were much higher than those produced with untreated fibers. The oxidation was reported to have increased oxygen percentage by approximately 20% with the addition of anhydride, quinine, ether, and ester functional groups. The oxidized layer could reduce conductivity through percolation pathways. The functional groups also should improve wetting by the polymer matrix. A strong bond between matrix and fiber could encase the fiber and serve as an insulating covering.

Finegan et al. [13] examined the mechanical properties of carbon nanofiber/polypropylene composites in an attempt to optimize carbon nanofiber surface treatment. The fiber–matrix adhesion was qualitatively studied by scanning electron microscopy (SEM), and the strength and stiffness of the composites were evaluated from tensile tests. One sample of nanofibers was oxidized in air at 450°C, and a second sample was oxidized with carbon dioxide in a tube furnace. Fiber–matrix adhesion was improved by moderately oxidizing the fibers in either air or carbon dioxide. The carbon dioxide oxidation was more effective as it increased the external surface area and the surface energy of the fibers. However, in the preparation of the nanocomposites, the fibers were either ball-milled or force-sieved, thereby possibly altering the fiber aspect ratios resulting in decreased mechanical testing results.

Cortes et al. [20] exposed carbon nanofibers to a series of chemical treatments in nitric acid prior to mixing with polypropylene. The oxidized fiber nanocomposites did not improve electrical properties of the polymer, did not produce significant changes in the mechanical properties of the composites, and showed a decrease in tensile strength. The nanocomposites produced had only 5 wt% vapor grown carbon fiber (VGCF) compositions. Higher fiber content may have led to increases in the mechanical properties of the composites.

In conventional carbon fiber reinforced composites, there have been a number of studies completed to generate strong adhesion between the fiber surface and matrix to improve stress transfer from the matrix to the reinforcing fibers. Continuous surface electrochemical oxidation has been the preferred method of fiber surface treatment to enhance interfacial bonding. Electrochemical treatments have been carried out in acid and alkaline aqueous solutions of ammonium sulfate, ammonium bicarbonate, sodium hydroxide, diammonium hydrogen phosphate, and nitric acid.

Anodic oxidation of fibers in electrolytes can produce a variety of chemical and physical changes in the fiber surface. Harvey et al. [21] examined surfaces of conventional carbon fibers by XPS after electrochemical treatment by galvanostatic and potentiostatic cell control under varying potential, reaction time, and electrolytes. They noted that the rise in interlaminar shear strength (ILSS) with surface treatment is not dependent on O-1s:C-1s ratios or the amount of carboxyl functionality on the surface, thereby supporting the view that mechanical keying of the resin to the fiber surface plays an important role in forming the resin–fiber bond.

Gulyas et al. [22] subjected polyacrylonitrile (PAN)-based carbon fibers to electrochemical oxidation under a wide variety of conditions, including varying electrolyte, electrolyte concentration, and applied voltage. The functional groups formed on the surface of the fibers were dependent on the type of electrolyte used, and the number of functional groups found on the fiber surface depended on electrolyte concentration and voltage. A close correlation was found between surface chemistry and fiber–matrix adhesion. The concentration of certain functional groups could be quantitatively related to ILSS.

Yue et al. [23] applied continuous electrochemical oxidation to high-strength PAN-based carbon fibers in 1% with weight potassium nitrate. Fiber weight loss increased with electrochemical oxidation. A large internal microporous surface area was generated due to the formation of acidic functions. XPS indicated that the concentration of oxygen within the outer 50 Å of the fibers increased on oxidation. XPS C-1s and O-1s spectra showed an increase in primarily carboxyl or lactone groups. The oxygen-rich surfaces in the microporous regions chemisorbed oxygen and water. The O-2s to C-2s peak separation increased in the valence band spectra as the extent of oxidation increased due to carbonyl group contribution.

In this study, carbon nanofibers were subjected to electrochemical oxidation in $0.1M$ nitric acid for varying times to modify the interface between the nanofibers and epoxide molecules in epoxide/nanofiber composites [24]. X-ray photoelectron spectroscopy was employed to characterize surfaces with regard to the content of carbon, oxygen, and nitrogen. The strength of adhesion between the fiber and an epoxy (thermoset) matrix was characterized by the tensile strength and modulus, and the electrical and thermal properties of the composites were investigated from the viewpoint of surface treatment of carbon nanofibers.

The nanofibers were electrochemically surface treated using nitric acid as an electrolyte in a concentration of 1 wt%. Approximately 15 g of nanofibers for each trial were packed into covered, porous plastic beakers and submerged in the acidic solution. Previous studies have shown that an amount of 12% in weight added to epoxy polymer led to maximum mechanical performance. We chose a graphite electrode submerged into the packed nanofibers, and a specific applied current was set at 0.1 Amps. The time of the treatments was 30 sec, 1 min, 2 min, 4 min, 8 min, and 15 min. Following treatment, the oxidized fibers were washed with distilled water until attaining a neutral pH and were dried in a vacuum oven at 100°C for 48 h.

The functionalized carbon nanofibers were characterized using X-ray photoelectron spectroscopy (XPS). The samples were oriented such that the axial direction was in the plane of the X-ray source and the analyzer detection slit. During all XPS experiments, the pressure inside the vacuum system was maintained at approximately Torr. A nonlinear least squares curve fitting program with a Gaussian–Lorentzian mix function and background subtraction was used to deconvolve the XPS peaks.

The carbon nanofibers were added to an epoxy resin matrix (EPON™ 862) forming a nanocomposite. Because bulk nanofibers are difficult to incorporate into resin matrices, the nanofibers were first dispersed in the epoxy resin. The resulting mixture was cured under pressure in a silicone mold.

Surface characterization of the nanofibers by XPS (Table 2.1) showed an increase in oxygen content from 2% in the as-received fibers to 8.6% in the fibers treated for 15 min. The N 1s peak was negligible in all cases. The increase in relative concentration of carbon oxygen complexes occurs because the outer layers of the fibers become increasingly porous. The fraction of carbon atoms

TABLE 2.1

Atomic Percentages of Nanofibers versus Electrochemical Treatment Time

Sample	Percent (%) Oxygen	Percent (%) Carbon
As-received	2.0	98.0
30-sec treatment	2.3	97.7
1-min treatment	2.8	97.2
2-min treatment	3.5	96.5
4-min treatment	4.2	95.8
8-min treatment	5.4	94.6
10-min treatment	7.6	92.4
12-min treatment	8.3	91.7
15-min treatment	8.6	91.4

TABLE 2.2

Nanocomposite Mechanical Properties with Relative Standard Deviation

Sample	Surface Treatment Time (min)	Mass %	Modulus (GPa)	STD	Strength (GPa)	STD
EPON 862		0	1.7730	0.0688	0.0976	0.0018
PR-24-PS	0	12	2.5783	0.1267	0.1196	0.0057
PR-24-PS-ET (0.5)	0.5	12	2.6898	0.5464	0.0901	0.0143
PR-24-PS-ET (1)	1	12	2.7399	0.1619	0.1037	0.0076
PR-24-PS-ET (2)	2	12	2.8014	0.1124	0.1072	0.0150
PR-24-PS-ET (4)	4	12	3.0261	0.3434	0.1017	0.0071
PR-24-PS-ET (8)	8	12	3.4915	0.0217	0.1163	0.0093
PR-24-PS-ET (10)	10	12	3.9535	0.0118	0.1263	0.0087
PR-24-PS-ET (12)	12	12	4.4856	0.0123	0.1462	0.0068
PR-24-PS-ET (15)	15	12	3.0690	0.3558	0.1041	0.0099

in the region which exist on the pore surfaces increases. These carbon atoms are the sites of oxidation, thereby increasing the relative amount of oxygen.

As shown in Table 2.2, a modest degree of electrochemical treatment improved the flexural modulus of the nanocomposite by enhancing the interfacial adhesion between the fiber and resin. The treatment time of 30 sec allowed for a 4.32% increase in flexural modulus. A maximum improvement of 74% occurred with an electrochemical treatment time of 12 min. Prolonged subjection to the electrochemical treatment beyond 8 min showed a drop in flexural modulus. This may be due to the oxide layer actually causing failure of the fiber–resin bond or possibly the additional treatment time may have damaged the fiber surface.

There was no significant change in the thermal conductivity results (Table 2.3). In addition, the electrical resistivity of the nanocomposites decreased with increasing electrochemical treatment time (Table 2.3). This

TABLE 2.3

Thermal and Electrical Properties of Nanocomposites as a Function of
Treatment Time

Sample	Surface Treatment Time (min)	Mass %	Resistivity (Ohm-cm)	K (W/m-K)	Density (g/cm³)
EPON 862		0	3.28E + 10	0.1768	1.198
PR-24-PS	0	12	1.58E + 00	0.3731	1.252
PR-24-PS-ET (0.5)	0.5	12	4.13E + 00	0.2992	1.237
PR-24-PS-ET (1)	1	12	5.16E + 01	0.3476	1.253
PR-24-PS-ET (2)	2	12	3.14E + 03	0.3303	1.246
PR-24-PS-ET (4)	4	12	5.57E + 06	0.3032	1.240
PR-24-PS-ET (8)	8	12	3.35E + 07	0.3290	1.235
PR-24-PS-ET (10)	10	12	5.56E + 10	0.3334	1.257
PR-24-PS-ET (12)	12	12	4.54E + 09	0.3493	1.234
PR-24-PS-ET (15)	15	12	3.69E + 10	0.3348	1.254

may be due to the increase in oxygen content causing the formation of a narrow insulating layer along the surface of the fibers.

The SEM examination of the fracture surface of various samples following mechanical testing has shown that at the interface between the nanofiber and epoxy resin, the matrix has only minimal adhesion in the form of covalent bonds (Figure 2.8 and Figure 2.9). An interfacial gap between the two phases is always present, and its location may vary from one sample to another (single arrows in Figure 2.8 and Figure 2.9).

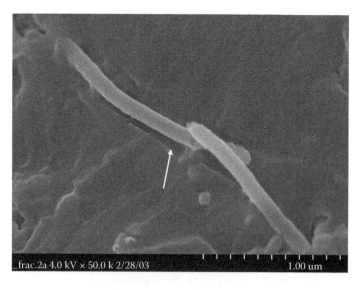

FIGURE 2.8
Untreated nanofibers–based nanocomposite fracture surface.

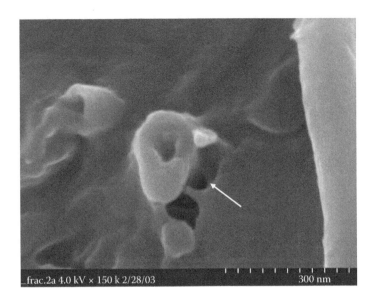

frac.2a 4.0 kV × 150 k 2/28/03   300 nm

FIGURE 2.9
Untreated nanofibers–based nanocomposite fracture surface.

The inability to create an effective interface between the carbon nanofiber and the neat resin prevents the transfer of mechanical loads between the two regions. This hindrance of load transference results in mechanical properties that are nominally better than those of the neat resin. To better understand the role of surface chemistry of the nanofiber, the effect of surface functionalization of the nanofiber on the mechanical properties of produced nanocomposites was studied.

The fracture surface of the surface-functionalized nanofiber-based nanocomposite is markedly different from that of a nonsurface-functionalized one. Along the length of the chemically modified nanofiber within the epoxy matrix, there is an apparent interface devoid of the gaps present in the pyrolytically stripped nanocomposite (Figure 2.10). The body of the nanofiber is actually covered with lighter regions of "strings" and bumps arranged in the form of "knife teeth." These bumps are made of resin residue. This region is more closely studied using high-resolution SEM and TEM techniques.

In Figure 2.11, it is shown that the surface of the functionalized nanocomposite is covered with masses of residual epoxy resin. This demonstrates that the nanocomposite did not fracture along the nanofiber/resin interface. Rather, the resin remained adhered to the nanofiber following fracture, and the mechanical failure took place within the resin phase of the nanocomposite. TEM analysis offers a more thorough explanation for the interaction between the modified surface of the nanofiber and the epoxy resin.

The black arrows in Figure 2.12 show the exposed ends of nanofibers and bodies of nanofibers within the resin matrix at the fracture site. The bright

FIGURE 2.10
Scanning electron microscopy (SEM) micrograph of a surface-functionalized nanocomposite fracture surface.

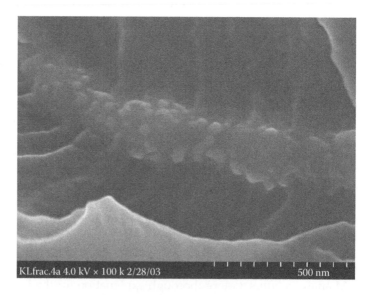

FIGURE 2.11
Scanning electron microscopy (SEM) micrograph of a surface-functionalized nanocomposite fracture surface.

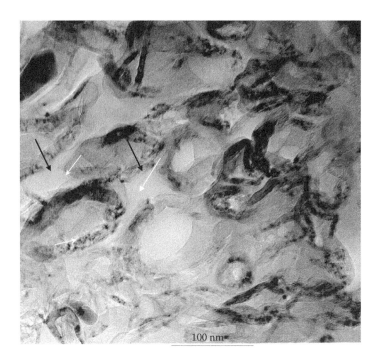

FIGURE 2.12
Transmission electron microscopy (TEM) micrograph of a surface-functionalized transverse fracture surface.

field micrograph shows a change in a gray scale to demonstrate differences in electronic density within the resin matrix. The epoxy resin areas do not have one consistent shade, however. The single white arrows in Figure 2.12 point to areas of higher electronic density than those of areas showing lighter contrast (black arrows in Figure 2.12). This difference in a gray scale may indicate some structural changes within the continuous resin matrix. A longitudinal representation offers a clearer understanding of the physical property gradient within the surface-functionalized carbon nanocomposite.

An analysis using high-resolution TEM (HR-TEM) allows for a more complete understanding of the changes that occur at the interface between the surface-functionalized carbon nanofibers and the continuous epoxy resin phase.

At high resolution, the individual graphene planes of a nanofiber can be seen along with the region of the epoxy region. Moving from the upper-left toward the lower-right of the micrograph, the interface with the nanofiber is crossed into the parallel layers within the nanofiber representing the wall of nanofiber. Further down the micrograph, another interface is crossed into the epoxy resin phase with very organized turbostratic carbon (circled areas in Figure 2.13 with a wavy, spaghetti-like look). In this region, there is a local molecular orientation of basic structural units in the form of columns and clusters with discontinuous but preferential molecular orientation parallel to

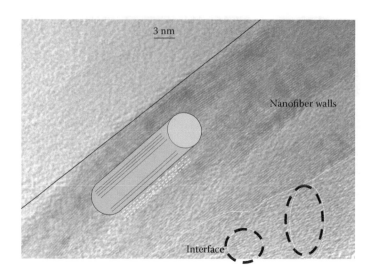

FIGURE 2.13
High-resolution transmission electron microscopy (TEM) showing an oriented interface between nanofiber and matrix.

the nanofiber axis. The polymer interphase is no longer amorphous but has gained a two-dimensional order similar to the microstructure of a carbonized PAN carbon fiber. Epoxy generally has a gravel-like appearance under HR-TEM, which indicates an amorphous structure. This change in physical properties demonstrates the presence of an "interphase." Due to the chemical interaction between the surface-functionalized nanofiber and the resin, a new material has been created with unique physical properties.

In principle, achieving high tensile strength composites with nanofibers as the reinforcement relies on factors including weight fraction, strength of the nanofibers, dispersion of the fibers, and strength of the interface. Assembly of these factors may be envisioned without difficulty; however, the latter factor—strength of the interfacial bond between nanofiber and matrix—is problematic due to the inert nature of the smooth hexagonal surface commonly presented by the nanofiber class of reinforcement.

Due to the exposed graphitic edge planes of nanofiber surfaces, it is possible to add functional groups. The addition of functional groups will greatly increase the ability of the nanofiber to bond to polar matrix materials such as epoxy. As a result of this surface modification, the mechanical properties of the nanocomposites were significantly enhanced, and the chemical properties such as heat capacity of the polymer were lowered. This improvement might be obtained by an internal reaction between the surface functional groups and polymer matrix. Transmission electron microscopy characterization shows that the postsynthesis surface treatment has contributed to the formation of a very dense and oriented interface between nanofiber and matrix.

The surface-treated nanofiber-based nanocomposites samples exhibit better mechanical properties than any of the nonsurface-treated nanofiber-based nanocomposites. The improvement in mechanical properties is due to the formation of gradients at interfaces and interphases between the nanofibers and epoxy resins. These interphases cause the polymeric nanocomposites to behave as a continuous phase in which the mechanical transport properties between parent individual ingredients (nanofibers and epoxy matrix) were enhanced.

There was a significant decrease in the relative content of graphitic carbon and an increase in the relative content of carbon bonded to oxygen containing functions with an increasing amount of electrochemical treatment. The increase in relative concentration of carbon oxygen complexes occurs because the outer layers of the fibers become increasingly porous, allowing for additional sites of oxidation. The increasing amounts of oxygen with electrochemical treatment corresponded to an increase in the flexural modulus of nanocomposites manufactured with the treated fibers. This indicates an improvement in interfacial adhesion between the fibers and the resin. The treated nanofibers had a negative impact on the heat capacity of the nanocomposites but did not affect the thermal diffusivity of the composites. The electrical resistivity of the nanocomposites decreased with increasing electrochemical treatment time due in part to the increased concentration of oxygen functional groups.

2.4 Nanometric Analysis

Although there are obstacles to directly measuring the mechanical properties of nanostructures due to their very small size, transmission electron microscopy coupled with electron energy loss spectra (EELS) can become an analytical tool for nanoscale measurements. The observed peaks in the plasmon energy spectrum are related to the intrinsic property of the material [25–28]. In this chapter, applications of plasmon spectroscopy are described to determine some mechanical properties of materials such as flexural modulus at the nanometer level. It is challenging to obtain such properties for interfaces and interphases regions.

EELS has a resolution of 1 to 10 nm and can be viewed as an analysis of the energy distribution of electrons that have interacted inelastically with the specimen.

Plasmons are longitudinal, wave-like oscillations of weakly bound electrons. The plasmon peak is the second-most dominant feature of energy loss spectrum after the zero loss peak. The energy lost by a beam electron when it generates a plasmon is equal to the Ep value. Typical Ep values are between 5 and 25 eV. Measurement of the Ep can give indirect microanalytical information. Plasmon-loss electrons are strongly forward scattered and interact with the electrons in the interior of the specimen. EELS relates the observed peaks in the plasmon energy spectrum to the intrinsic property of the material.

2.4.1 Analytical Approach in Plasmon Energy– Mechanical Property Correlations [25]

The volume plasmon energy, Ep, varies with the valence electron density, n, of a material due to single-electron excitations [30,31] approximately as

$$Ep = [(hw_p^f)^2 + E_g^2]^{0.5}$$

where $w_p^f = [ne^2 / (\varepsilon_o m)]^{0.5}$ is the free electron plasma frequency, E_g is the bandgap energy, e is the electron charge, $\varepsilon_o m$ is the permittivity of vacuum, and m is the electron mass.

Ep is influenced by a material's composition, bonding, and band structure. Although low-energy interband transitions can decrease the plasma frequency by increasing the effective mass of valence electrons, for all semiconductors and some insulators, Eg2 << Ep2, so the "free-electron" approach is often a good approximation. Ym also increases with the electron density as described by the equation for Ep above, because the bulk modulus, which is closely related to the elastic modulus, is proportional to the square root of n [26]. After considering a set of assumptions, the elastic moduli would be the following:

$$Ym = 25.32(1 - 2v) \times \left(\frac{me}{h^2}\right)^2 E_p^2$$

where v is the Poisson's ratio.

These moduli are similarly related to the square of the plasmon peak energy.

2.5 Experimental Data on Interphase Regions

For polyaromatic solids with sp2 hybridization, the low loss area exhibits two main plasmon peaks, which correspond to the excitation of the π and $\pi + \sigma$ electrons. The results presented here focus specifically on the Ep value for the bulk plasmon peak—that is, the $\pi + \sigma$ electron-related peak (Figure 2.14).

For this current study, only the value of plasmon peaks, which corresponds to the excitation $\pi + \sigma$ electrons, was reported. Figure 2.12 and Figure 2.13

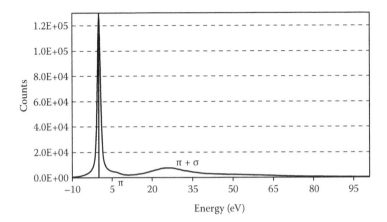

FIGURE 2.14
Electron energy loss spectra (EELS) characterization of a polyaromatic solid.

show that there is a change in the structural gradient moving away from the interface–interphase regions, and it vanishes when far from the nanofiber surface. The EELS techniques were carried out at various locations within a 200 nm distance from the nanofibers' surface with an Ep measurement every 20 nm. For each location, on average, 12 measurements were recorded and plotted (Figure 2.15). It was observed that the Ep value decreases with increasing distance from the surface of the nanofibers. Taking into account the above equations, we find that the mechanical properties (i.e., the bulk, elastic, and shear moduli of the interface and interphase regions) are much higher than the original polymer properties.

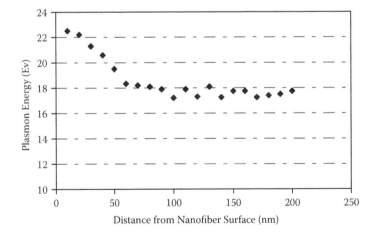

FIGURE 2.15
Interfacial electron energy loss spectra (EELS) analysis of a surface-functionalized nanocomposite.

2.5.1 Nanocomposite Flow Behavior

Numerical models were established to predict the full potential of novel nanocomposites [32,33]. From a fluid dynamics point of view, nanocomposites have a distinctive characteristic—nanometer-sized particles are involved. Nanoparticles can create a pressure drop when flowing through the passages [34]. The possibility of clogging micrometric channels remains a challenge. For instance, this issue is magnified in the case of resin transfer molding (RTM) and other composite processes because of a high resin viscosity and a low preform permeability. Solving such problems is also considered a challenge, because the simulation and prediction of the characteristics of multiphase flows is very complex [35,36]. The continuous phase (fluid) and the dispersed phase (solid) can be described by a Eulerian model. In the case of a RTM scenario, a two-dimensional simulation model based on the Eulerian multiphase approach for the flow of nanocomposites around a carbon microfiber matrix is introduced to investigate and predict the flow characteristics.

A nanocomposite is infiltrated into a two-dimensional carbon microfiber matrix as shown in Figure 2.16a. A unit cell is taken as a representative unit cell for the system as shown in Figure 2.16b, which illustrates that one-half of each side-by-side microfiber is only taken for symmetrical consideration, and the third microfiber, which is in a staggered orientation with the other two microfibers, is completely considered.

The derivation of the conservation equations is performed by using the mixture theory approach [37–42].

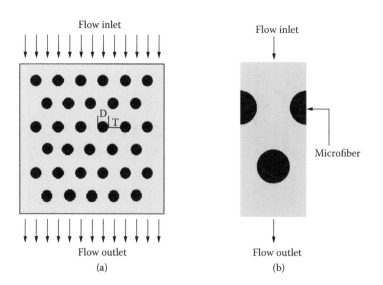

FIGURE 2.16
(a) Carbon microfibers matrix configuration. (b) Model's unit cell.

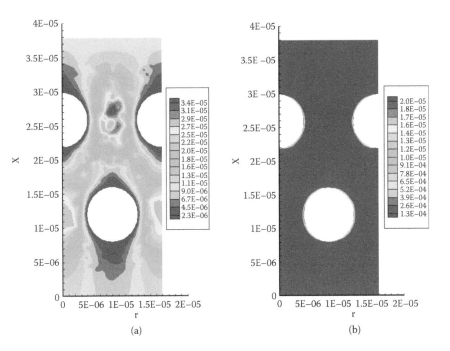

FIGURE 2.17
(a) Cell Reynolds number. (b) Wall's shear stress contours for neat resin, respectively.

The predicted cell Reynolds number contours for the flow of neat resin are presented in Figure 2.17a. The figure shows that the cell Reynolds number has minimum values around all microfiber walls as a result of high friction. The friction decreases gradually toward the center of the cell at the radial direction between the side-by-side microfibers, where microfiber walls–fluid interaction diminished, while this friction decreases gradually around the third microfiber. Similarly, the shear stress is much higher at the microfiber walls–fluid interface (Figure 2.17b). The wall shear stress was estimated with a thickness of 0.6 μm.

As shown in Figure 2.18, the cell Reynolds number decreases as the carbon particle volume load ratio increases in the resin. This trend could be explained by an increase in nanocomposite effective viscosity. Similarly, the magnitude of the wall shear stresses increases by increasing the particle load ratio (Figure 2.19). The relation between the load ratio and the maximum shear stress around the microfibers for all cases is illustrated in Figure 2.20, which shows a quadratic relation between them.

As was expected during the flow process, the interfacial fluid layers around the microfiber walls exhibit high friction. This friction tends to reduce the flow velocity, which may cause it to stick on the microfiber walls, and after some time, perhaps the flow passages will be blocked. To prevent any potential sticking of the adjacent fluid layers on the microfiber walls during the

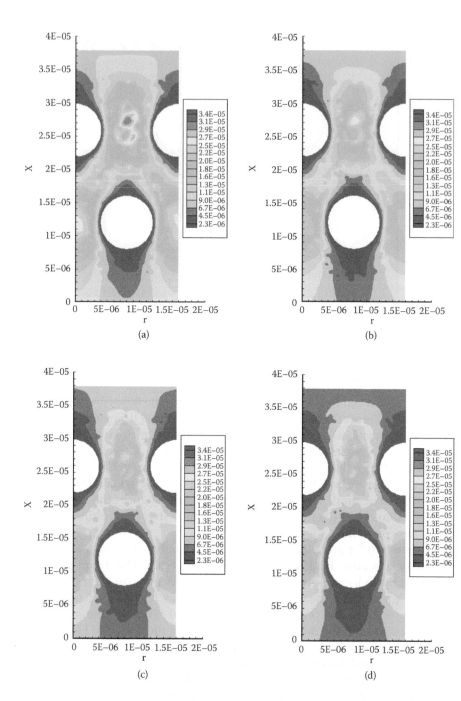

FIGURE 2.18
Cell Reynolds number contours for volume ratios: (a) 0.5%, (b) 1%, (c) 1.5%, and (d) 2.0%.

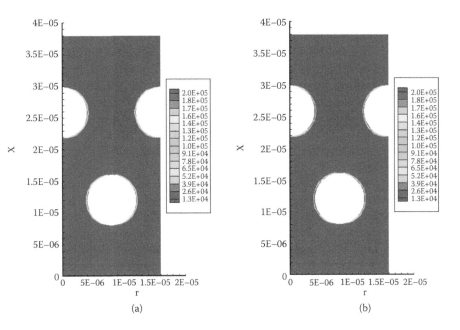

FIGURE 2.19
Wall shear stress contours for nanoparticle volume ratios: (a) 0.5% and (b) 2%.

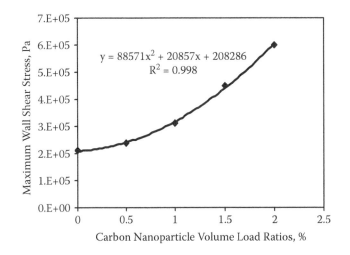

$$y = 88571x^2 + 20857x + 208286$$
$$R^2 = 0.998$$

FIGURE 2.20
The relation between volume ratio and maximum walls shear stress.

flow process, an energy imbalance technique could be created between the carbon microfiber walls and the fluid flow. The temperature-induced surface stress will contribute to thermocapillary flow within the liquid layer [43]. These thermocapillary effects give rise to various gravity-independent phenomena, including convective flows, interface distortions, as well as interface rupture. Thermocapillary flows are driven by the imbalance of tangential stress on the interface caused by temperature dependence of surface tension [44]. This tangential stress is called Marangoni shear stress. It is a combination of a surface tension and a temperature gradient, which offers a wealth of possible responses.

After introducing several combinations of the surface tension temperature coefficient and the temperature difference between fluid flow and carbon microfiber walls, it has been found that the interfacial fluid layers around the microfibers resulted in higher velocities when these two parameters were 0.4 N/m.K and 70°C, respectively [45–47]. These higher velocities allow the interfacial fluid layers the potential to flow more smoothly around the microfiber walls, relative to their motion, before applying the energy imbalance technique. Thus, the likelihood of sticking is less presumed.

The predicted cell Reynolds number contours for the flow with different volume load ratios after introducing the energy imbalance technique are shown in Figure 2.21.

In general, one can see that applying the proposed energy imbalance technique creates convective currents around the carbon microfiber walls, which causes vortices around them. These vortices force the flow to move away from the carbon microfiber walls. From the figures, we can also see that these vortices are stronger in the case of low loading ratios of the carbon nanoparticles in the resin, which means that in the case of higher load ratios, higher temperature differences may be needed [45].

Carbon nanofibers have been successfully integrated with an epoxy resin matrix to improve the physical properties of the neat resin. A successful integration of two independent phases is imperative to the effective translation of superior properties of the carbon nanofiber filler constituent into a physically superior carbon nanocomposite. Through the creation of a superior resin matrix, the foundation of the next generation of advanced composites has been laid with the introduction of the first carbon nanocomposites.

The properties of the next generation of advanced composites can be tailored according to degree and manner of improvement through the introduction of specific carbon nanofibers. Nanofibers offer improvements to the thermal, electrical, and mechanical properties of neat epoxy resin. In efforts to maximize the thermal and electrical conductivity of the constituent nanofibers, they may be heat treated. Following heat treatment, the graphene planes within the nanofiber align, allowing a higher degree of order. Increasing order provides an easier path for transmission of thermal and electrical energy translating into increased thermal diffusivity and conductivity.

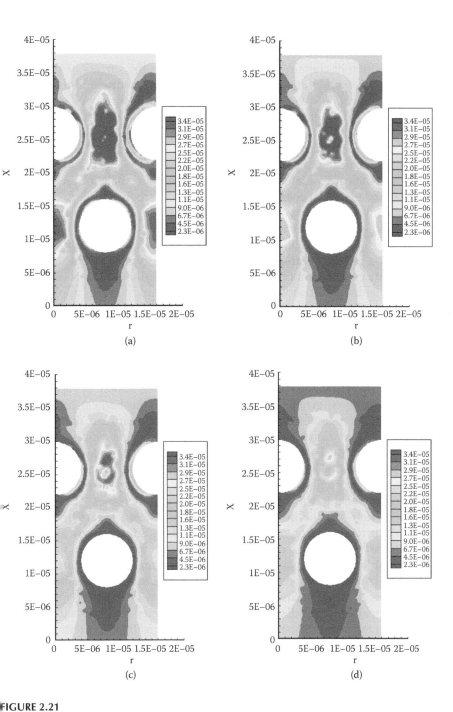

FIGURE 2.21
Cell Reynolds number contours after applying the energy imbalance technique for nanoparticle volume ratios: (a) 0.5%, (b) 1%, (c) 1.5%, and (d) 2%.

Surface functionalization of carbon nanofibers adversely affects their thermal and electrical properties.

The addition of chemical functional groups on the surface of the carbon nanofiber allows them to physically and chemically adhere to the continuous resin matrix. The chemical interaction with the continuous epoxy resin matrix results in the creation of a third independent phase. This region within the nanocomposite has been referred to as the "interphase." This new, novel phase must be studied in more depth to better understand its properties. Through chemical and physical bonding, the surface-functionalized nanofibers become encapsulated, thereby preventing the efficient transference of thermal and electrical energy between adjacent carbon nanofibers. The ability to chemically and physically interact with the epoxy region is beneficial to the mechanical properties of a carbon nanocomposite.

The primary focus behind most advanced composites is the ability to maximize a material's strength while decreasing its weight. The addition of low-density carbon nanofibers to epoxy resin has allowed for its mechanical improvement without adding weight. As a mechanical load is applied to the nanocomposite, the pressure is felt by the continuous resin matrix. If there is a high degree of interaction between the two phases, the load is then transferred to the discontinuous nanofiber phase possessing excellent mechanical properties. This effective transfer between the two phases is seen in the surface-functionalized nanocomposites and to a lesser extent in the pyrolytically stripped nanocomposites. The heat-treated nanofibers are stripped of their functional groups, and even though the mechanical properties of the constituent nanofibers are improved with heat treatment, their ability to adhere to the continuous resin phase is restricted. This causes mechanical performance of the heat-treated nanofiber composites to be inferior to that of either pyrolytically stripped or surface-functionalized nanocomposites. As the amount of nanofiber added as the nanofiber loading rate is increased, the performance of the resultant nanocomposite also increases. This is seen in thermal, electrical, and mechanical properties. There appears to be a point at which further addition of nanofibers provides minimal physical improvement to the neat resin and causes an adverse effect to the nanocomposite due to the inability to create a homogeneous nanocomposite. Through the use of nondestructive methods, it may be concluded that the obtained correlations between the experimental and calculated average elastic moduli and plasmon peak energy indicate that plasmon energies can potentially be used to determine the local elastic moduli of technologically important materials. Further exploration of the "interphase" will provide information pertaining to the resin's chemical reactivity to certain chemical functional groups, allowing for the creation of more predictable nanocomposite properties.

The flow characteristics of carbon nanocomposite around a staggered carbon microfiber matrix have been investigated. Observations of the incident illustrate interactions between the microfiber sidewalls and the interfacial fluid layers. These interactions have a tendency to reduce the flow velocity, causing an attraction of the flow to the microfiber sidewalls. After some time, the flow passages will be blocked. To circumvent this tendency, an energy imbalance technique was established between the nanocomposite flow and microfiber walls. As a result, convective currents around the carbon microfiber walls have been produced causing vortices around them. The forces from these vortices cause the flow to move away from the carbon microfiber walls, preventing the interfacial fluid layers from sticking on the microfiber walls [45–47].

References

1. Sittig, Marshall, ed., *Carbon and Graphite Fibers,* Noyes Data Corporation, 1980.
2. Strong, A. Brent, *Fundamentals of Composites Manufacturing: Materials, Methods, and Applications,* Society of Manufacturing Engineers, Michigan, 1989.
3. Iijima, S., "Helical Microtubules of Graphitic Carbon," *Nature,* 354 (7), pp. 56–58, 1991.
4. Saito, R. et al., *Physical Properties of Carbon Nanotubes,* Imperial College Press, Singapore, 1998.
5. Lafdi, Khalid, "Carbon Nanotechnology," Winter 2003, University of Dayton, Ohio.
6. Callister, W.D. Jr., *Materials Science and Engineering: An Introduction,* John Wiley & Sons, New York, 2000.
7. Thostenson, E. et al., "Advances in the Science and Technology of Carbon Nanotubes and Their Composites: A Review," *Composites Science and Technology,* 61, pp. 1899–1912, 2001.
8. Endo, M., Kim, Y.A., Hayashi, T., Yanagisawa, T., Muramatsu, H., Ezaka, M., Terrones, H., Terrones, M., and Dresselhaus, M.S., "Microstructural Changes Induced in Stacked-Cup Carbon Nanofibers by Heat Treatment." *Carbon,* 41, pp. 1941–1947, 2003.
9. Lim, S., Yoon, S.-H., Mochida, I., and Chi, J.-H., "Surface Modification of Carbon Nanofiber with High Degree of Graphitization," *Journal of Physical Chemistry B,* 108, pp. 1533–1536, 2004.
10. Katayama, T., Araki, H., and Yoshino, K., "Multiwalled Carbon Nanotubes with Bamboo-Like Structure and Effects of Heat Treatment," *Journal of Applied Physics,* 91 (10), pp. 6675–6678, 2002.
11. Kiselev, N.A., Sloan, J., Zakharov, D.N., Kukovitskii, E.F., Hutchinson, J.L., Hammer, J., and Kotosonov, A.S., "Carbon Nanotubes from Polyethylene Precursors: Structure and Structural Changes Caused by Thermal and Chemical Treatment Revealed by HREM," *Carbon,* 36, pp. 1149–1157, 1998.
12. Xu, J., Donohoe, J.P., and Pittman, Jr., C.U., "Preparation, Electrical and Mechanical Properties of Vapor Grown Carbon Fiber (VGCF)/Vinyl Ester Composites," *Composites Part A,* 35, pp. 693–701, 2004.
13. Finegan, I.C., Tibbetts, G.G., Glasgow, D.G., Ting, J.-M., and Lake, M.L., "Surface Treatments for Improving the Mechanical Properties of Carbon Nanofiber/Thermoplastic Composites," *Journal of Materials Science,* 38, pp. 3485–3490, 2003.

14. Kuriger, R.J., Alam, M.K., Anderson, D.P., and Jacobsen, R.L., "Processing and Characterization of Aligned Vapor Grown Carbon Fiber Reinforced Polypropylene," *Composites Part A*, 33, pp. 53–62, 2002.

15. Ma, H., Zeng, J., Realff, M.L., Kumar, S., and Schiraldi, D.A., "Processing, Structure, and Properties of Fibers from Polyester/Carbon Nanofiber Composites," *Composites Science and Technology*, 63, pp. 1617–1628, 2003.

16. Matzek, M.D., "Polymeric Carbon Nanocomposites: Physical Properties and Osteoblast Adhesion Studies." Master's thesis, University of Dayton, Ohio, 2004.

17. Toebes, Marjolein L., et al., "Impact of the Structure and Reactivity of Nickel Particles on the Catalytic Growth of Carbon Nanofibers," *Catalysis Today*, 76, pp. 33–42, 2002.

18. Bubert, H., Ai, X., Haiber, S., Heintze, M., Brueser, V., Pasch, E., Brandl, W., and Marginean, G., "Basic Analytical Investigation of Plasma-Chemically Modified Carbon Fibers," *Spectrochimica Acta Part B*, 57, pp. 1601–1610, 2002.

19. Xu, J., Donohoe, J.P., and Pittman, Jr., C.U., "Preparation, Electrical and Mechanical Properties of Vapor Grown Carbon Fiber (VGCF)/Vinyl Ester Composites," *Composites Part A*, 35, pp. 693–701, 2004.

20. Cortes, P., Lozano, K., Barrera, E.V., and Bonilla-Rios, J., "Effects of Nanofiber Treatments on the Properties of Vapor-Grown Carbon Fiber Reinforced Polymer Composites," *Journal of Applied Polymer Science*, 89, pp. 2527–2534, 2003.

21. Harvey, J., Kozlowski, C., and Sherwood, P.M.A., "X-ray Photoelectron Spectroscopic Studies of Carbon Fibre Surfaces," *Journal of Materials Science*, 22, pp. 1585–1596, 1987.

22. Gulyas, J., Foldes, E., Lazar, A., and Pukanszky, B., "Electrochemical Oxidation of Carbon Fibres: Surface Chemistry and Adhesion," *Composites Part A*, 32, pp. 353–360, 2001.

23. Yue, Z.R., Jiang, W., Wang, L., Gardner, S.D., and Pittman, Jr., C.U., "Surface Characterization of Electrochemically Oxidized Carbon Fibers," *Carbon*, 37, pp. 1785–1796, 1999.

24. Lafdi, K., Fox, W., Matzek, M., and Yildiz, E., "Effect of Carbon Nanofiber-Matrix Adhesion on Polymeric Nanocomposite Properties (Part II)," *Journal of Nanomaterials*, 2008.

25. Oleshko, V. et al., "Electron Microscopy and Scanning Microanalysis," *Encyclopedia of Analytical Chemistry*, Wiley & Sons, Chichester, UK, pp. 9088–9120, 2000.

26. Gilman, J.J., "Plasmons at Shock Fronts," *Phil Mag B*, 79, pp. 643–654, 1999.

27. Thomas, G., *Transmission Electron Microscopy of Metals*, Wiley, New York, 1962.

28. Thomas, G., "The Impact of Electron Microscopy on Materials Research. Microstructural Design and Tailoring of Advanced Materials." In *Impact of Electron and Scanning Probe Microscopy on Materials Research*, Rickerby, D.G., Valdre, G., and Valdre, U., Eds., pp. 1–40. Wiley, New York, 1998.

29. Thomas, G., and Goringe, M.J., *Transmission Electron Microscopy of Materials*, Wiley, New York, 1979.

30. Egerton, R.F., *Electron Energy Loss Spectroscopy in the Electron Microscope*, 2nd edition, Plenum Press, New York, 1996.

31. Raether, H., *Excitation of Plasmons and Interband Transitions by Electrons*, *Springer Tracts in Modern Physics*, 88, Springer-Verlag, Berlin, 1980.

32. Yang, R., and Chen, G., "Thermal Conductivity Modeling of Periodic Two-Dimensional Nanocomposites," *Physical Review B*, 69, 195316, pp. 1–10, 2004.

33. Valavala, K., and Odegard, M., "Modeling Techniques for Determination of Mechanical Properties of Polymer Nanocomposites," *Reviews on Advanced Materials Science*, 9, pp. 34–44, 2005.
34. Khanafer, K., Vafai, K., and Lightstone, M., "Buoyancy-Driven Heat Transfer Enhancement in a Two-Dimensional Enclosure Utilizing Nanofluids," *International Journal of Heat and Mass Transfer*, 46, pp. 3639–3653, 2003.
35. Kleinstreuer, C., *Two-Phase Flow*, Taylor & Frances, New York, 2003.
36. Wachem, B., and Almstedt, A., "Methods for Multiphase Computational Fluid Dynamics," *Chemical Engineering Journal*, 96, pp. 81–98, 2003.
37. Bowen, M., *Theory of Mixtures*, Academic Press, New York, 1976.
38. Gidaspow, D., *Multiphase Flow and Fluidization: Continuum and Kinetic Theory Descriptions*, Academic Press, San Diego, CA, 1994.
39. Ding, J., and Gidaspow, D., "A Bubbling Fluidization Model Using Kinetic Theory of Granular Flow," *American Institute of Chemical Engineers Journal*, 36, pp. 523–538, 1990.
40. Drew, D., and Lahey, R., *Particulate Two-Phase Flow*, Butterworth-Heinemann, Boston, MA, 1993.
41. Vasquez, S., and Ivanov, V., "A Phase Coupled Method for Solving Multiphase Problems on Unstructured Meshes." In *Proceedings of ASME FEDSM'00: ASME 2000 Fluids Engineering Division Summer Meeting*, Boston, MA, 2000.
42. Weiss, J., Maruszewski, J., and Smith, W., "Implicit Solution of Preconditioned Navier-Stokes Equations Using Algebraic Multigrid," *AIAA Journal*, 37, pp. 29–36, 1999.
43. Schwartz, L., "On the Asymptotic Analysis of Surface-Stress-Driven Thin-Layer Flow," *Journal of Engineering Mathematics*, 39, pp. 171–188, 2001.
44. Jiang, Y., and Floryan, J., "Effect of Heat Transfer at the Interface on Thermocapillary Convection in Adjacent Phase," *ASME Journal of Heat Transfer*, 125, pp. 190–194, 2003.
45. Elgafy, A., and Lafdi, K., "Engineering Solution in Monitoring Nanoparticle-Fluid Flow during Nanocomposites Processing," *Journal of Nanoparticle Research*, 9 (3), pp. 441–454, 2007.
46. Elgafy, A., and Lafdi, K., "Carbon Nanofluids Flow Behavior in Novel Composites," *Journal of Microfluidic Nanofluidic*, 2 (5), pp. 425–433, 2006.
47. Elgafy, A., and Lafdi, K., "Carbon Nanoparticle-Filled Polymer Flow in the Fabrication of Novel Fiber Composites," *Carbon*, 44 (9, August), pp. 1682–1689, 2006.

3

Rheology in Polymer/Clay Nanocomposites: Mesoscale Structure Development and Soft Glassy Dynamics

Masami Okamoto

Toyota Technological Institute, Japan

CONTENTS

3.1 Introduction

Over the last few years, the utility of inorganic nanoscale particles as filler to enhance the polymer performance has been established. Of particular interest is recently developed nanocomposite technology consisting of a polymer and organically modified layered filler (organo-clay), because they often exhibit remarkably improved mechanical and various other materials properties as compared with those of virgin polymer or conventional composite (micro-/macrocomposites) [1–6]. These concurrent property improvements are well beyond what can generally be achieved through the micro-/macrocomposites preparation.

The synthetic strategy and molecular design was first explored by Toyota group with nylon 6 as the matrix polymer [7]. This new class of material is now being introduced in structural applications, such as gas barrier film and other load-bearing applications [5]. Polymer/clay nanocomposites (PCNs) and their self-assembly behaviors have recently been approached to produce nanoscale polymeric materials [1–6]. Additionally, these nanocomposites have been proposed as model systems to examine polymer structure and dynamics in confined environments [8–10].

In order to understand the processability of these materials (i.e., the final stage of any polymeric material), one must understand the detailed rheological behavior of these materials in the molten state. Understanding the rheological properties of PCN melts is not only important in gaining a fundamental knowledge of the processability but is also helpful in understanding the structure–property relationships in these materials. Although rheological measurement is an indirect probe, it is a well-established approach to probe the interaction between nanofiller and polymer matrix and the time-dependent structure development. In addition, more clear nanoscale and mesoscale structure development of the systems could be provided when combined with X-ray/light-scattering experiments and electron microscopy.

The original mesoscale structure in PCNs consists of randomly oriented exfoliated layers or tactoids of layers. This randomly distributed nanofiller forms a "organo-clay network" structure that is mediated by polymer chains and organo-clay–organo-clay interactions, responsible for the linear viscoelastic response observed in PCN melts. This mesostructure, which is intrinsically in a disordered metastable state and out of equilibrium, and offers an apt analogy to soft colloidal glasses and gels, was extensively discussed [11,13–27].

In this chapter, we survey mesostructure development in PCN melts with a primary focus on flow behavior and an analogy to soft glassy dynamics [12].

3.2 Linear Viscoelastic Properties

Dynamic oscillatory shear measurements of polymeric materials are generally performed by applying a time-dependent strain of $\gamma(t) = \gamma_o \sin(\omega t)$, and the resultant shear stress is $\sigma(t) = \gamma_o[G' \sin(\omega t) + G'' \cos(\omega t)]$, with G' and G'' being the storage and loss modulus, respectively.

Generally, the rheology of polymer melts strongly depends on the temperature at which measurement is carried out. It is well known that for thermorheological simplicity, isotherms of storage modulus ($G'(\omega)$), loss modulus ($G''(\omega)$), and complex viscosity ($|\eta^*|(\omega)$) can be superimposed by horizontal shifts along the frequency axis:

$$b_T G'(a_T \omega, T_{ref}) = b_T G' (\omega, T)$$

$$b_T G''(a_T \omega, T_{ref}) = b_T G'' (\omega, T)$$

$$|\eta^*|(a_T \omega, T_{ref}) = |\eta^*| (\omega, T)$$

where a_T and b_T are the frequency and vertical shift factors, respectively; and T_{ref} is the reference temperature. All isotherms measured for pure polymer and for various PCNs can be superimposed along the frequency axis.

In the case of polymer samples, it is expected, at the temperatures and frequencies at which the rheological measurements were carried out, that the polymer chains should be fully relaxed and exhibit characteristic homo-polymer-like terminal flow behavior (i.e., curves can be expressed by a power law of $G'(\omega) \propto \omega^2$ and $G''(\omega) \propto \omega$).

The rheological properties of *in situ* polymerized nanocomposites with end-tethered polymer chains were first described by Krisnamoorti and Giannelis [13]. The flow behavior of poly(E-caprolactone) (PCL)– and nylon 6–based nanocomposites differed extremely from that of the corresponding neat matrices, whereas the thermorheological properties of the nanocomposites were entirely determined by that behavior of matrices [13]. The slope of $G'(\omega)$ and $G''(\omega)$ versus the $a_T\omega$ is much smaller than 2 and 1, respectively. Values of 2 and 1 are expected for linear monodispersed polymer melts and are a large deviation, and especially in the presence of a very small amount of layered silicate, loading may be due to the formation of network structure in the molten state. However, such nanocomposites based on the *in situ* polymerization technique exhibit fairly broad molar mass distribution of the polymer matrix, which hides the structurally relevant information and impedes interpretation of the results.

To date, the melt state linear dynamic oscillatory shear properties of various kinds of PCNs have been examined for a wide range of polymer matrices, including nylon 6 with various matrix molecular weights [14], PS [15], polystyrene (PS)-polyisoprene (PI) block copolymers [16,17], PCL [18], polypropylene (PP) [11,19–22], polylactide (PLA) [23,24], and poly(butylene succinate) (PBS) [25,26]. In the linear viscoelastic regime, a big change in the terminal (low-frequency) region from a liquid-like response to a solid-like response was observed for all PCNs ($G'(\omega) \sim G''(\omega) \propto \omega^0$), ascribed to the formation of a volume spanning mesoscale organo-clay network above the mechanical percolation threshold [22]. The terminal rheology is sensitive to organo-clay loading and the extent of exfoliation/intercalation in the polymer matrix.

A typical example of the linear dynamic viscoelastic master curves for the neat polylactide (PLA) and various PLA-based nanocomposites (PLACNs) with different organo-clay loading [24] is shown in Figure 3.1. The linear dynamic viscoelastic master curves were generated by applying the time-temperature superposition principle and shifted to a common temperature T_{ref} using both frequency shift factor a_T and modulus shift factor b_T. The moduli of the PLACNs increase with increasing clay loading at all frequencies ω. At high ωs, the qualitative behavior of $G'(\omega)$ and $G''(\omega)$ is essentially the same and is unaffected by frequencies. However, at low frequencies, $G'(\omega)$ and $G'(\omega)$ increase monotonically with increasing organo-clay content. In the low-frequency region, the curves can be expressed by power law of $G'(\omega) \propto \omega^2$

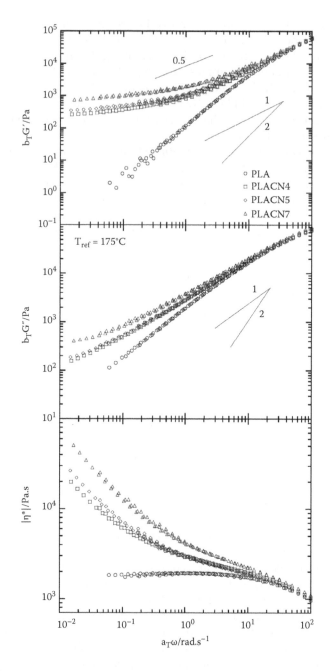

FIGURE 3.1
Reduced frequency dependence of storage modulus, loss modulus, and complex viscosity of neat polylactide (PLA) and various PLA-based nanocomposites (PLACNs) under dynamic oscillatory shear measurement. (Reprinted from Sinha Ray S, Yamada K, Okamoto M, Ueda K, *Polymer* 44, 6631. Copyright 2003, Elsevier Science. With permission.)

TABLE 3.1

Terminal Slopes of G' and G'' versus $a_T\omega$ for
Polylactide (PLA) and Various PLA-Based
Nanocomposites (PLACNs)

System	Organo-Clay/wt%	G'	G''	E_a [a]/kJmol^{-1}
PLA	0	1.6	0.9	170 ± 5
PLACN4	4.0	0.2	0.5	225 ± 10
PLACN5	5.0	0.18	0.4	225 ± 5
PLACN7	7.0	0.17	0.32	230 ± 8

[a] Flow activation energy obtained from an Arrhenius fit of
master curves.
Source: Reprinted from Sinha Ray S, Yamada K, Okamoto M,
Ueda K, *Polymer* 44, 6631, Copyright 2003, Elsevier
Science. (With permission.)

and $G''(\omega) \propto \omega$ for neat PLA, suggesting this is similar to those of the narrow molecular weight (M_w) distribution homopolymer melts. On the other hand, for $a_T\omega < 5$ rad.s^{-1}, the viscoelastic response (particularly $G'(\omega)$) for all the nanocomposites displays significantly diminished frequency dependence as compared with the matrices. In fact, for all PLACNs, $G'(\omega)$ becomes nearly independent at low $a_T\omega$ and exceeds $G''(\omega)$, characteristic of materials exhibiting a pseudo-solid-like behavior. The terminal zone slopes values of both neat PLA and PLACNs are estimated at the lower $a_T\omega$ region (<10 rad.s^{-1}) and are presented in Table 3.1. The lower slope values and the higher absolute values of the dynamic moduli indicate the formation of an "organo-clay network" structure in the PLACNs in the molten state [16,24]. This mesoscale structure is partially supported by transmission electron microscopic (TEM) observation [24]. The correlation length between clay particles (ξ_{clay}) is smaller than the average value of the particle length (L_{clay}) for PLACNs, suggesting the formation of highly geometric constraints. A plausible explanation and its model are presented by Ren et al. [16] (Figure 3.2). The individual or stacked silicate layers are incapable of freely rotating; hence, by imposing small ω, the relaxation of the structure is prevented almost completely. This type of prevented relaxation leads to the presence of the volume-spanning mesoscale network. The effective volume occupied by the clay platelets is much larger than may be calculated from their volume alone. At the same time, the pseudo-solid-like behavior is observed in all PCNs, despite the fact that the thermodynamic interaction between those polymers and organo-clay is quite different [11,13–27]. In addition, the correlation between rheological response and the confinement of polymer chains into silicate nanogalleries is reported [28]. The confinement is directly concerned with rheological properties and enhanced modulus.

The temperature dependence frequency shift factor (a_T, Williams-Landel-Ferry (WLF) type [29]) used to generate the master curves shown in Figure 3.1 is shown in Figure 3.3. The dependence of the frequency shift factors on the

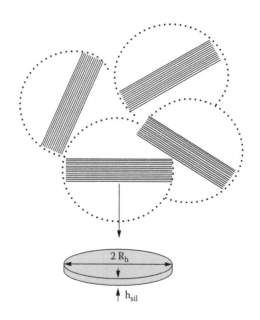

FIGURE 3.2
Percolation of hydrodynamic volumes of silicate layers at low concentration. Stacked silicate layers and their interaction with each other, resulting in complete relaxation of the nanocomposite melts. (Reprinted from Ren J, Silva AS, Krishnamoorti R, *Macromolecules* 33, 3739. Copyright 2000, American Chemical Society. With permission.)

silicate loading suggests that the temperature-dependent relaxation process observed in the viscoelastic measurements is somehow affected by the presence of the silicate layers (225 kJ/mol for PLACN4 and 170 kJ/mol for neat PLA, see Table 3.1). In the case of nylon 6–based nanocomposites, where the hydrogen bonding is already formed to the silicate surface [30], the system exhibits a large flow activation energy almost one order higher in magnitude compared with that of neat nylon 6 (450 kJ/mol and 1900 kJ/mol for nylon 6–based nanocomposites with 1.6% and 3.7% clay loading, respectively, and 350 kJ/mol for neat nylon 6) (see Figure 3.10) [31].

The shift factor b_T shows large deviation from a simple density effect, and it would be expected that the values would not vary far from unity ($b_T = \rho T / \rho_{ref} T_{ref}$, where ρ and ρ_f are the densities at T and T_{ref}, respectively) [29]. One possible explanation is a network structural change occurring in PLACNs during measurement (shear process). The reconstituting of the organo-clay network probably supports for PCN melts under weak shear flow (terminal zone), thereby leading to the increase in the absolute values of $G'(\omega)$ and $G''(\omega)$.

The subsequent reconstitution of organo-clay domains into a mesoscale network is also argued by simple Brownian motion, where the rotary diffusivity (D_{r0}) of a circular disk of diameter (d) ($= L_{clay}/2$) is [32,33]

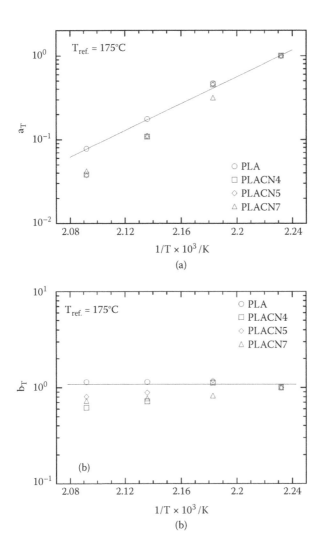

FIGURE 3.3
(a) Frequency shift factors a_T and (b) modulus shift factor b_T as a function of temperature. (Reprinted from Sinha Ray S, Yamada K, Okamoto M, Ueda K, *Polymer* 44, 6631. Copyright 2003, Elsevier Science. With permission.)

$$D_{r0} = \frac{3k_B T}{4\eta_0 d^3} \qquad (3.1)$$

where η_0 is the polymer matrix viscosity, k_B is the Boltzmann constant, and T is the temperature.

The rotational relaxation time (t_D) is given by

$$t_D \sim \frac{(\pi/2)^2}{D_{r0}} \tag{3.2}$$

A quarter period is provided to complete rotational relaxation. This time scale is useful to discuss the reconstituting of the network via the rotation of the organo-clays as compared with an experimental time scale. For PLACN4, the estimated time scale for rotational Brownian motion is about 2×10^3 s at 175°C. This value is larger than the experimental time scale ($1/a_T\omega \sim 10^2$ s), suggesting the major driving force for the structural evolution is not simple Brownian relaxation of the organo-clays. The extensive argument of the reorganization of organo-clay domains in light of the rotational Brownian motion has been presented in several papers [11,19,33,34].

However, in many intercalated nanocomposites (i.e., complete exfoliation is not feasible), the organo-clay domains consist of the penetrated polymer chains into nanogalleries and stacked silicate layers [35]. Irrespective of the intercalated nanocomposite structure, the opposite conclusion was also reported with respect to the time scale between Brownian relaxation and experimental relaxation [33]. In this regard, we have to clarify the mesoscale network structure. For this reason, the stress recovery following prolonged large-amplitude oscillatory shear (LAOS) [19,33] and the stress response upon start-up of steady shear or small-amplitude oscillatory shear (SAOS) [11,22] were examined to contrast the aging behavior of as-processed (unsheared) and presheared (leading to parallel alignment of the silicate layers) nanocomposites.

3.3 Nonlinear Shear Response

The start-up of shear and SAOS experiments offer insight into mesoscale network structural changes.

Solomon et al. conducted nonlinear reversing shear flow experiments to study the disorientation kinetics of flow-aligned organo-clay domains in a PP matrix during the annealing period between deformations. The transient stress in start-up of steady shear scaled with the applied strain (not shear rate) (Figure 3.4) [19]. This transient nonlinear rheology has non-Brownian style. They showed the disorientation kinetics faster than that predicted by Brownian motion and the constant WLF shift factor for the PP matrix and nanocomposites, and they concluded that the strong attraction between the silicate layers led to the rapid disorientation upon cessation of shear [19]. They suggested that the stacked silicate layers into larger organo-clay domains assemble into a heterogeneous organo-clay network as a result of attractive interaction (van der Waals force). The organo-clay network formed under quiescent conditions is easily perturbed by deformation. Upon the cessation

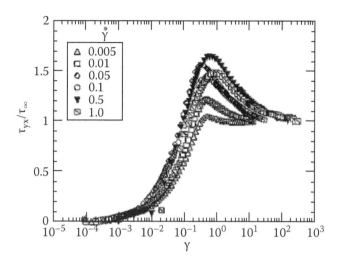

FIGURE 3.4
Onset of stress overshoot in start-up of steady shear depends on the strain applied to the poly-propylene (PP)-based nanocomposite at 180°C. (Reprinted from Solomon MJ, Almusallam AS, Seefeld KF, Somwangthanaroj S, Varadan P, *Macromolecules* 34, 1864. Copyright 2001, American Chemical Society. With permission.)

of flow, attractive interparticle interactions promote reconstitution of the network. The reconstituting network, which more completely re-forms as the rest time increases, is the initial state upon which deformation is oriented and ruptured, thereby giving rise to an overshoot in the stress during the flow reversal experiments [19].

For viscosity overshoot, Treece et al. reported that the flow ruptures the organo-clay network in the as-processed material and flow-aligns its domains to a degree that depends on the magnitude of the deformation [11]. The morphology in both as-processed and presheared samples changes significantly during annealing. This result supports the large increases in the magnitude of the viscosity overshoot. The strong dependence on annealing time for the as-processed sample clearly indicates that it is out of equilibrium [11,28]. This behavior is an "aging" phenomenon characteristic of systems (i.e., microgel pastes, nematic polymer, aqueous hectrite–type clay [Laponite] dispersion). Those are classified as soft colloidal glasses and are far from equilibrium (Figure 3.5) [36].

3.4 Analogy to Soft Colloids

Ren et al. [33] discussed the analogy to the dynamics of soft colloidal glasses, exhibiting yield stress, thixotropy, and slow stress recovery under

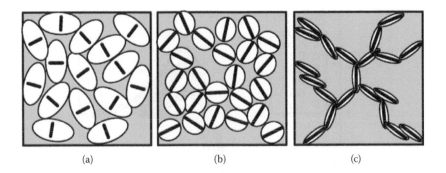

(a)   (b)   (c)

FIGURE 3.5
(a) Repulsive colloidal glass, (b) attractive glass, and (c) gel. Each thick line represents a Laponite disk, and a white ellipsoid indicates the range of electrostatic repulsions. For (a), long-range electrostatic repulsions dominate. In (b), attractive interactions affect the spatial distribution, but repulsive interactions still play the predominant role in the slow dynamics of the system. In (c), attractive interactions play a dominant role; a percolated network forms, which gives the system its elasticity. (Reprinted from Tanaka H, Meunier J, Bonn D, *Phys Rev. E* 69, 031404. Copyright 2004, American Physical Society. With permission.)

deformation [12,36]. A glass is homogenous on interparticle structure with elasticity derived from the caging effect, which is characterized by long-range electrostatic repulsion and short-range attraction [36]. In the colloidal solid, the storage modulus and complex viscosity show logarithmic dependence time, like PCN melts. For example, in an aqueous Laponite dispersion, the structure reorganization responsible for the rheological behavior depends upon the ionic strength of the dispersion, which balances the attractive and repulsive forces between clay domains.

On the other hand, a gel also shows the soft glassy dynamics, which is associated with metastable, structural heterogeneity on mesoscale length, caused by a percolation infinite (volume spanning) network. PCNs have a structural hierarchy without long-range repulsive interaction, leaving short-range attractive forces (van der Waals forces) between polymer and clay because of the surface modification of clay by cationic intercalant. For this reason, PCN melts were classified as attractive colloidal gels [11]. This picture supports other experimental results on transient start-up viscosity under shear flow [22,33].

3.5 Reversibility of Network Formation Process

The central discussion in PCN melts is structure with elasticity. The idea is an organo-clay network with a few weight percent (wt%). An important question is how long the building of the solid-like network continues during annealing (i.e., reversibility of the network formation process).

In an aqueous dilute suspension of smectite swelling clays such as hectorite and montmorillonite, the existence of a long-range structural correlation (~100 nm) was proposed via an *in situ* small-angle neutron scattering (SANS) experiment of the smectite suspension under the quiescent state [37]. This is a worthy hypothesis for a new interpretation of network formation and complex rheological properties. Such magnitude of the structural length in the clay suspensions is suitable for a Rayleigh scattering experiment on the basis of the different polarizabilities between clay particles and medium.

Okamoto et al. constructed a rheo-optical device, small-angle light-scattering apparatus under shear flow (Rheo-SALS), which enables us to perform time-resolved measurements of light intensity scattered from the internal structure developed under shear flow [38]. They examined long-range correlation and its evolution in the organo-hectorite/styrene (3.5/96.5 vol/vol) suspensions (dense condition: volume fraction $\phi \geq 0.01$ [39]) under shear flow. In this region, the power law relation for both zero-ω limiting value of modulus $G'(\omega \to 0)$ and fluidizing stress (σ_{fl}) were given by

$$G'(\omega \to 0) \sim \sigma_{fl} \propto \phi^{4 \pm 0.2} \qquad (3.3)$$

The ratio of $G''(\omega)/G'(\omega)(= \tan \delta)$ exhibits independence of ϕ and remains a constant value of about 0.1. This behavior relates to self-similarity of the linked network structure in the gelation region as reported by Winter [40].

The hectorite $(Na_{0.33}(Mg_{2.67}Li_{0.33})Si_4O_{10}(OH)_2)$ was modified with quarternized hexadecyl, octadecyl ammonium chloride $([(C_{16}H_{33})_{0.5}(C_{18}H_{37})_{1.5}N^+(C H_3)_2]Cl^-)$ [38]. The estimated individual silicate layer under very dilute conditions ($\phi \ll 0.01$) was 60 nm in length scale. Four different $\dot{\gamma}$ were selected to confirm the time dependence of the internal structural change under different steady shear flow via stress rheometer. The time development of transient viscosity ($\eta(\dot{\gamma}:t)$) under constant shear rate ($\dot{\gamma}$) for the first run at 25°C is shown in Figure 3.6. Under $\dot{\gamma} = 0.1$ and 1.0 s^{-1}, $\eta(\dot{\gamma}:t)$ gradually increases with time and exhibits the power law, $\eta(\dot{\gamma}:t) \propto t^{\upsilon}$ with $\upsilon = 0.5$ for 0.1 s^{-1} and $\upsilon = 0.2$ for 1.0 s^{-1}, whereas under $\dot{\gamma} = 10$ and 10^2 s^{-1} (yielding regime due to the strong shear stress), $\eta(\dot{\gamma}:t)$ decreases, and power law decay, $\eta(\dot{\gamma}:t) \propto t^{-\upsilon}$ with $\upsilon = 0.17$ is exhibited for both cases (Figure 3.6). The quite different viscosity profiles under constant $\dot{\gamma}$ runs were reported.

The time variation of the mean-square density fluctuation $<\eta^2>$, the mean-square anisotropy $<\delta^2>$, and the relevant value of long-range correlation distance (ξ_η and ξ_δ) upon imposition/cessation of steady shear flow at both low (= 0.5 s^{-1}) and high $\dot{\gamma}$ (= 60 s^{-1}) were discussed (Figure 3.7). Before flowing, the organo-clay network in length scale was 450 nm estimated from ξ_η value, suggesting the existence of a large-scale flocculated network structure due to the edge-to-edge interaction [41]. In addition, the boundary between networks is of importance in predicting the complex flow dynamics. The suppression of $\langle \eta^2 \rangle$ is observed during flow

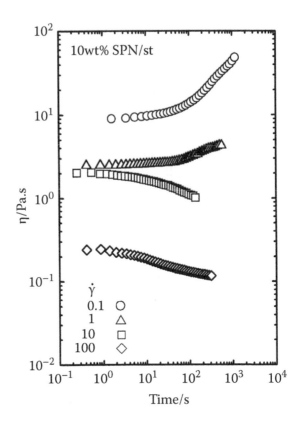

FIGURE 3.6
Time variation of shear viscosity for organo-hectorite/styrene (3.5/96.5 vol/vol) suspension with four shear rate. (Reprinted from Okamoto M, Sato H, Taguchi H, Kotaka T, *Nippon Rheology Gakkaishi* 28, 201. Copyright 2000, The Society of Rheology, Japan. With permission.)

in both cases as compared with the initial quiescent state (before shearing). Presumably, this is due to the hydrodynamic forces, which promote alignment of the organo-clay networks because of the large dimensions of the networks.

Under $\dot{\gamma} = 0.5$ s^{-1} (corresponding to unabated viscosity regime), the gradual decreasing of $\langle \eta^2 \rangle$ and increasing of ξ_η upon imposition of shear are observed. Simultaneously, the value of $\langle \delta^2 \rangle$ is jumped upward and then remains almost constant during flow, where ξ_δ also keeps a constant value of 400 nm, indicating that the anisotropy in the system is developed. Upon cessation of shear at 750 s, on the contrary, $\langle \eta^2 \rangle$ increases with time and finally levels off at 1300 s with ξ_η of 250 nm. On the other hand, $\langle \delta^2 \rangle$ does not become zero and developed orientation of the network with ξ_δ of 690 nm in the shear field is stable, even upon cessation of shear. The anisotropy gradually develops with time, forming relatively stable oriented networks, which are stable even after the cessation of the shear.

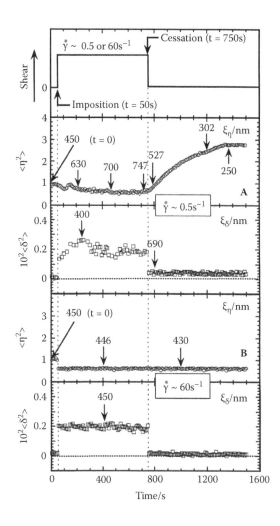

FIGURE 3.7
Time variation of $<\_^2>$ and $<\_^2>$ upon imposition/cessation of steady shear under low (= 0.5 s^{-1}) and high $\dot{\gamma}$ (= 60 s^{-1}) conditions. The Debye–Bueche equation is applicable in anisotropic shear flow field for dense suspension [42]. (Reprinted from Okamoto M, Sato H, Taguchi H, Kotaka T, *Nippon Rheology Gakkaishi* 28, 201. Copyright 2000, The Society of Rheology, Japan. With permission.)

In the case of under $\dot{\gamma}$ = 60 s^{-1}, $\langle\eta^2\rangle$ slightly decreases upon imposition of shear, but ξ_η does not change and agrees with the value of the initial quiescent state during flow. For the anisotropy, during shear flow, $\langle\delta^2\rangle$ stably appears with ξ_δ of 450 nm; upon cessation of shear, it disappears suddenly, suggesting that such an aligned, orientated network structure in the high shear field is labile (not stable) as compared with that under weak shear flow. Furthermore, following the large deformation, the slippage and rotation take place at the interface between network domains, and the resulting shear thinning behavior is observed in this condition [39]. This feature may resemble the rotation of the grains in block copolymer melt under uniaxial deformation [43]. In PCN melts, however, the network structure is easily destroyed (ruptured) by deformation due to the polymer chain entanglement as discussed by Solomon et al. [19].

The structure disorder creates energy barriers that prevent reorganizing/reconstituting networks into states of lower free energy. In dynamics for a typical soft glassy material, a slow degree of freedom is taken into account [12,44]. Jamming is a common property of complex fluids. In the presence of stress, the viscosity is given by the distribution of relaxation times of the networks of "slow mode."

Imposition of shear is considered to change the energy landscape and allows for the system to access new metastable states [45]. For this reason, the imposition of weak shear ($\dot{\gamma}$ = 0.5 s^{-1}) (much smaller than yield stress) is considered a shear rejuvenating condition ($\eta(\dot{\gamma}:t) \propto t^{0.5}$), meaning that the longest relaxation time of the slow mode decreases in time. Accordingly, the initial network structure grows with time, accompanied by an increase of ξ_n.

Upon cessation of shear, aging begins anew, because flow alters the energy landscape, new metastable states are now accessible, and the system evolves spontaneously, accompanied by an increase of ξ_δ and decrease of ξ_n. This suggests that aging of the system (i.e., the reconstituting in between networks) starts, and a steady state is finally reached.

In the case of large deformation ($\dot{\gamma}$ = 60 s^{-1}), those energy barriers become greater the longer a system is aged, such that the longest relaxation time continuously increases. As a result, upon cessation of shear, thermal motion alone is insufficient to mediate complete structural relaxation. Therefore, the network may become trapped in a higher energy state (almost constant value of ξ_n). Brownian forces alone are unable to change the energy barriers created by such an oriented organo-clay network structure, because the estimated rotational Brownian motion of the hectrite platelets is about 10^{-2} s at 25°C. Such discussion on energy landscape appears to be entirely valid for the experimental results as well as isotropic particle dispersion in a Newtonian fluid matrix [44].

3.6 Alignment of Silicate Layers in Networks

The organo-clay platelets orient in both shear and elongational flow fields. A second question is how the platelets are oriented during the flow direction. Lele et al. [46] reported the *in situ* Rheo-X-ray investigation of flow-induced orientation in syndiotactic PP/layered silicate nanocomposite melt. The clay platelets rapidly oriented and remain at constant orientation in the long time regime (~1500 s).

The orientation of silicate layers and nylon 6-base nanocomposite (N6CN) using *ex situ* small-angle X-ray scattering (SAXS) is examined [33]. The clay layers, due to their higher aspect ratio, were predominantly oriented in a "parallel" orientation (with layer normals along the velocity gradient direction) at different times following LAOS.

Kojima et al. [47] found three regions of different orientations in the injection-molded bar as a function of depth. Near the middle of the sample, where the shear forces are minimal, the clay platelets are oriented randomly, and the nylon 6 crystallites are perpendicular to the silicate layers. In the surface region, shear stresses are very high, so both the clay layers and the nylon 6 crystallites are parallel to the surface. In the intermediate region, the clay layers, presumably due to their higher aspect ratio, still orient parallel to the surface, and the nylon 6 crystallites assume an orientation perpendicular to the silicate. Medellin-Rodriguez et al. [48] reported that the molten N6CN samples showed planar orientation of silicate layers along the flow direction, which is strongly dependent on shear time as well as clay loading, reaching a maximally orienting level after being sheared for 15 min with $\dot{\gamma} = 60$ s^{-1}.

In contrast, the orientation occurs by the "normal" to the clay surface aligning the flow direction through vorticity during shear. Okamoto et al. conducted the transmission electron microscopic (TEM) observation for the sheared N6CN3.7 (clay loading = 3.7 wt%) with $\dot{\gamma} = 0.0006$ s^{-1} for 1000 s [49]. The edges of the silicate layers laying along the z-axis (marked with the arrows [A]) or parallel alignment of the silicate edges to the shear direction (x-axis) (marked with the arrows [B]) rather than random orientation in the matrix is observed, but in fact, one cannot see these faces in this plane (Figure 3.8). Here, it should be emphasized that the planar orientation of the silicate faces along the x–z plane does not take place prominently. For the case of rapid shear flow, the commonly applicable conjecture of the planar orientation of the silicate faces along the shear direction first demonstrated to be true by Kojima et al. [47].

In uniaxial elongational flow (converging low) for a PP-based nanocomposite (PPCN4) (clay loading = 4 wt%), the formation of a *house-of-cards* structure is found by TEM analysis [50,51]. The perpendicular (not parallel) alignment of clay platelets with large anisotropy toward the flow direction might sound unlikely, but this could be the case, especially under an elongational flow field, in which the extentional flow rate is the square of the converging flow rate along the thickness direction, if the assumption of *affine* deformation without volume change is valid. Obviously, under such conditions, the energy dissipation rate due to viscous resistance between the platelet surface and the matrix polymer is minimal when the platelets are aligned perpendicular to the flow direction.

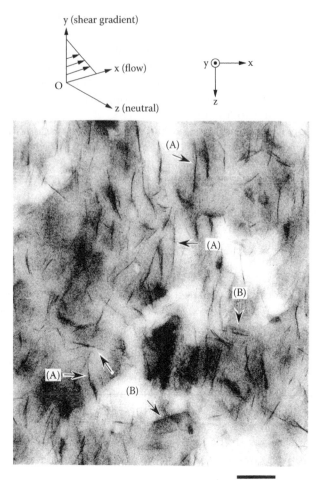

FIGURE 3.8
Transmission electron micrograph (TEM) in the x–z plane showing N6CN (clay loading = 3.7 wt%) sheared at 225°C with $\dot{\gamma}$ = 0.0006 s^{-1} for 1000 s. The x-, y-, and z-axes correspond respectively to flow, shear gradient, and neutral direction. (Reprinted from Okamoto M, *Rapra Review Report* 163. Copyright 2003, Rapra Technology Ltd., London. With permission.)

Figure 3.9 shows double logarithmic plots of transient elongational viscosity $\eta_E(\dot{\varepsilon}_0;t)$ against time t observed for N6CN3.7 and PPCN4 with different Hencky strain rates $\dot{\varepsilon}_0$ ranging from 0.001 s^{-1} to 1.0 s^{-1}. The solid curve represents time development of threefold shear viscosity, $3\eta_0(\dot{\gamma};t)$, at 225°C with a constant shear rate ($\dot{\gamma}$ = 0.001 s^{-1}). First, the extended Trouton rule, $3\eta_0(\dot{\gamma};t) \cong \eta_E(\dot{\varepsilon}_0;t)$ [52], as well as an empirical Cox–Merz relation ($\eta(\dot{\gamma}) = |\eta^*|(\omega)$) [53], fails for both N6CN3.7 and PPCN4 melts, as opposed to the melt of ordinary homopolymers. In $\eta_E(\dot{\varepsilon}_0;t)$ at any $\dot{\varepsilon}_0$, the N6CN3.7 melt shows a weak tendency of *strain-induced hardening* as compared to that

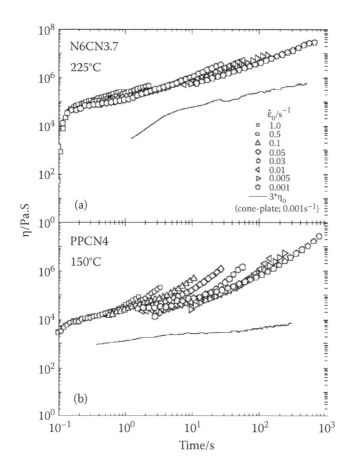

FIGURE 3.9
Time variation of elongational viscosity $\eta_E(\dot{\varepsilon}_0;t)$ for (a) N6CN3.7 melt at 225°C and for (b) PPCN4 at 150°C. The solid line shows three times the shear viscosity, $3\eta_E(\dot{\gamma};t)$, taken at a low shear rate $\dot{\gamma} = 0.001$ s^{-1} on a cone-plate rheometer. (Reprinted from Okamoto M, *Rapra Review Report* 163. Copyright 2003, Rapra Technology Ltd., London. With permission.)

of the PPCN4 melt. A strong behavior of strain-induced hardening for the PPCN4 melt was originated from an aging phenomenon characteristic of reconstituting the networks through the perpendicular alignment of the silicate platelets to the stretching direction.

From TEM observation (see Figure 3.8), the N6CN3.7 forms a fine dispersion of the silicate platelets of about 100 nm in L_{clay}, 3 nm thickness in d_{clay}, and ξ_{clay} of about 20 to 30 nm between them. The ξ_{clay} value is one order of magnitude lower than the value of L_{clay}, suggesting the formation of a rigid network domain structure of the dispersed clay platelets in end-tethered polymer chains (see flow activation energy in Figure 3.10). This suggests that both slow ($\dot{\gamma} = 0.001$ s^{-1}) and rapid ($\dot{\gamma} = 1.0$ s^{-1}) elongational flow rates are

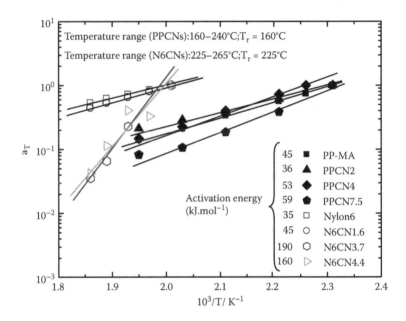

FIGURE 3.10
Frequency shift factors a_T as a function of temperature. Flow activation energy of neat PP and nylon 6, PPCNs and N6CNs with different contents of clay are shown.

unable to erase the energy barriers created by the *in situ* polymerization condition. Accordingly, the longest relaxation time in the network may remain constant. This tendency was also observed in the PPCN7.5 melt having a higher content of clay (= 7.5 wt%) [31].

SANS is useful in determining the orientation of the organo-clay under shear because of contrast matching to clay (montmorillonite) in D_2O [34]. In aqueous dispersions of hectite (3 wt%) and poly(ethylene oxide) (PEO) (2 wt%), the platelets were oriented in the flow direction with the surface normal in the neutral direction (Figure 3.11) [34]. It is quite possible that the dispersed organo-clay platelets attain not only parallel alignment but also perpendicular or even transverse alignment during shear and elongational flow fields.

The aqueous clay (kaolinite) suspensions have been investigated both in the quiescent state [54] and under shear flow using SANS [55]. Some 20 years ago, van Olphen [51] pointed out that the electrostatic attraction between the layers of natural clay in aqueous suspension arises from higher polar force in the medium. The intriguing features such as yield stress thixotropy and rheopexy exhibited in aqueous suspensions of natural clay minerals may be taken as a reference to the present PCNs. More detailed surveys on various types of experiments can also be found in the literature [56–63].

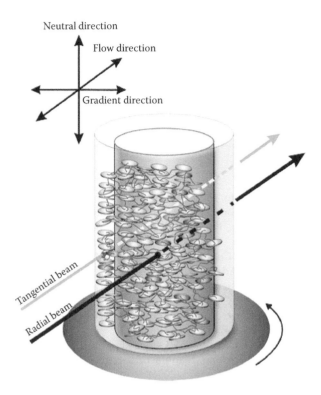

Neutral direction

Flow direction

Gradient direction

Tangential beam

Radial beam

FIGURE 3.11
Couette-type shear cell for small-angle neutron scattering (SANS) and model for real space orientation of oriented clay platelets in the cell. The reference coordinate frame is anchored in the tangential beam. (Reprinted from Scmidt G, Nakatani AI, Butler PD, Karim A, Han CC, *Macromolecules* 33, 7219. Copyright 2000, American Chemical Society. With permission.)

3.7 Summary

For an improved understanding of the soft glassy dynamics in PCN melts, in this review, we described some recent results concerning the mesoscale organo-clay network structure and its reversibility in the light of melt rheometry with a combination of scattering experiments and electron microscopy. We studied the dynamics in PCN melts; however, it is difficult to discuss the intrinsic feature of the networks.

Although our experimental results are still weak evidence to discuss the mesoscale structure development in PCN, many papers ignore the existence of percolated organo-clay networks and their intrinsically metastable states and out-of-equilibrium condition. Once a percolated network is formed, the networks retard crystalline capability and enhance thermal stability as well as modulus [64,65].

For the future, for designing high-performance PCN materials and their processing, the correlation between the mesoscale network structure and macroscopic properties will be probed via an innovative methodology such as three-dimensional TEM and fast scanning Fourier transform infrared (FTIR) imaging.

References

1. Sinha Ray S, Okamoto M, "Polymer/Layered Silicate Nanocomposites: A Review from Preparation to Processing" *Prog. Polym. Sci.*, 28, 1539–1641 (2003).
2. Vaia RA, Wagner HD, "Framework for Nanocomposites" *Materials Today*, 7, 32–37 (2004).
3. Gao F, "Clay/Polymer Composites: The Story" *Materials Today*, 7, 50–55 (2004).
4. Okamoto M, "Recent Advances in Polymer/Layered Silicate Nanocomposites: An Overview from Science to Technology" *Mater. Sci. Tech.*, 22, 7, 756–779 (2006).
5. Okada A, Usuki A, "Twenty Years of Polymer-Clay Nanocomposites" *Macromol. Mater. Eng.*, 291, 1449–1476 (2006).
6. Hussain F, Hojjati M, Okamoto M, Gorga RE, "Review Paper: Polymer–Matrix Nanocomposites, Processing, Manufacturing, and Application: An Overview" *J. Composite Mater.*, 40, 1511–1575 (2006).
7. Usuki A, Kojima Y, Okada A, Fukushima Y, Kurauchi T, Kamigaito O, *J. Mater. Res.*, 8, 1174 (1993).
8. Vaia RA, Giannelis EP, *Macromolecules*, 30, 8000 (1997).
9. Krishnamoorti R, Vaia RA, Giannelis EP, *Chem. Mater.*, 8, 1728 (1996).
10. Rao Y, Pochan JM, *Macromolecules*, 40, 290 (2007).
11. Treece MA, Oberhauser JP, *Macromolecules*, 40, 571 (2007).
12. Sollich P, Lequeux F, Hebraud P, Cate ME, *Phys. Rev. Lett.*, 78, 2020 (1997).
13. Krishnamoorti R, Giannelis EP, *Macromolecules*, 30, 4097 (1997).
14. Fornes TD, Yoon PJ, Keskkula H, Paul DR, *Polymer*, 42, 9929 (2001).
15. Hoffman B, Dietrich C, Thomann R, Friedrich C, Mulhaupt R, *Macromol. Rapid Commun.*, 21, 57 (2000).
16. Ren J, Silva AS, Krishnamoorti R, *Macromolecules*, 33, 3739 (2000).
17. Mitchell CA, Krishnamoorti R, *J. Polym. Sci. Part B Polym. Phys.*, 40, 1434 (2002).
18. Lepoittevin B, Devalckenaere M, Pantoustier N, Alexandre M, Kubies D, Calberg C, Jerome R, Dubois P, *Polymer*, 43, 1111 (2002).
19. Solomon MJ, Almusallam AS, Seefeld KF, Somwangthanaroj S, Varadan P, *Macromolecules*, 34, 1864 (2001).
20. Galgali G, Ramesh C, Lele A, *Macromolecules*, 34, 852 (2001).
21. Lele A, Mackley M, Galgali G, Ramesh C, *J. Rheol.*, 46, 1091 (2002).
22. Treece MA, Oberhauser JP, *Polymer*, 48, 1083 (2007).
23. Sinha Ray, S, Maiti, P, Okamoto, M, Yamada, K, Ueda, K, *Macromolecules*, 35, 3104 (2002).
24. Sinha Ray S, Yamada K, Okamoto M, Ueda K, *Polymer*, 44, 6631 (2003).
25. Sinha Ray S, Okamoto K, Okamoto M, *Macromolecules*, 36, 2355 (2003).

26. Okamoto K, Sinha Ray S, Okamoto M, *J. Polym. Sci. Part B: Polym. Phys.*, 41B, 3160 (2003).
27. Krishnamoorti R, Yurekli K, *Current Opinion in Colloid Interface Sci.* 6, 464 (2001).
28. Maiti P, Nam PH, Okamoto M, Kotaka T, Hasegawa N, Usuki A, *Macromolecules*, 35, 2042 (2002).
29. Williams ML, Landel RF, Ferry JD, *J. Amer. Chem. Soc.*, 77, 3701 (1955).
30. Maiti P, Okamoto M, *Macromole. Mater. Eng.*, 288, 440 (2003).
31. Nam, PH, Master Thesis, Toyota Technological Institute (2001).
32. Brenner H, *Int. J. Multiphase Flow*, 1, 195 (1974).
33. Ren J, Casanueva BF, Mitchell CA, Krishnamoorti R, *Macromolecules*, 36, 4188 (2003).
34. Scmidt G, Nakatani AI, Butler PD, Karim A, Han CC, *Macromolecules*, 33, 7219 (2000).
35. Saito T, Okamoto M, Hiroi R, Yamamoto M, Shiroi T, *Polymer*, 48, 4143 (2007).
36. Tanaka H, Meunier J, Bonn D, *Phys Rev. E*, 69, 031404 (2004).
37. Mourchid A, Delville A, Lambard J, Lecolier E, Levitz P, *Langmuir*, 11, 1942 (1995).
38. Okamoto M, Sato H, Taguchi H, Kotaka T, *Nippon Rheology Gakkaishi*, 28, 201 (2000).
39. Okamoto M, Taguchi H, Sato H, Kotaka T, Tatayama H, *Langmuir*, 16, 4055 (2000).
40. Winter HH, Chambon F, *J. Rheol.*, 30, 367 (1986).
41. Sinha Ray S, Okamoto K, Okamoto M, *Macromolecules*, 36, 2355 (2003).
42. Hsiao BS, Stein RS, Deutscher K, Winter HH, *J. Polym. Phys.*, 28, 1571 (1990).
43. Kobori Y, Kwon YK, Okamoto M, Kotaka T, *Macromolecules*, 36, 1656 (2003).
44. Bonn D, Tanase S, Abou B, Tanaka H, Meunier J, *Phys. Rev. Lett.*, 89, 015701 (1992).
45. Lacks D, *Phys Rev. E*, 64, 51508 (2001).
46. Lele A, Mackley M, Galgali G, Ramesh C, *J. Rheol.*, 46, 1091 (2002).
47. Kojima Y, Usuki A, Kawasumi M, Okada A, Kurauchi T, Kamigaito O, Kaji K, *J. Polym. Sci. Part B: Polym. Phys.*, 33, 1039 (1995).
48. Medellin-Rodriguez FJ, Burger C, Hsiao BS, Chu B, Vaia RA, Phillips S, *Polymer*, 42, 9015 (2001).
49. Okamoto M, "Polymer/Layered Silicate Nanocomposites," Rapra Review Report 163, 166 pp, Rapra Technology Ltd., London (2003).
50. Okamoto M, Nam PH, Maiti P, Kotaka T, Hasegawa N, Usuki A, *Nano. Lett.*, 1, 295 (2001).
51. van Olphen H, *An Introduction to Clay Colloid Chemistry*, Wiley: New York (1977).
52. Trouton FT, *Proc. Roy. Soc.*, A77, 426 (1906).
53. Cox WP, Merz EH, *J. Polym. Sci.*, 28, 619 (1958).
54. Pignon F, Magnin A, Piau JM, *J. Rheol.*, 42, 1349 (1998).
55. Jogun SM, Zukoski CF, *J. Rheol.*, 43, 847 (1999).
56. Roe R, *Methods of X-ray and Neutron Scattering in Polymer Science.* New York: Oxford University Press, p.199 (2000).
57. Bafna A, Beaucage G, Mirabella F, Skillas G, Sukumaran S, *J. Polym. Sci., Part B: Polym. Phys.*, 39, 2923 (2001).
58. Koo CM, Kim SO, Chung IJ, *Macromolecules*, 36, 2748 (2003).

59. Yalcin B, Valladares D, Cakmak M, *Polymer*, 44, 6913 (2003).
60. Bafna A, Beaucage G, Mirabella F, Mehta S, *Polymer*, 44, 1103 (2003).
61. Yalcin B, Cakmak M, *Polymer*, 45, 2691 (2004).
62. Loo LS, Gleason KK, *Polymer*, 45, 5933 (2004).
63. Kim JH, Koo CM, Choi YS, Wang KH, Chung IJ, *Polymer*, 45, 7719 (2004).
64. Wang K, Liang S, Deng J, Yang H, Zhang Q, Fu Q, Dong X, Wang D, Han CC, *Polymer*, 47, 7131 (2006).
65. Rao Y, Pochan JM, *Macromolecules*, 40, 290 (2007).

4

Polymer/Graphite Nanocomposites

Guohua Chen and Weifeng Zhao

Huaqiao University, China

CONTENTS

4.1 Introduction

Modification of polymers by adding inorganic fillers as a second phase has become ubiquitous in polymeric systems. It is a common practice to improve polymer materials' properties such as stiffness, toughness, heat distortion temperature, and mold shrinkage to satisfy applications. The addition of filler materials is also an effective way to develop novel polymer materials with notable properties to meet new industry requirements and to replace some of the existing materials with property limitations. Polymer composites containing inorganic additives including $CaCO_3$, clay, alumina trihydrate, carbon or glass fibers, and carbon blacks, have been manufactured for a wide range of applications such as household goods, packaging, sporting goods, aerospace components, and automobiles. These composites always combine the advantages of corrosion resistance, light weight, and ease of processing from polymers and other functional performance such as electrical conductivity from fillers.

During the last decades, intense research on polymer/graphite composites has been promoted to develop polymer materials for applications where electrical conductivity is often required [1–10]. Because naturally abundant graphite is a well-known material for its mechanical, electrical, and thermal properties, it offers great opportunity for making multifunctional composites in a cost-effective way. In particular, the good electrical conductivity of 10^4 S/cm at ambient temperature makes graphite an ideal candidate to incorporate conduction to polymer composites. Graphite powders have been extensively incorporated into polymers that are either insulating or inherently electrical conducting [11–13]. By changing the volume fraction of graphite filler, both the mechanical property and conductivity of polymer matrix can be notably modified. The electrical conductivity of polymer composites varies depending on the conductivities of graphite and filler concentration in polymer matrixes. Moreover, not only the filler concentration but also the graphite particle size or shape have a notable influence on the electrical conductivity of graphite/polymer composites [14–16]. These electro-conductive polymeric composites can find their applications in many fields, such as electromagnetic interference shielding, electrostatic discharging, lightning-protected aircraft composite panels, battery electrodes, fuel cell bipolar plates, electronic pressure-sensitive switches or sensors, and other functional applications.

In addition to excellent electrical conductivity, graphite is a good solid lubricant and pencil material which has been selected to enhance the tribological properties of polymer materials such as rubbers [17,18]. Solid-lubricant-filled rubber composites with high wear resistance and low friction coefficient are favored to be used as sealing materials in radial lip seals of shafts, valve shaft seals, reciprocating piston, and piston rod seals [19,20].

Tremendous developments have been achieved in the manufacture and application of conductive polymer composites by blending conductive components with polymer matrices since early 1970s, when graphite began to be incorporated into polymer composites as a conductive ingredient [1,2]. People find that the conductive properties of electrically conducting polymer composites are governed by the conduction network developed by the conductive fillers. It is greatly affected by many factors, including the nature of the polymers, the types of conductive fillers, the processing parameters, the temperature, and other related factors [21]. Several theories have been proposed to interpret the electrical conductivity—namely, percolation theory, mean-field theory, and excluded volume theory. However, the conventional electrical conductive fillers whose size are of the micrometer scale commonly lead

TABLE 4.1

Properties and Cost of Carbon Materials

Carbon Materials	Electrical Resistivity (S/cm)	Thermal Conductivity (W/m K)	Tensile Modulus	Young's Modulus	True Density (g/cm³)	Cost ($/lb)
Graphite nanoplatelet[a,b,c]	~2.0 × 10⁴	~3000	1000 GPa	1060 GPa	~2.25	<5
Carbon nanotube[b,d]	10⁻²~10⁵	1700~6000	270GPa~1 TPa	>1.2 TPa	~2.10	>45000
Carbon fiber[b]	~10⁴	500~1000	200~700 GPa	—	1.7~2.1	>15
Carbon black[e]	~10²	—	—	—	~1.8	~1.5

[a] Thostenson, E., Li, C., and Chou, T. [22].
[b] Fukushima, H., Drzal, L.T. et al. [23].
[c] Stankovich, S., Dikin, D.A. et al. [24].
[d] Hussain, F., Hojjati, M. et al. [25].
[e] Krueger, Q.J. and King, J.A. [26].

to material redundancy and mechanical detriment when composites become sufficiently conductive. In recent years, there has been a strong emphasis placed on nanostructured polymer composites, where at least one of the filler dimensions is of nanometer order (less than 100 nm). Compared with the conventional composites, nanoscale additives are more efficient in reinforcement because of their larger surface area for a given volume. The transition of graphite thickness from micro- to nanoscale will yield a dramatic increase in the surface area/volume ratio by three orders of magnitude, logically. This results in graphite nanoplatelets of high aspect ratio and large surface area which are favored for the improvement of reinforcement efficiency and minimizing the needed filler concentration. Thus, in order to utilize the layered graphite filler efficiently, the layers of graphite are expected to be exfoliated as sufficiently as possible and dispersed throughout polymer matrix properly, in addition to obtaining good adhesion at the filler–matrix interface, which also plays a crucial role in determining the mechanical performance of the nanocomposite.

A graphite nanoplatelet combines low cost as well as excellent mechanical, electrical, and thermal properties that may rival the values of carbon nanotubes (Table 4.1). It can be an ideal alternative to the carbon black particles, carbon fibers, and nanotubes most commonly used in recent polymer modifications. There is little doubt that the polymer materials incorporated with graphite nanoplatelets are superior to the traditional polymer composites filled with micron-sized natural graphite powders. They have tremendous potential to replace commercially available conductive polymer composites. The focus of this review is to highlight the most recent findings regarding graphite nanostructures, processing, characterization, material properties, and challenges and potential applications for polymer/graphite nanocomposites.

TABLE 4.2

Properties of Graphite

Properties at Room Temperature	In the Basal Plane	Across the Basal Plane
Electrical resistivity (Ωcm)	4×10^{-5}	6×10^{-3}
Thermal conductivity (W/mK)	250	80
Coefficient of thermal expansion (K^{-1})	-1.2×10^{-6}	25.9×10^{-6}
Young's modulus (GPa)	1060	35.6

Source: Reprinted from Li J., Sham M.L., Kim J.K., and Marom G., *Comp. Sci. Techn.* 2006, 67(2): 296–305. Copyright 2006, Elsevier Ltd. (With permission.)

4.2 Nanostructured Graphite

As a layered material, the surface area/volume ratio of graphite can be expressed as $2/t + 4/l$, where t is the thickness and l is the side length of a four-sided filler plane [22]. The key point to utilizing graphite as a nanoreinforcement is the ability to create graphite platelets with thickness in the nanoscale range based on their unique layered structure. Thus far, graphite nanoplatelets are mainly obtained from sulfuric acid–intercalated graphite and graphite oxide.

4.2.1 Graphite Intercalation Compounds

Graphite is the crystalline of carbon that has a layered structure. In graphite, the 2s, $2p_x$, $2p_y$ electrons of the carbon atom form three sp^2 hybridized orbitals, and the $2p_z$ electron forms a delocalized orbital of π symmetry. Within each layer, the sp^2 hybridized orbital overlaps another one, leading to the formation of σ-bonding between carbon atoms. Through the σ-bond, the carbon atoms are arranged in a hexagonal pattern on a layer plane. The layers, on the other hand, are held together in a staggered sequence denoted as ···ABAB··· by weak van der Waals forces that are related to the loosely bound π-electrons of high mobility. As a result, graphite has different physical properties for in-plane and c-axis crystallographic directions. The highly anisotropic properties of graphite are shown in Table 4.2 [27].

Being different from silicate clay minerals, graphite does not bear any net charge. There are no reactive ion groups on the graphite layers; thus, it is impossible to intercalate other substances into the graphite galleries by ion exchange reactions, as do layered silicates. However graphite can react with certain atoms, molecules, and ions by allowing these guest species to reside between the interplanar spaces of the graphite lattice, forming graphite intercalation compounds (GICs). The guest substances are called intercalate. The mechanism of the intercalation is interpreted by a charge transfer between the intercalate and the carbon layers. It is believed that the delocalized π-bonds can gain electrons from or lose electrons to the intercalate, shifting the position of the Fermi energy from that in pure graphite. Thus, GIC tends to be more conductive electrically than

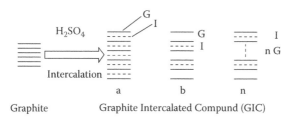

FIGURE 4.1
Stage 1(a), 2(b), and n(n) of graphite bisulfate.

pristine graphite. Graphite compounds intercalated by electron donors such as alkali metals are known as the donor-type GICs, whereas compounds formed by the intercalation of molecular species acting as electron acceptors such as halogens, halide, and acids, are viewed as acceptor-type GICs. However, it should be emphasized that the interaction between the carbon atoms and the intercalates is not scientifically true ionic bonding as is that in the totally ionic solids where simple ions are involved, although the electron transfers really exist in GICs. In fact, the degree of the ionicity in GICs is very low, and many of the intercalates retain their molecular form in the graphite lattice.

In the GIC, only a fraction of the interlayer spacing is filled with intercalate, and the carbon and intercalate layers stack on top of one another in a periodic sequence called the staging phenomenon. The definite number of carbon layers between two intercalate layers defines the "stage" of the GICs which is variable from 1 to n, depending on the particular intercalate and intercalation situations. The stage can be identified by X-ray diffraction (XRD) and by the intercalation isotherms, and it always decreases with an increasing intercalate concentration.

Graphite forms intercalation compounds with a large number of acids. The acids involved include nitric acid, sulfuric acid, perchloric acid, and selenic acid. Graphite bisulfate (graphite-H_2SO_4) is the most commonly used GIC to prepare graphite nanoplatelets for the fabrication of polymer/graphite nanocomposites.

Graphite bisulfate is typically prepared by spontaneous chemical interaction of graphite with a mixture of concentrated sulfuric acid and fuming nitric acid serving as oxidizing agent. The stacking sequence varies depending on the intercalation situation as shown in Figure 4.1. The reaction taking place between graphite and concentrated sulfuric acid can be expressed as [24,25]

$$n \text{ (graphite)} + nH_2SO4 + n/2[O] \rightarrow n \text{ (graphite} \cdot HSO_4) + n/2H_2O$$

$$(O) \text{ ------Oxidant; (graphite} \cdot HSO_4) \text{------ GIC}$$

4.2.2 Exfoliation of Graphite

GICs are uniquely capable of huge irreversible expansion. Some of the GICs can be exfoliated irreversibly by as many as several hundred times upon

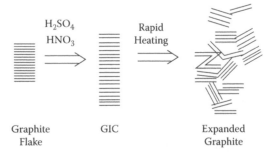

Graphite   GIC   Expanded
Flake   Graphite

FIGURE 4.2
Making expanded graphite through the HNO_3-H_2SO_4 route. (Reprinted from Chen, G.H., Wu, D.J., Weng, W.G., He, B., and Yan, W.L., *Polym. Int.*, 2001, 50(9): 980–985. Copyright 2001, Society of Chemical Industry. With permission.)

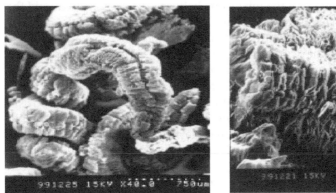
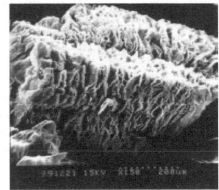

FIGURE 4.3
Scanning electron micrograph of expanded graphite. (Reprinted from Chen, G.H., Wu, D.J., Weng, W.G., He, B., and Yan, W.L., *Polym. Int.*, 2001, 50(9): 980–985. Copyright 2001, Society of Chemical Industry. With permission.)

heating. The origin of the exfoliation lies in the vaporization of intercalate at a high temperature, leading to a much larger elongation of graphite flake along the c-axis than in the in-plane direction. In principle, any intercalate can be used for graphite exfoliation. However, the surface area of exfoliated graphite varies from intercalate to intercalate. The most common choice of GIC is graphite bisulfate obtained through the HNO_3-H_2SO_4 route as mentioned earlier. After suffering an extensive thermal shock where the heating is excessive, the graphite bisulfate expands irreversibly by up to hundreds of times along the c-axis (Figure 4.2), resulting in a wormlike shape material with a low density and a high temperature resistance. The puffed-up product is known as exfoliated or expanded graphite (EG), as shown in Figure 4.3. The expansion ratio, defined as the reciprocal of the apparent density, can yield as high as 300. The typical structure of EG is an incompact and porous wormlike shape with many pores of different sizes, 10 nm to 10 µm, structured with parallel boards, which desultorily collapsed and deformed, possessing the apparent density of about 0.004 g/cm³ [30].

EG consists of a large number of delaminated graphite sheets, most of which have a thickness in nanometers, and have been proposed as a good graphite nanoplatelet precursor to develop polymer/graphite nanocomposites since the late 1990s [31]. Due to their high expansion ratio, the galleries of EG can be easily intercalated by monomers or polymers to achieve nanocomposites. After intercalation compound, the EG particles are distributed into the polymer matrix in the form of nanosheets with thicknesses varying from about 10 to 50 nm. The resulting composites have low percolation thresholds. Many works have been conducted on polymer/EG nanocomposites such as poly(methyl methacrylate) (PMMA)/EG [32], poly(styrene-co-acrylonitrile)/ EG [33], polystyrene (PS)/EG [34], and PS-PMMA/EG [35].

A crucial aspect in the production of polymer nanocomposites is the filler dispersion, which should be as homogeneous as possible. Aggregates of nanomaterials are proved to degrade the mechanical properties. Because the graphite sheets on EG interlock with each other firmly, it is not easy to disperse these graphite nanoplatelets effectively via blending the EG with polymers directly. Thus, in an attempt to enhance the nanodispersion of graphite in a polymer matrix, a simple but effective procedure was developed by Chen et al. to prepare individual graphite nanoplatelets from EG [36,37]. The EG particles were further fragmented by ultrasonication in an alcohol solution. The scanning electron micrograph (SEM) of Figure 4.4 reveals that the resulting graphite nanosheets (GNs) have thicknesses ranging from 30 to 80 nm and possess a high aspect ratio (width-to-thickness) of around 100 to 500, which is an advantage in forming the conducting network in a polymer matrix. The powder had an apparent density of 0.015 g/cm³, much smaller than that of

FIGURE 4.4
Scanning electron micrograph of graphite nanosheets: (a) lower magnification and (b) higher magnification. (Reprinted from Chen, L., Lu, L., Wu, D.J., and Chen, G.H., *Polym. Comp.*, 2007, 28: 493–498. Copyright 2007, Society of Plastics Engineers. With permission.)

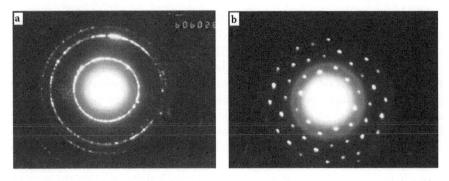

FIGURE 4.5
Selected area diffraction on graphite nanosheet (a) and conventional graphite (b). (Reprinted from Chen, G.H., Weng, W.G., Wu, D.J., and Wu, C.L., *Eur. Polym. J.*, 2003, 39: 2329–2335. Copyright 2003, Elsevier Ltd. With permission.)

the original flake graphite, 2.25 g/cm³ [38]. GNs offers people a new choice for fabricating electrically conductive polymer/graphite nanocomposites.

Selected area diffraction (SAD) conducted on a graphite nanosheet is very different from the SAD patterns of conventional graphite (Figure 4.5). According to the SAD patterns, the GNs crystals have been broken down into smaller ones by the thermal shock and the subsequent ultrasonication.

4.2.3 Graphene

Graphene is the one-atom-thick layer of graphite with carbon atoms arranged in a six-numbered-ring plane. Graphene sheets are believed to offer extraordinary electrical, thermal, and mechanical properties, and are expected to find a variety of applications.

Two routes are available for the preparation of graphene, one is from highly oriented pyrolytic graphite (HOPG), and the other is from graphite oxide (GO). In 2004, A.K. Geim originally peeled carbon layers from HOPG using "Scotch" tape and then captured the monolayers released in acetone on a Si Wafer with a SiO₂ layer on its top [39]. This reported method known as micromechanical cleavage led to the discovery of graphene sheets and become a strategy currently used in most experimental studies of graphene. However, this approach is severely limited by very low yield and, thus, is unsuitable for large-scale use (e.g., for fabricating polymer nanocomposites). Starting from GO is a promising methodology to prepare bulk quantities of graphene. By treating graphite with a mixture of concentrated sulfuric acid, sodium nitrate, and potassium of permanganate, natural graphite can be chemically converted into hydrophilic GO. Due to the presence of hydroxyl and epoxide on their basal planes and carbonyl and carboxyl groups located on the edge atoms, GO is water dispersible, but loses electrical conductivity. It has a layered structure similar to original graphite. When being sonicated in aqueous media, GO tends to exfoliate into single graphene oxide layers

FIGURE 4.6
Atomic force microscopy images of graphene on a silicon substrate, cast from a dilute dispersion in water. (Reprinted from Li, D., Muller, M.B., Gilje, S., Kaner, R.B., and Wallace, G.G., *Nat. Nanotechnol.* 2008, 3: 101–105. Copyright 2008, Nature Publishing Group. With permission.)

readily. After a subsequent chemical reduction, the electrically insulating graphene oxide layers are converted back to conducting graphene, presumably owing to the restoration of the graphitic network of the sp^2 bond during the reaction [40–42]. The reducing agents include hydrazine and its derivatives, such as dimethylhydrazine. Figure 4.6 shows an atomic force microscopy (AFM) image of a conducting graphene sheet produced via chemical conversion of GO on a silicon wafer [43].

4.3 Polymer/Graphite Nanocomposites

Polymer nanocomposites are of research focus both scientifically and commercially owing to their attractive potentials for enhancement of polymer properties. Improvements either in mechanical or functional properties are expected, provided that the nanofillers with a high aspect ratio can be well dispersed in the polymer, except for a strong interfacial interaction between them. Apart from the prototypical nanofillers, including layered silicates and carbon nanotubes, graphite nanoplatelets with thicknesses in nanometers are also among the leading nanoscale fillers in the research of nanocomposite systems. Similar to the silicate clay minerals, exfoliation of graphite layers is needed to achieve nanostructures. Thanks to the expandable GIC, such as graphite bisulfate, expanded graphite can be easily obtained from them. Expanded graphite offers great opportunities to incorporate graphite nano-platelets in polymer matrixes. In fact, in the early 1990s, expanded graphite had been developed and was proposed to be used as reinforcement in polymer systems. Since then, significant efforts have been made to optimize the processing method in order to achieve good exfoliation and dispersion of graphite and strong graphite matrix adhesion at the interface for improved strength and conductivity. Studies have been conducted intensely on exfoliated nanocomposites using graphite particles of various dimensions and a wide range of polymers [44,45]. Researchers have suggested that polymer/graphite nanocomposites possess many advantages compared to conventional microcomposites. The remarkable improvements in their properties are subjected to the high aspect ratio and the increased relative surface area of graphite nanostructures. The use of exfoliated graphite opens up many application sectors where electromagnetic shielding, high thermal conductivity, gas barrier resistance, and low flammability are needed [46–51].

In the following section, the fabrication methodologies, characterization techniques, novel properties, and potential applications of polymer/graphite nanocomposites will be discussed.

4.3.1 Manufacturing and Processing

For the manufacturing of polymer/graphite nanocomposites, the first step will be choosing a proper blending method to reach a satisfactory dispersion of graphite nanolayers throughout the matrix phase. Once the graphite is exfoliated, many methods utilized to fabricate and process conventional composites also work well for the nanocomposite counterparts. Based on the exfoliation of graphite, various polymer/graphite nanocomposites, thermoplastic polymers in particular, can be successfully fabricated by incorporating exfoliated graphite into an organic using conventional processing techniques such as extrusion, injection molding, and pressing.

Depending on the state of the organic during mixing, the main methods can be classified into three classes: *in situ* polymerization, solution compounding, and melt blending. Generally, *in situ* compounding methods give more satisfactory dispersion of fillers and stronger interactions between the reinforcements and polymeric phase. Thus, composites fabricated by *in situ* processing have better mechanical properties and a lower percolation threshold than those of the composites made by melt mixing and other *ex situ* fabrication methods. The three kinds of techniques will be discussed in detail.

4.3.1.1 In Situ *Polymerization*

In situ polymerization is a composite method where the monomer (or oligomer) is polymerized in the presence of fillers. The polymerization of monomer can be carried out either in a solution (called solution polymerization) or in the monomer system (called bulk polymerization). In the beginning, this approach was applied in manufacturing conventional polymer composite filled with graphite fiber [52,53], and then it was extended to fabricate polymer/EG nanocomposites in the 1990s [54]. Fabricating EG nanocomposites by *in situ* polymerization involves inhabitancy or "intercalation" of monomer into the larger EG interplanar spacing and a subsequent polymerization. The polymerization can be initiated by heat or radiation, and by a suitable initiator. The spaces include the pores and the enlarged sheet galleries of EG, as EG is of a worm-like structure with many pores in large sizes built by delaminated graphite nanosheets. The pores can be readily inhabited by monomers in a liquid state. Moreover, due to the high expansion ratio and the polar groups on the EG layers, the galleries of EG can be easily "intercalated" through physical adsorption [55]. Additionally, it is worth emphasizing that the "intercalation" mentioned here is very different from that used in the case of GIC because the original interplanar spacing of graphite crystal has been increased into much larger irregular voids embedded into each delaminated nanosheet by the evaporation of sulfate intercalate when GIC is subjected to heating. After polymerization, the interlinked nanosheets are separated into isolated nanoplatelets, and they reassemble, forming a dispersed phase in the matrix, resulting in a final composite filled with graphite nanoplatelets.

In situ polymerization is a convenient method that had been used widely for the preparation of polymer nanocomposites using EG particles as fillers. Researchers have used both thermoplastics and thermosets. Celzard et al. added ground EG flakes with an average diameter of 10 μm and an average thickness of 100 nm into epoxy resin to investigate the conductive behavior of the composite films [56]. It was only 1.3 vol% graphite additive that was needed to reach the percolation threshold. Hu et al. succesfully fabricated electrically conducting nylon-6/graphite nanocomposites via intercalation polymerization of ε-caprolactam in the presence of expanded graphite [55]. They found that the nylon-6 transit from an electrical insulator to an electrical semiconductor at 0.75 vol% of filler content, which is much lower than the conventional conducting composites. Chen et al. conducted *in situ* polymerization of styrene with EG in a sealed vessel at 150°C [58]. The obtained composites have a percolation threshold 1.8 wt% at which point the composite conductivity increased sharply. It exhibited excellent electrical conductivity of 10^{-2} Scm^{-1} when the graphite content was 2.8 ~ 3.0 wt%. Ou et al. carried out studies on the conductive mechanism of PS and nylon-6 nanocomposites with EG and managed to interpret the conductive phenomena using a model where a primary particle composed of a graphite particle, a compact-adsorbed layer, and a wrapping shell is assigned as the basic conductive unit in the composites [59]. Via *in situ* polymerization, EG particles have also been successfully incorporated into poly(methyl methacrylate) (PMMA), poly(vinyl chloride) (PVC), poly(styrene-methyl methacrylate), aromatic polydisulfide, and poly(methyl acrylic acid). The resulting polymer/graphite nanocomposites display a markedly low percolation threshold and high electrical conductivity due to the high aspect ratio (ratio of diameter to thickness) of graphite nanosheets in polymer matrix [60–63]. Transmission electron microscopy has shown good exfoliation of EG along with homogeneous graphite nanoplatelets dispersion in the final composites using this methodology.

Although great success has been achieved for polymer nanocomposites with EG, effective dispersion of graphite nanoplatelets from EG could hardly be achieved, especially for rubbers, because EG is somewhat partially exfoliated graphite in which the graphite subsheets with thicknesses in nanoscale are still interlinked with each other [64]. Pan et al. take advantage of this honeycombed structure to carry out *in situ* polymerization of styrene and acrylonitrile monomers inside the pores of EG, which have an expansion ratio as low as 30 [65]. SEM disclosed that the EG fillers maintain their honeycomb structures in the finished composites, contributing excellent conductivity to the matrix. To overcome this disadvantage in the dispersing of EG, Chen et al. sonicated EG to make separate graphite nanosheets and dispersed in PMMA and polystyrene via *in situ* polymerization [66–68]. As an individual graphite nanosheet still contains thinner sheets with spaces between them, intercalations of monomers consequently take place facilitating the exfoliation and dispersion of graphite. Meng et al. fabricated poly (4,4 prime-oxybis (benzene) disulfide)/graphite nanosheet composite via *in situ* ring-opening polymerization of cyclo(4,4'-oxybis (benzene)disulfide) oligomers with graphite nanosheets at 200°C [69]. The graphite nanosheets were intercalated by the oligomers and were dispersed well in a foliated state into the as-prepared composite as indicated by TEM micrograph. The intercalation polymerization process for forming a polystyrene/nanographite composite is shown schematically in Figure 4.7. The high aspect ratio characteristic and satisfactory dispersion of graphite nanosheets give remarkable advantages in establishing a conducting network in a polymer matrix, resulting in enhanced conductivities compared with conventional graphite powder. Being attracted by the high-diameter-to-thickness ratio and dispersion convenience of graphite nanosheets using *in*

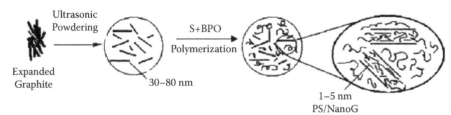

FIGURE 4.7
The process for the formation of polystyrene/graphite nanocomposite starting from expanded graphite. (Reproduced from Chen, G.H., Wu, C.L., Weng, W.G., Wu, D.J., and Yan, W.L., *Polymer*, 2003, 44(6): 1781–1784. Copyright 2007, Elsevier Ltd. With permission.)

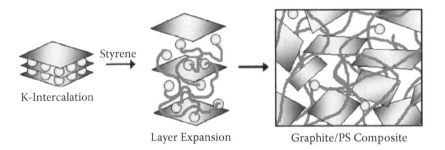

FIGURE 4.8
In situ polymerization of styrene in the potassium intercalated graphite galleries. (Reprinted from Kim, H., Hahn, H.T., Viculis, L.M., Gilje, S., and Kaner, R.B., *Carbon*, 2007, 45(7): 1578–1582. Copyright 2007, Elsevier Ltd. WIth permission.)

situ polymerization, intensive studies on polymer/graphite nanocomposite have been carried out with this new nanostructure as a conductive modifier. Different studies have suggested that the isolated graphite nanosheet is an excellent nanofiller for the fabrication of a polymer/graphite nanocomposite with high conductivity [70–72].

Apart from graphite bisulfate, potassium intercalated graphite is also a capable graphite nanoplatelet precursor which has been used in polymer/ graphite nanocomposite preparation. Gong et al. [73] carried out anionic polymerization of styrene in a tetrahydrofuran (THF) solution containing K-THF-GIC as the catalyst and filler at room temperature for 3 h. They obtained polystyrene/graphite nanocomposites with higher glass transition and thermal degradation temperatures compared with neat polystyrene. Their percolation threshold in conductivity is lower than 8.2 wt%. In another study [74], polystyrene/graphite nanocomposite was synthesized from styrene vapor and potassium intercalated graphite of a different stage. In their work, stage I, II, III, IV K-GIC compounds KC_8, KC_{24}, KC_{36}, and KC_{48} were prepared by heating foliated graphite with potassium in a sealed, evacuated Pyrex tube in a helium atmosphere at 220 for 3 days. When being exposed to saturated styrene vapor, the potassium intercalation complexes began to expand slowly as styrene penetrated into the K-GIC interlayer galleries. After intercalation, the monomers take part in anionic polymerization in the gallery spaces of K-GIC. During the polymerization, the spacing expands farther and the graphite layers are separated, as shown schematically in Figure 4.8. The resulting nanocomposites in the form of powder have good filler dispersion. This is an interesting way to achieve good exfoliation and dispersion of graphite nanoplatelets in polymer matrix.

In situ polymerization is an efficient way to improve the exfoliation and dispersion of foliated graphite in the matrix. Consequently, polymer/graphite nanocomposites with low percolation threshold and high conductivity can be achieved. However, this technique certainly has some disadvantages. For example, the compounding processing involves a polymerization of monomers forming macromolecular chains, and the active center responsible for the chain growth can be easily terminated by some reactive species such as radicals and ions derived from treated graphite, leading to polymers of low molecular weight and of broadened molecular-weight distribution, which is negative for the mechanical properties of the final composites. In addition, it is obvious that this compounding method is unworkable when the monomers are gaseous, such as in the cases of polyethylene, polystyrene, and other commercial resins. Moreover, it is usually electrical energy consuming, as heating is usually needed to promote reactions and to exfoliate and disperse graphite nanoplatelets into the polymer matrix during processing. These limitations may be the critical reasons why this method seems unsuitable for the production of polymer/graphite nanocomposites on an industrial scale.

4.3.1.2 Solution Compounding

Solution compounding is a blending technique that filler dispersion is accomplished in a solvent medium where polymers are dissolved. Once the mixing is completed, the solvent is removed, leaving bulk polymer material containing the fillers, and a melting molding is usually followed to give the hybrid products a regular shape. Obviously, the solution mixing method is restricted to areas where the polymer matrix is dissoluble in a certain solvent.

Being facilitated by the solution, the polymer macromolecules are able to encroach upon the pores of EG and "intercalate" into the enlarged layer spaces of the nanosheet through physical adsorption [75]. This process is known as "solution intercalation," and the polymers can be either thermoplastics or elastomers. Wong et al. [76] reported that only 1.0 wt% EG content was needed to satisfy the percolation threshold of PMMA/EG nanocomposite using chloroform as solvent. The EG embedded in matrix almost maintains the same morphology of the original EG particles as revealed by SEM. This phenomenon was also observed in Shen's work [77]. In their synthesis, EG was predispersed into xylene, and then the suspension was added into the xylene solution of maleic anhydride grafted polypropylene drop by drop while the solvent was kept in a refluxing state at 130°C. SEM images of the obtained *g*PP/EG composite show that the internal structure of EG in the matrix was identical to that of the start EG despite the graphite board thickness becoming somewhat less after compounding. Solution intercalation is also very welcome for elastomer matrices. George et al. [78] dissolved rubber-grade ethylene vinyl acetate (EVA) containing 60% vinyl acetate in THF solution with the addition of dicumyl peroxide and triallyl cyanurate as curing agent and co-cross-linker, respectively. After addition of the EG/THF suspension, the mixture was dried and hot pressed for a certain time for curing. TEM images show a fine dispersion of graphite nanoplatelets in EVA at a low filler concentration, whereas agglomerations are observed at high filler loading (8 phr). Using this method, nanocomposite synthesis has also been conducted with GNs. Zhang et al. [79] prepared NBR/

graphite nanocomposite by blending the rubber latex with GNs in a water medium giving a much finer filler dispersion of GNs in rubber matrix and an enhancement of interface adhesion between the graphite sheets and NBR matrix compared with the composites fabricated by directly blending GNs into NBR in a mixer. Chen et al. [80] chose hexane as a solvent to mix liquid silicon rubber (SR) with GNs. In their synthesis, the uniform dispersion of GNs into SR was accomplished by a combination of mechanical stirrer and ultrasonic vibration.

More recently, using the solution compounding approach, graphene-based composite materials have been generated for the first time, offering a notable way to harness the distinctive properties of the single carbon sheet. Stankovich et al. [81] incorporated individual graphene sheets into polystyrene using aprotic solvent N,N-dimethylformamide (DMF). In order to obtain a well-dispersed graphene suspension in DMF where polystyrene matrix can be dissolved, the graphite oxide was first chemically pretreated with phenyl isocyanate to increase their compatibility with the organics. After dissolving polystyrene into the DMF suspension, chemical reduction was carried out to increase the electrical conductivity of the graphene oxide sheets, and then the solution was precipitated by adding the suspension into methanol. The coagulation of graphene during the reduction step can be prevented by the polymers in solution.

Although the graphite nanoplatelets in EG can hardly be separated into isolated nanosheets in blending processing, consequently leading to serious filler agglomerations at a higher filler loading, solution intercalation is unquestionably a feasible blending technique to fabricate polymer nanocomposites once the graphite is exfoliated. The main limitation of this approach is that this technique usually requires large amounts of solvent and high temperatures to dissolve the polymers and, consequently, introduces serious environmental pollution.

4.3.1.3 Melt Blending

Melt blending is the most commonly used compounding technique in the plastics industry. This traditional blending approach which is usually carried out using a twin-screw extruder, injection molding, twin-roller, and internal mixer typically is also capable for manufacturing graphite nanocomposites with either thermoplastics or thermoelastomers. Research has suggested that melt mixing conducted with Masterbatch always gives more intensive flake delamination of EG and better nanosheets dispersion than direct mixing [82–84]. In addition, as EG is soft and has a loose structure, when being mixed with polymers, for example with rubbers such as NBR which usually have a much higher viscosity on twin-roller, the EG worms can be broken down into smaller pieces or compressed into a tighter form by the harsh shear forces, subsequently decreasing the reinforcement efficiency of EG [82,85].

Chen et al. [86] modified GNs with acrylonitrile-styrene (AS) resin to introduce GNs into polymers. The as-prepared GNs were first coated with AS
resin by coagulation of AS in its 2-butanone solution with GNs presence. The
designed Masterbatch, containing 70 to 80 wt% of GN fillers was then blended
with AS and high-density polyethylene (HDPE), respectively, by a single-
screw extruder to prepare the target nanocomposites. The results suggested
that the dispersion state of GNs in the final composites was strongly affected
by the miscibility of the AS-coated GNs with the polymer matrix. Good compatibility between the Masterbatch and polymer matrix led to much finer
GNs dispersion. When the polymeric phases in composites are immiscible,
such as with the HDPE/EVA system in which the two polymer matrices have
great differences in crystallinity, the GNs tend to locate in a certain polymer
phase where the interfacial tension between them is low [87].

Except for polymer/graphite nanocomposites made from GO, EG, and GNs,
all of which had been pretreated chemically and thermally before blending, graphite reinforced polymer nanocomposites have also been achieved
utilizing as-received graphite without any modification via melt extrusion
as reported by Torkelson et al. [88,89]. They accomplished polypropylene/
graphite nanocomposites by mixing polypropylene pellets and graphite particles in a modified twin-screw device with 25 mm of screw diameter and 26
of L/D. The results showed that the graphite in the extrusion products had a
thickness less than 10 nm.

Melt blending is a promising technique to volume-produce polymer nanocomposites industrially. This method is the most preferred for many industry
applications. But ensuring a constant homogeneous dispersion of graphite
nanoplatelets in a polymer by using the existing melt-blending devices is
still a big challenge. Efforts are needed to improve the processability of the
polymer/graphite nanosystems. Developing advanced devices and modifying graphite surface properties are recommended for future works.

4.3.2 Characterization Techniques

Characterization tolls undoubtedly have a significant role in facilitating the
studies of polymer nanocomposites. Powerful techniques for characterization are crucial for getting information from bulk materials and for compre-
hending the natural properties of polymer nanocomposites. The structural
information of the final graphite composites can be revealed through several
techniques, including wide-angle X-ray diffraction (WAXD), small-angle
X-ray scattering (SAXS), scanning electron microscopy (SEM), and transmission electron microscopy (TEM).

Both SEM and TEM have been used extensively to investigate the dispersion
and exfoliation states of graphite particles embedded in polymers. The SEM provides morphologies of the surfaces of the compounds, from which the interface
situation between graphite and polymer resins can be observed clearly. Because
the fractured surface of composite samples is less conductive, a golden coating

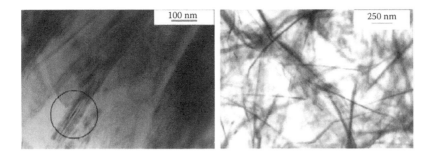

FIGURE 4.9
Transmission electron micrograph images of polymer/graphite nanocomposites. (Reprinted from Chen, G.H., Wu, D.J., Weng, W.G., and Wu, C.L., *Carbon*, 2003, 41(3): 619–621. Copyright 2002, Elsevier Science Ltd. With permission.)

is always needed before carrying out the SEM. Compared with SEM, TEM is more effective at showing the internal structures of composites along with spatial distribution of different phases. From TEM images, both the dispersion and exfoliation of the nanostructured graphite can be observed through direct visualization as presented in Figure 4.9. The dark line observed on the micrograph at a higher magnification corresponds to the graphite sheets with thickness about 10 nm. The circled area shows the subtle structures of thinner sheets with thickness of about 2 to 5 nm inside EG. The conductive network formed by the nanosheets in polymer matrix can also be seen clearly from the TEM image.

X-ray diffraction including WAXD and SAXS is the most common technique to probe the nanostructures of composites due to its ease of use and availability. It allows a quantitive understanding of the interlayer space change of layered materials when intercalation occurs [90]. The principle for the characterization is that increasing exfoliation of a layered material is associated with a decrease of diffraction peak intensity scanning from the repeated layers. For example, in nanocomposites containing fully exfoliated layered montmorillonite, the sharp (001) diffraction peak of montmorillonite disappeared in the XRD pattern after intercalation [91]. Zhang et al. [79] reported the WAXD analysis of NBR/GNs nanocomposite using the solution intercalation technique. Torkelson et al. [89] used WAXD to characterize the nanoscale dispersion of graphite in polypropylene matrix.

For thermal characterization (crystalline behaviors, glass transition and degradation temperatures, and rheological properties) and to study the compatibility between different polymeric phases of the graphite nanocomposites, traditional techniques including differential scanning calorimeter (DSC), thermogravimetric analysis (TGA), thermomechanical analysis (TMA), dynamic mechanical analysis (DMA), and rheometry are always workable. The following are examples. Katbab et al. incorporated EG into thermoplastic vulcanizate via melt compounding and carried out studies on the melt rheological behavior of the resulting compounds using parallel plate rheometry [92]. Feng et al. employed DMA to reveal the effect of mixing methods on the storage modulus of silicone rubber/EG composite [93]. In Pan's study, TGA was applied to test the thermal stability of poly(styrene-co-acrylonitrile)/EG nanocomposites [65]. Chen et al. used DSC to analyze the isothermal and nonisothermal crystallization and melting behaviors of nylon-6/GNs nanocomposites [94].

4.3.3 Properties and Potential Applications

Graphite is a rigid material with sound electrical conductivity. By the introduction of nanostructured graphite particles, the performances of polymers can be modified dramatically. The mechanical, thermal, and electrical properties of polymer/graphite nanocomposites are of research focus.

4.3.3.1 Electrical Properties

Graphite has a similar geometry to silicate clays. In addition, it possesses electrical conductivity, which is absent in clay minerals. When graphite powers are incorporated into polymers properly, they are capable of conveying interesting electrical properties to insulated polymeric materials. In the early 1970s, graphite had been compounded into polymer matrix as an electrically conductive filler [95].

In order to give a detailed description of the conduction behaviors of electrically conductive polymer composites, insulator–conductor transition, which is fundamental, has been studied extensively. The mechanism of electrical conduction of polymer composite is largely associated with a formation of a continuous conductive filler network throughout the insulating polymer matrix. When graphite fillers are added into the insulating polymers, the conductivity of the composites usually increases gradually with increasing filler loading. At the beginning, with a small volume fraction of graphite fillers, the resistivity of the composite is close to that of neat polymer. As the volume filler concentration reaches a certain value, a continuous conductive network is built up for the first time, and the polymer composites begin to transit from insulator to conductor showing a sharp decrease in resistivity by several orders of magnitude, as shown typically in Figure 4.10. This critical volume fraction of the conducting additive is referred to as percolation threshold P_c, and the transition characteristic of conductive polymer composites is known as percolation phenomena. Near the percolation threshold, which corresponds to a relatively narrow filler loading range, a small increase in filler concentration will result in a dramatic increase in conductivity.

The percolation phenomenon in conducting composites consisting of insulating matrices and dispersed conducting fillers is, indeed, a nonlinear behavior of conductivity as a function of the filler concentration. The nonlinear conduction in the polymer/graphite composite has also attracted attention [96–99].

The percolation threshold P_c is a critical parameter to determine the level of electrical conductivity of a complex disordered system based on polymers.

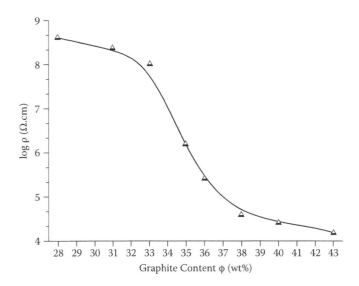

FIGURE 4.10
Resistivity of graphite-filled high-density polyethylene as a function of filler content. (Reprinted from Chen, G.H., Chen, X.F., Wang, H.Q., and Wu, D.J., *J. Appl. Polym. Sci.*, 2007, 103: 3470–3475. Copyright 2007, Elsevier Science Ltd. With permission.)

An important aspect in the conductive polymer composite production is the P_c, which should be minimized. Considering this another way, it is preferred to allow the composite to fulfill its electrical requirements at a filler fraction as low as possible because a heavy additive loading always leads to difficulty in mixture processing and is a detriment to mechanical properties. Therefore, many efforts have been devoted to decreasing the percolation threshold of graphite filler concentration in polymeric matrices, which are mainly based on the exfoliation and fragmentation of graphite particles, the optimization of mixing techniques, as well as the modification of graphite surface properties.

Exfoliation of graphite is an effective way to improve the reinforcement efficiency of graphite. The exfoliation allows the graphite platelets to provide the maximum efficiency for a given volume percentage. Polymer composites reinforced by exfoliated graphite always have a much lower P_c than that of the conventional composites incorporated by unexfoliated graphite powders. Wong et al. [100] reported that in high-density polyethylene (HDPE)/EG nanocomposites, fabricated by melt blending, their transition from an insulator to a conductor occurred at about 3.0 wt% of EG. It was significantly lower than 5.0 wt% of the untreated graphite reinforced composites that were manufactured under the same processing conditions. The advances of EG in forming a conductive network in polymer matrix ignite tremendous innovative ideas of developing novel polymeric materials with much lower P_c. In the past years, EG has been successfully incorporated into various polymeric systems with different compounding techniques, mainly including polyethylene [101,102], polypropylene

[103,104], polyamide [105], polystyrene [59,106], PMMA [31,76,107], and nylon [57,59]. GNs also possess remarkable advantages in forming conductive networks due to their high aspect ratio between the diameter and thickness. The critical percolation transition can be much more easily attained with only a slight amount of GNs filled. For example, the P_c values for HDPE composites filled with GNs, 7500, and 2000 mesh graphite were 6.0, 17, and 19 vol%, respectively, as reported by Chen et al. [108]. The advances of GNs as conductive nanofiller were also revealed in a silicon rubber system, where the P_c of GNs-reinforced nanocomposites is far less than that of the composites incorporated with 8000 and 2000 mesh graphite [38]. Researchers have shown that both *in situ* polymerization and solution intercalation are more beneficial in distributing exfoliated graphite (EG and GNs) into the polymer matrix. And, thus, composite materials fabricated by these two methods exhibit a much lower P_c than that of the corresponding composites prepared by direct melt mixing. One example is the PE/maleic anhydride grafted PE (g-PE)/EG composite system as reported by Shen et al. [80]. In addition, composites prepared by melting blending via Masterbatch usually exhibit higher electrical conductivity compared with those manufactured by direct melt mixing [82,109]. The conducting properties of the polymer/graphite composites can be interpreted theoretically by certain theories such as percolation theory, tunneling theory, and effective media theory. Chen et al. [110] prepared unsaturated polyester resin/GNs conducting composites with a low percolation threshold of 0.64 vol% via *in situ* polymerization under the application of ultrasonic irradiation. They found that the low critical volume concentration could be successfully interpreted by mean-field theory and excluded volume theory, which indicated that the morphologies of the GNs with a high aspect ratio play an active role in minimizing the P_c value.

As mentioned above, the mechanism of electrical conduction in a composite is the formation of a continuous conductive network formed by the fillers throughout the insulating polymer matrix. The conductive network can be altered or destroyed by some external fields such as temperature, pressure, and solvent [111], leading to a change in the overall conductivity of the composites. Li et al. investigated the electrical response of PMMA/EG conductive nanocomposite upon exposure to organic solvent vapors [112]. It was found that the electrical increased as the composite under atmosphere of tetrahydrofuran, DMF, chloroform, and acetone, and a positive vapor coefficient (PVC) effect was observed in the case of chloroform. Chen et al. [80] demonstrated that the silicon rubber nanocomposites containing 1.36 vol% GNs near the percolation threshold exhibit a super positive pressure coefficient (PPC) effect with good repeatability in the finger-pressure range from 0 to 0.8 MPa as shown in Figure 4.11. The experimental results can be explained well using the tunneling conduction model, suggesting that under external pressure, the charge carrier transport in silicon rubber/GNs composites is mainly controlled by the tunneling-conduction mechanism. In some cases, the resistivity decreases with increasing pressure, which is known as the

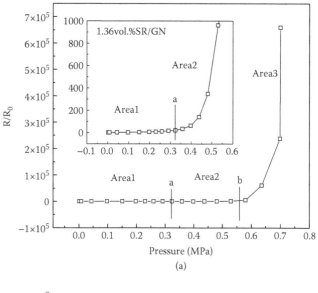

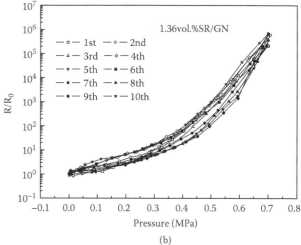

FIGURE 4.11

Relative resistance of silicon/GNs composite with 1.36 vol% GNs as a function of applied pressure (a) and its repeatable piezoresistive behaviors under cyclic compression (b). (Reprinted from Chen, L., Chen, G.H., and Lu, L., *Adv. Funct. Mater.*, 2007, 17(6): 898–904.)

negative pressure coefficient (NPC) effect, for example, in the polyethylene/GNs and polypropylene/EG systems as reported by Chen et al. and Wong et al., respectively [113,114]. When the pressure was fixed at a certain value, the resistivity of the composites varied as a function of time. This is closely related to the deformation of the polymer matrix caused by the creep of polymer chains under a fixed pressure [113,115].

Due to the temperature-dependent expansion of polymer matrix, conductive polymer composites consequently show a sharp increase in electrical resistivity at a transition temperature [116]. This phenomenon is defined as positive temperature (PTC). GNs possess high aspect ratios and thus gain great advantages over conventional graphite in forming conducting networks in polymer matrices. Due to the less needed GNs loading in the vicinity of the critical percolation point, the conductive networks in polymer matrix are much easier to disconnect when the matrix undergoes expansion at a certain temperature. Thus, the HDPE/GNs nanocomposites exhibit higher PTC intensity which is defined as the ratio of peak resistivity to the resistivity at room temperature, compared with that of the HDPE composites reinforced by traditional graphite powder of 2000 mesh [117].

These sensitive properties of polymer/graphite nanocomposites are crucial for a wide range of potential applications for pressure measurement and as a transducer for gas or liquid concentration. The composites as PTC materials have various technological applications such as current and temperature sensors, as well as detection and control of parameters related to temperature.

Polymer/graphite nanocomposites hold great potential applications as conductive materials. In order to meet applications such as electromagnetic shielding and electrostatic dissipation where high electrical conductivity is the most critical requirement, significant efforts have been directed toward improving the electrical conductivity of graphite fillers. Kim et al. [118] treated graphite nanoplatelets with bromination vapor and incorporated the modified fillers into epoxy resin. It was found that bromination treatment increased the absolute values of the electrical conductivity of the final composites at a filler content higher than the percolation threshold. Enhancement of the conductivity of graphite nanoplatelets was also conducted successfully by treating graphite using UV/ozone, resulting in improvement of the electrical conductivity of their epoxy-based composites [119,120].

Chen et al. [121] coated GNs with silver nanoparticles by electroless deposition and obtained silver-coated GNs with excellent conductivity up to 10^6 Sm^{-1}, which is equivalent to the conductivity of silver powder.

4.3.3.2 Thermal and Mechanical Properties

Both the thermal and mechanical behaviors of polymer/graphite nanocomposites are crucial properties for industrial applications and attracted research focus. The introduction of graphite particles into polymers certainly has a significant influence on the mechanical properties of the final composites. The thermal properties are very important for obtaining optimum material processing conditions and for analyzing heat transport in materials during practical applications. Fukushima et al. [50] investigated the thermal conductivities of nanocomposites of nylon-6, nylon-66, and HDPE incorporated by EG through melt extrusion and found that composites with EG up to 20 vol% exhibited thermal conductivities more than 4 $Wm^{-1}K^{-1}$, which is significantly higher than the control polymer matrix. Feng et al. [93] paid special attention to the thermal conductivity of silicon rubber/EG composites. In their work, the thermal conductivities of silicon rubber/EG nanocomposites increased with increasing EG content, and the solution intercalation method was proven to be better than the conventional melt mixing method in improving the thermal conductivity. Also, a higher storage modulus was observed for the solution-intercalated composites. Meng et al. reported that the tensile strength of NBR increased by 78% at a loading of only 5.0 wt% EG, and the increase in storage modulus was more than 90% [122]. Zhang and Mai [79] incorporated GNs into NBR latex to achieve rubber nanocomposites of greatly increased elastic modulus and tensile strength along with super thermal conductivity. Increases of tensile strength, modulus, and thermal conductivity were also observed in thermoplastic polymers, except for enhanced thermal degradation stability [73,123]. For example, with 4.0 wt% EG addition in EVA, the tensile strength and modulus at 100% elongation were increased by 35% and 150%, respectively, and the temperature of maximum degradation rate shifted to a higher temperature by 14°C [78]. The interfacial interaction between graphite nanoplatelets and matrix impacts the glass transition temperature (Tg) of polymer/graphite nanocomposites due to the suppression of the mobility of the macromolecular segment near the surface [120]. One example is the 5°C of Tg improvement with the addition of nanostructured graphite into polystyrene [73]. The flammability of polymeric materials is another important issue, as many disasters are caused by fire. Chuang et al. [51] investigated the influence of EG on the smoke formation and flammability of ethylene-propylene-diene terpolymer. They demonstrated that EG is an effective flame-retardant and smoke-suppressant additive for polymer compounds.

4.4 Summary

Recent achievements have demonstrated that graphite nanoplatelets are novel nanofillers to provide both dramatic reinforcement and multifunction to various polymers either thermoplastic or thermoset. The potential of polymer/graphite nanocomposites is tremendous and promising, undoubtedly as replacements for some of the current market-available products and in opening new applications through their outstanding properties. On the other hand, people still face many critical issues. One of the biggest challenges is to develop efficient processing–manufacturing technologies for complete exfoliation and dispersion of graphite layers in polymer matrix, especially in terms of industrial quantity and value for commercialization. Currently, it is difficult to achieve nanocomposites containing graphite of approved separation and dispersion using the traditional melt blending method. Thus, it is necessary to pay special attention to designing novel blending devices for nanocomposite manufacturing at the industry scale. Furthermore, from the intercalation mechanism viewpoint, in graphite the distance between adjacent carbon layers is only 0.335 nm. This interplanar space cannot be intercalated by polymer macromoleculars, which also limits the exfoliation and separation of graphite during melt mix processing. Fortunately, this limitation can be overcome by the emergence of graphene, the monolayer of graphite at atomic scale, which has a great deal of promise as a future conductive graphitic nanofiller for exploiting high-performance polymer materials. Although their large-scale production seems decades away.

Graphite is innocuous and biocompatible. It has been widely used as a lubricant and an electrically conductive material in many areas such as aerospace and ionic battery. Moreover, GO has been incorporated into silicon rubber to fabricate nanocomposites with novel anticoagulative properties, and good mechanical strength, biocompatibility, and longtime blood compatibility were observed [124]. In fact, the excellent biocompatible properties and high inertness make graphite materials attractive in constructing biomaterials, for instance, in medical dressings and heart valves [125,126].

Clearly, industry and a diverse range of other sectors related to polymer materials will benefit significantly from the introduction of nanostructured graphite, which is safe enough provided excellence techniques for the composition are available. However, the change in graphite scale from micrometer to nanometer poses new problems in processing techniques. Complete exfoliation of the graphite plate is the biggest challenge. Additionally, a homogeneous dispersion of graphite nanoplatelets in a polymer by using traditional compounding techniques is very difficult due to their strong tendency to agglomerate. To explore such attractive polymer nanocomposites and fully realize their novel properties, in-depth investigations surrounding fundamental issues including compounding techniques, compatibilities between graphite and polymers, and surface modifications are critical.

Acknowledgment

The authors are thankful for the support of the National Natural Science Foundation of China (NO.50373015, 20574025).

References

1. Ohe, K. and Naito, Y., *Jpn. J. Appl. Phys.*, 1971, 10: 99–108.
2. Squire, E., *Br. Plast. Rubber*, 1984, 35: 37–38.
3. Delvigs, P., *Polym. Compos.*, 1989, 10(2): 134–139.

4. Zhou, J. and Lucas, J.P., *J. Thermoplast. Compos. Mater.*,1996, 9(4): 316–328.
5. Busick, D.N., Spontak, R.J., and Balik, C.M., *Polymer*, 1999, 40(22): 6023–6029.
6. Zheng, Q., Song, Y., Wu, G., and Yi, X. *J. Polym. Sci. Part B*, 2001, 39(15): 2833–2842.
7. Zheng, W.G., Wong, S.C., and Sue, H.J., *Polymer*, 2002, 43(25): 6767–6773.
8. Krupa, I., Novak, I., and Chodak, I. *Synth. Met.*, 2004, 145(2–3): 245–252.
9. Li, J. and Kim, J.K. *Compos. Sci. Technol.*, 2007, 67(10): 2114–2120.
10. Du, X.S., Yu, Z.Z., Dasari, A., Ma, J., Mo, M.S., Meng, Y.Z., and Mai, Y.W. *Chem. Mater.*, 2008, 20(6): 2066–2068.
11. Bourdo, S.E. and Viswanathan, T. *Carbon*, 2005, 43(14): 2983–2988.
12. Bourdo, S., Li, Z.R., Biris, A.S., Watanabe, F., Viswanathan, T., and Pavel, I. *Adv. Funct. Mater.*, 2008, 18(3): 432–440.
13. Li, W., Johnson, C.L., and Wang, H.L., *Polymer*, 2004, 45(14): 4769–4775.
14. Gilje, S., Hughes, J., Beliciu, M., Choi, O., Hahn, H.T., and Kaner, R.B. *In: International SAMPE Symposium and Exhibition (Proceedings), v51, SAMPE 2006 Conference Proceedings — 2006*, p. 15.
15. Heo, S.I., Yun, J.C., Oh, K.S., and Han, K.S., *Adv. Compos. Mater. Off. J. Jpn. Soc. Compos. Mater.*, 2006, 15(1): 115–126.
16. Nagata, Kazuya, Iwabuki, Hitoshi, Nigo, and Hideyuki. *Compos. Interf.*, 1999, 6(5): 483–495.
17. Shiro, N., Katsuzo, O., and Toshio, K., *Trans. Jpn. Soc. Mech. Eng. Part C.*, 1994, 60(572): 1376–1381.
18. Yang, J., Tian, M., Jia, Q.X, Zhang, L.Q., and Li, X.L., *J. Appl. Polym. Sci.*, 2006, 102: 4007–4015.
19. Stair, W.K. *Handbook of Lubrication (Tribology), Vol. II: Dynamic Seals*, Booser, E.R., Ed., CRC Press: Boca Raton, FL, 1983, pp. 581–622.
20. Schweitz, J.A. and Ahman, L., *Friction and Wear of Polymer Composites*, Fredrich, K., Ed., Elsevier: Amsterdam, 1986, Chapter 9.
21. Zhang, W., Dehghani-Sanij, A.A., and Blackburn, R.S., *J. Mater. Sci.*, 2007, 42 (10): 3408–3418.
22. Thostenson, E., Li, C., and Chou, T., *Comp. Sci. Techn.*, 2005, 65: 491–516.
23. Fukushima, H., Drzal, L.T., Rook, B.P., and Rich, M.J., *J. Therm. Anal. Calor.*, 2006, 85(1): 235–238.
24. Stankovich, S., Dikin, D.A., Dommett, G.B., Kohlhaas, K.M., Zimney. E.J., Stach, E.A., Piner, R.D., Nguyen, S.T., and Ruoff, R.S., *Nature*, 2006, 442(20): 282–286.
25. Hussain, F., Hojjati, M., Okamoto, M., and Gorga, R.E., *J. Compos. Mater.*, 2006, 40 (17): 511–575.
26. Krueger, Q.J. and King, J.A., *Adv. Polym. Technol.*, 2003, 22(2): 96–111.
27. Li, J., Sham, M.L., Kim, J.K., and Marom, G., *Comp. Sci. Techn.*, 2006, 67(2): 296–305.
28. Nishimura, T., *MaKromol. Chem., Rapid. Commun.* 1980, 1: 573–579.
29. Shioyama, H. and Fujii, R., *Carbon*, 1987, 25(6): 771–774.
30. Chen, G.H., Weng, W.G., Wu, D.J., and Wu, C.L., *Eur. Polym. J.*, 2003, 39: 2329–2335.
31. Celzard, A., Mcrae, E., Mareche, J.F., Furdin, G., Dufort, M., and Deleuze, C. *Phys. Chem. Solids*, 1996, 57: 715–718.
32. Zheng, W. and Wong, S.C., *Compos. Sci. Technol.*, 2003, 63: 225–235.
33. Wang, W.P., Pan, C.Y., and Wu, J.S. *J. Phys. Chem. Solids*, 2005, 66(10): 1695–1700.
34. Xiao, P., Xiao, M., and Gong, K., *Polymer*, 2001, 42: 4813–4816.

35. Chen, G.H., Wu, D.J., Weng, W.G., and Yan, W.L., *J. Appl. Polym. Sci.*, 2001, 82: 2506–2513.
36. Chen, G.H., Wu, D.J., Weng, W.G., and Wu. C.L., *Carbon*, 2003, 41(3): 619–621.
37. Chen, G.H., Wu, C.L., Weng, W.G., Wu, D.J., and Yan, W.L., *Polymer*, 2003, 44(6): 1781–1784.
38. Chen, G.H., Weng, W.G., Wu, D.J., and Wu, C.L., *Eur. Polym. J.*, 2003, 39: 2329–2335.
39. Novoselov, K.S., Geim, A.K., Morozov, S.V., Jiang, D., Zhang, Y., Dubonos, S.V., Grigorie-va, I.V., and Firsov, A.A., *Science*, 2004, 306: 666–669.
40. Stankovich, S., Dikin, D.A., Piner, R.D., Kohlhaas, K.A., Kleinhammes, A., Jia, Y., Wu, Y., Nguyen, S.T., and Ruoff, R.S., *Carbon*, 2007, 45: 1558–1565.
41. Gilje, S., Han, S., Wang, M., Wang, K.L., and Kaner, R.B., *Nano. Lett.*, 2007, 7(11): 3394–3398.
42. Stankovich, S., Dikinl, D.A., Dommett, G.B., Kohlhaas, K.M., Zimney, E.J., Stach, E.A., Piner, R.D., Nguyen, S.T., and Ruoff, R.S., *Nature*, 2006, 442(20): 282–286.
43. Li, D., Muller, M.B., Gilje, S., Kaner, R.B., and Wallace, G.G., *Nat. Nanotechnol.*, 2008, 3: 101–105.
44. Kalaitzidou, K., Fukushima, H., and Drzal, L.T., *Compos. Sci. Technol.*, 2007, 67(10): 2045–2051.
45. Yang, J., Tian, M., Jia, Q.X, Zhang, L.Q., and Li, X.L., *J. Appl. Polym. Sci.*, 2006, 102: 4007–4015.
46. Pramanik, P.K., Khastgir, D., and Saha, T.N., *J. Elastomers Plast.*, 1991, 23(4): 345–361.
47. Foy, J.V. and Lindt, J.T. *Polym. Compos.*, 1987, 8(6): 419–426.
48. Hussain, M., Choa, Y.H., and Niihara, K., *Scripta. Mater.*, 2001, 44(8–9): 1203–1206.
49. Todorova, Z., El-Tantawy, F., Dishovsky, N., and Dimitrov, R., *J. Appl. Polym. Sci.*, 2007, 103: 2158–2165.
50. Fukushima, H., Drzal, L.T., Rook, B.P., and Rich, M.J., *J. Therm. Anal. Calor.*, 2006, 85(1): 235–238.
51. Chuang, T.H., Chern, C.K., and Guo, W.J., *J. Polym. Res.*, 1997, 4(3): 153–158.
52. Litt, M.H. and Brinkmann, A.W., *J. Elastoplast.*, 1973 (5): 153–160.
53. Serafini, T.T., Delvigs, P., and Vannucci, R.D., *In: Society of the Plastics Industry, Reinforced Plastics/Composites Institute, Annual Conference, Proceedings—1974*, p. 6.
54. Chung, D.D.L., Composites of *in situ* exfoliated graphite. U.S. patent, 4946892, 1990.
55. Martin, C.A., Sandler, J.K.W., Shaffer, M.S.P., Schwarz, M.K., Bauhofer, W., Schulte, K., and Windle, A.H., *Comp. Sci. Techn.*, 2004, 64(15): 2309–2316.
56. Celzard, A., McRae, E., Mareche, J.F., Furdin, G., Dufort, M., and Deleuze, C., *J. Phys. Chem. Solids*, 1996, 57(6–8): 715–718.
57. Pan,Y.X., Yu, Z.Z., Ou, Y.C., and Hu, G.H., *J. Polym. Sci. Part B*, 2000, 38(12): 1626–1633.
58. Chen, G.H., Wu, D.J., Weng, W.G., He, B., and Yan, W.L., *Polym. Int.*, 2001, 50: 980–985.
59. Zou, J.F., Yu, Z.Z., Pan, Y.X., Fang, X.P., and Ou, Y.C., *J. Polym. Sci. Part B*, 2002, 40(10): 954–963.
60. Chen, G.H., Wu, D.J., Weng, W.G., and Yan, W.L., *Polym. Eng. Sci.*, 2001, 41(12): 2148–2154.
61. Chen, G.H., Wu, D.J., Weng, W.G., and Yan, W.L., *J. Appl. Polym. Sci.*, 2001, 82: 2506–2513.

62. Song, L.N., Xiao, M., and Meng, Y.Z., *Compos. Sci. Technol.*, 2006, 66(13): 2156–2162.
63. Li, L.W., Luo, Y.L., and Li, Z.Q., *Smart. Mater. Struct.*, 2007, 16(5): 1570–1574.
64. Weng, W.G., Chen, G.H., Wu, D.J., Chen, X.F., Lu, J.R., and Wang, P.P., *J. Polym. Sci., Part B*, 2004, 42: 2844–2856.
65. Zheng, G.H., Wu, J.D., Wang, W.P., and Pan, C.Y., *Carbon*, 2004, 42(14): 2839–2847.
66. Chen, G.H., Wu, D.J., Weng, W.G., and Wu, C.L., *Carbon*, 2003, 41: 619–621.
67. Chen, G.H., Weng, W.G., Wu, D.J., and Wu, C.L., *Eur. Polym. J.*, 2003, 39(12): 2329–2335.
68. Chen, G.H., Wu, C.L., Weng, W.G., Wu, D.J., and Yan, W.L., *Polymer*, 2003, 44(6): 1781–1784.
69. Du, X.S., Xiao, M., Meng, Y.Z., and Hay, A.S., *Polymer*, 2004, 45(19): 6713–6718.
70. Chen, G.H., Weng, W.G., Wu, D.J., and Wu, C.L., *J. Polym. Sci. Part B*, 2004, 42: 155–167.
71. Mo, Z.L., Zuo, D.D., Chen, H., Sun, Y.X., and Zhang, P. *Chinese J. Inorg. Chem.*, 2007, 23(2): 265–269.
72. Mo, Z.L., Sun, Y.X., Chen, H., Zhang, P., Zuo, D.D., Liu, Y.Z., and Li, H.J., *Polymer*, 2005, 46(26): 12670–12676.
73. Xiao, M., Sun, L., Liu, J., Li, Y., and Gong, K., *Polymer*, 2002, 43(8): 2245–2248.
74. Kim, H., Hahn, H.T., Viculis, L.M., Gilje, S., and Kaner, R.B., *Carbon*, 2007, 45(7): 1578–1582.
75. Kotov, N.A., Dekany, I., and Fendler, J.H., *Adv. Mater.*, 1996, 8: 637–641.
76. Zheng, W., Wong, S.H., and Sueb, H.J., *Polymer*, 2002, 73: 6767–6773.
77. Shen, J.W., Chen, X.M., and Huang, W.Y., *J. Appl. Polym. Sci.*, 2003, 88: 1864–1869.
78. George, J.J. and Bhowmick, A.K., *J. Mater. Sci.*, 2008, 43(1): 702–708.
79. Yang, J., Tian, M., Jia, Q.X., Shi, J.H., Zhang, L.Q., Lim, S.H., Yu, Z.Z., and Mai, Y.W., *Acta. Mater.*, 2007, 55: 6372–6382.
80. Chen, L., Chen, G.H., and Lu, L., *Adv. Funct. Mater.*, 2007, 17(6): 898–904.
81. Stankovich, S., Dikinl, D.A., Dommett, G.B., Kohlhaas, K.M., Zimney, E.J., Stach, E.A., Piner, R.D., Nguyen, S.T., and Ruoff, R.S., *Nature*, 2006, 442(20): 282–286.
82. Shen, J.W., Huang, W.Y., Zuo, S.W., and Hou, J., *J. Appl. Polym. Sci.*, 2005, 97(1): 51–59.
83. She, Y.H., Chen, G.H., and Wu, D.J., *Polym. Int.*, 2007, 56: 679–685.
84. Li, Y.C. and Chen, G.H., *Polym. Eng. Sci.*, 2007, 47: 882–888.
85. Yang, J., Tian, M., Jia, Q.X., Zhang, L.Q., and Li, X.L., *J. Appl. Polym. Sci.*, 2006, 102: 4007–4015.
86. Chen, G.H., Chen, X.F., Wang, H.Q., and Wu, D.J., *J. Appl. Polym. Sci.*, 2007, 103: 3470–3475.
87. Chen, G.H., Lu, J.R., and Wu, D.J., *Mater. Chem. Phys.*, 2007, 104: 240–243.
88. Furgiuele, N., Lebovitz, A.H., Khait, K., and Torkelson, J.M., *Macromolecules*, 2000, 33(2): 225–228.
89. Wakabayashi, K., Pierre, C., Dikin, D.A., Ruoff, R.S., Ramanathan, T., Brinson, L.C., and Torkelson, J.M., *Macromolecules*, 2008, 41: 1905–1908.
90. Yano, K., Usuki, A., Okada, A., Kurauchi, T., and Kamigaito, O., *Polym. Prep.* (Jpn), 1991, 32(1): 65.
91. Chen, G.H., Chen, X.Q., Lin, Z.Y., and Ye, W.L., *J. Mater. Sci. Lett.*, 1999, 18: 1761–1763.

92. Katbab, A.A., Hrymak, A.N., and Kasmadjian, K., *J. Appl. Polym. Sci.*, 2008, 107: 3425–3433.
93. Mu, Q. and Feng, S., *Thermochim. Acta.*, 2007, 462: 70–75.
94. Weng, W.G., Chen, G.H., and Wu, D.J., *Polymer*, 2003, 44: 8119–8132.
95. Ohe, K. and Naito, Y., *Jpn. J. Appl. Phys.* 1971, 10: 99–108.
96. Lin, H.F., Lu, W., and Chen, G.H., *Physica B*, 2007, 400: 229–236.
97. Chen, G.H., Weng, W.G., Wu, D.J., and Wu, C.L., *J. Polym. Sci., Part B: Polym. Phys.*, 2004, 42: 155–167.
98. Kannarpady, G.K., Mohan, B., and Bhattacharyya, A., *J. Appl. Polym. Sci.*, 2007, 106(1): 293–298.
99. Srivastava, N.K., Sachdev, V.K., and Mehra, R.M. *J. Appl. Polym. Sci.*, 2007, 104(3): 2027–2033.
100. Zheng, W., Lu, X.H., and Wong, S.C., *J. Appl. Polym. Sci.*, 2003, 91(5): 2781–2788.
101. Yang, Y.F. and Liu, M.J. *China Plast.*, 2002, 16(10): 46–48.
102. Weng, W.G., Chen, G.H., Wu, D.J., and Yan, W.L., *Compos. Interface.*, 2004, 11(12): 131–143.
103. Semko, L.S., Popov, R.E., and Chernysh, I.G., *Plasticheskie Massy*, 1996, 6: 22–25.
104. Chen, X.M., Shen, J.W., and Huang, W.Y., *J. Mater. Sci. Lett.*, 2002, 21(3): 213–214.
105. Uhl, F.M., Yao, Q., Nakajima, H., Manias, E., and Wilkie, C.A., *Polym. Degrad. Stabil.* 2005, 89(1): 70–84.
106. Wang, W.P. and Pan, C.Y., *J. Polym. Sci., Part A: Polym. Chem.*, 2003, 41(17): 2715–2721.
107. Wang, W.P., Liu, Y., Li, X.X., and You, Y.Z. *J. Appl. Polym. Sci.*, 2006, 100(2): 1427–1431.
108. Lu, J.R., Weng, W.G., Chen, X.F., Wu, D.J., Wu, C.L., and Chen, G.H., *Adv. Funct. Mater.*, 2005, 15: 1358–1363.
109. Zhang, M., Li, D.J., Wu, D.F., Yan, C.H., Lu, P., and Qiu, G.M., *J. Appl. Polym. Sci.*, 2008, 108(3): 1482–1489.
110. Lu, W., Lin, H.F., Wu, D.J., and Chen, G.H., *Polymer*, 2006, 47: 4440–4444.
111. Lundberg, B. and Sundqvist, B. *J. Appl. Phys.*, 1986, 60(3): 1074–1079.
112. Li, L.W., Luo, Y.L., and Li, Z.Q., *Smart. Mater. Struct.*, 2007, 16(5): 1570–1574.
113. Qu, S.Y. and Wong, S.C. *Compos. Sci. Technol.* 2007, 67(2): 231–237.
114. Lu, J.R., Chen, X.F., Lu, W., and Chen, G.H., *Eur. Polym. J.*, 2006, 42(5): 1015–1021.
115. Chen, G.H., Lu, J.R., Lu, W., Wu, D.J., and Wu, C.L., *Polym. Int.*, 2005, 54: 1689–1693.
116. Fournier, J., Boiteux, G., and Seyter, G., *J. Mater. Sci. Lett.*, 1997, 16: 1677–1679.
117. Chen, G.H., Lu, J.R., and Wu, D.J., *J. Mater. Sci.*, 2005, 40(18): 5041–5043.
118. Li, J., Vaisman, L., Marom, G., and Kim, J.K. *Carbon*, 2007, 45(4): 744–750.
119. Li, J., Kim, J.K., Sham, M.L., *Scripta Mater.*, 2005, 53(2): 235–240.
120. Li, J., Sham, M.L., Kim, J.K., and Marom, G. *Compos. Sci. Technol.*, 2007, 67(2): 296–305.
121. Sun, W.F., Chen, G.H., and Zheng, L.L., *Scripta Mater.*, 2008, 59: 1031–1034.
122. Liu, D.W., Du, X.S., and Meng, Y.Z., *Polym. Polym. Compos.*, 2005, 13(8): 815–821.
123. Uhl, F.M. and Wilkie, C.A., *Polym. Degrad. Stabil.*, 2002, 76(1): 111–122.
124. Chen, Y.H., Zhou, N.L., Meng, N., Hang, Y.X., Gao, N.X., Li, L., Zhang, J., Wei, S.H., and Shen, J., *Funct. Mater.*, 2007, 38(3): 438–440.

5

Polymer Nanocomposite Flammability and Flame Retardancy

Alexander B. Morgan
University of Dayton Research Institute, Ohio

CONTENTS

5.1 Introduction

Polymer nanocomposites as a field of materials science and chemical research has greatly expanded over the last 15 years since the first papers on polymer nanocomposites were published, but to some extent the field goes back farther than this. The initial papers from Toyota Research and Development (R&D) about their polyamide-6 nanocomposites in the early 1990s actually refer to these materials as "hybrids" [1,2], while the first use of the term "nanocomposite" was used in early papers by Giannelis [3,4] and Pinnavaia [5,6] in the mid-1990s. Regardless of the term used, polymer nanocomposite research expanded greatly past this point once it was realized that an inorganic particle of nanoscale dimensions could be dispersed into a polymer matrix to achieve enhancements in properties above and beyond that expected for a traditional filled system. There are numerous reviews and books on the subject of polymer nanocomposites today and what they can do [7–12], including the book that this chapter appears in, but the feature that has really driven the field of polymer nanocomposite research is that these materials are multifunctional. The nanoscale particles do more than just reinforce the polymer at the nanoscale, they completely change and often enhance the properties of the entire composite in the areas of thermal, electrical, mechanical, gas-barrier, optical, and flammability properties. This last property, flammability, is the area of emphasis of this chapter, which will explain how nanocomposites enhance flammability performance, how nanocomposites can provide improvements in flame retardancy, and where research in this field is likely headed in the next 5 to 10 years.

Before discussing the future of these materials or how they work, some background of polymer nanocomposite flammability is needed. Although the term "nanocomposite" has been a more recent definition in polymeric materials research, the use of nanoparticles in polymers goes back much farther in time. For example, carbon blacks (nanoscale in size) have long been used in tires, and organically treated layered silicates (clays) have long been used in paints as a rheology modifier. The difference is that these earlier materials were not understood, as such, to be nanocomposites (polymer plus nanoscale particle systems), but they showed properties that appear in polymer nanocomposite materials today. Some of the earliest work of enhanced thermal stability of a polymer plus nanoparticle (layered silicate) hybrid was done by Blumstein in the 1960s, where significant enhancements in thermal stability were noted in a high montmorillonite content, low polymer content system [13–22]. The general property shown for these clay–polymer hybrids was significant increases in onsets of polymer decomposition temperature. Blumstein hypothesized that the polymer intercalated in between the clay plates (the gallery region) was being protected from thermal decomposition by the clay plates. These results hinted at the improved flammability performance of polymer nanocomposite materials. Later, patents in the 1980s showed that small additions of clays to polymers would yield some antidripping effects for polymers exposed to flames [23,24]. These early results were largely forgotten until Gianellis in collaboration with Gilman and other researchers at the U.S. National Institute of Standards and Technology (NIST) [25,26] found that polymer–clay hybrids rich in the polymer phase showed enhanced char formation, antidripping behavior in vertical flame tests, and significant reductions in heat release by cone calorimeter [27,28]. With these results, materials clearly identified as polymer nanocomposites began to show that significant reductions in flammability were another feature of polymer nanocomposite enhanced properties. Not only were polymer nanocomposites showing increases in thermal stability, but once aflame, hey showed that they burned slower and with less heat release, making them potential fire-safe materials for new applications. From this moment on (late 1999, early 2000), research into polymer nanocomposite flammability really expanded and "took off," with major efforts being led by the NIST, under the leadership of Dr. Jeff Gilman, as well as efforts in academia with Professor Charles Wilkie (Marquette University, Milwaukee, Wisconsin) expanding upon chemical studies of polymer nanocomposite flammability, and the combination of other flame retardants with polymer nanocomposites being pioneered by Professor Serge Bourbigot (University of Lille, France) and Professor Giovanni Camino (University of Torino, Italy). This phenomenon of enhanced flammability for polymer nanocomposites has now been shown in many polymer systems and with just about all types of nanofillers, from layered silicates to carbon nanotubes/nanofibers to colloidal nanoparticles. Most of the work in this field can be found in greater detail in recent reviews and books on the subject [29,30], but this chapter will attempt to summarize the field as it stands today and explain how polymer nanocomposites reduce flammability and can be used as flame-retardant materials.

5.2 Polymer Nanocomposite Flammability

As previously mentioned, polymer nanocomposites show enhanced flammability performance over traditional composites or even "microcomposites," which are polymers filled with micron-sized fillers. Indeed, the precursors of the nanoparticles which make up polymer nanocomposite structures are typically micron-sized particles, and if these particles are not dispersed into the polymer matrix, then a microcomposite is formed rather than a nanocomposite. Specifically, the nanoparticles are agglomerated into micron-sized particles that often must be broken apart in (or before) the polymer processing step before the nanoparticles can form a nanocomposite. For layered silicates such as clays, these primary particles are called "tactoids," and if the tactoids are not broken up, then no reduction of polymer flammability will result. Instead, the clay agglomerate acts as a traditional filler, and the only reduction in flammability obtained is one caused by replacement of flammable polymer with nonflammable filler. In effect, a microcomposite only reduces flammability by diluting the total amount of fuel present to burn. The point of mentioning microcomposite versus nanocomposite is that it must be understood that anti-drip behavior and flammability reductions are only found with polymer nanocomposites, not microcomposites, and therefore the researcher, material scientist, or flame-retardant polymer formulator must pay attention to how the nanocomposite is synthesized. One cannot just add any type of nanoparticle to a polymer matrix and expect a nanocomposite to form.

Careful attention to nanoparticle interface design, polymer processing conditions, and desired nanocomposite structure must occur for a successful nanocomposite to be made. This chapter will not focus greatly on nanocomposite synthesis, as that is covered in other chapters of this book, but some comments on nanoparticle dispersion, interface, and processing will be provided in this chapter to show how these factors correlate to observed flammability effects and the mechanism of flammability reduction for polymer nanocomposites.

Flammability of polymer nanocomposites will be discussed in the following sections by grouping nanocomposites into three categories: layered materials, rods/tubes, and colloidal nanocomposites.

5.2.1 Layered Nanocomposite Flammability

Of all the polymer nanocomposites studied, nanocomposites based upon layered nanoscale platelets are the most well studied and published class of nanocomposites to date. This includes studies on flammability effects in specific polymers and the mechanism of flammability reduction. This class of nanocomposites includes natural layered silicates (clays) such as montmorillonite, as well as synthetic layered silicates such as fluorinated synthetic mica. It also includes work with double-layered hydroxides and a small amount of work on graphite and graphite-oxide systems [31–34]. Within this category, nanocomposites are typically defined by their nanoparticle dispersion structure, involving the following general definitions:

- *Microcomposite*: Like that described in the previous section, a microcomposite is a structure in which the layered nanoparticles have not deagglomerated and remain in their original tactoids. This is sometimes referred to as an immiscible nanocomposite due to the fact that the layered material (nanoplatelets) is not miscible in the polymer matrix and phase separates to give micron-sized agglomerates.

- *Intercalated*: For this system, polymer chains have entered the interlayer (gallery) spacing between the layered structure (silicate, double layered hydroxide, graphite, or other material), or "intercalated" into the gallery area of the layered material. Primary tactoids have not broken up completely (longer range order is still maintained), but each layered material gallery area now has some polymer separating each layer, although not enough to completely disperse the layered material evenly through the matrix. Intercalation is usually the first step on the way to exfoliation, and sometimes there are polymer nanocomposites that show both intercalated and exfoliated structures.

- *Exfoliated*: This category is used to describe individual layered plates that are now fully surrounded by polymer, and sometimes long-range order relative to the original tactoids has been lost as well.

Therefore, exfoliated nanocomposites can be put into two categories—ordered and disordered. The exfoliated structure (either in ordered or disordered form) is the hardest to obtain in a polymer nanocomposite and also has been somewhat subjective in its definition [35–37]. Some would argue that two- to three-layer intercalated stacks throughout the polymer matrix is a sign of exfoliation, while others would argue that only single layers throughout the polymer matrix are truly exfoliated.

A schematic of these general structures is shown in Figure 5.1, with the straight lines representing a layered structure seen edge on, and the wavy lines representing polymer chains.

For any significant flammability reduction to occur, one of the intercalated, intercalated/exfoliated, or ordered/disordered exfoliated nanocomposite structures must be present. If a microcomposite is present, as mentioned at the start of this section, the only flammability reduction that will occur will be a simple dilution effect from the amount of inorganic filler added. This effect is typically additive, so, for example, if 5 wt% of a montmorillonite clay was added to a polymer that yielded a microcomposite, one would typically see a reduction of heat release in that same microcomposite of about 5% compared to the base polymer. For the nanocomposites however, heat release rates as measured by the cone calorimeter are dramatically reduced by anywhere from 20% to 70% depending upon nanoparticle loading level, polymer type, and sample thickness [30]. The cone calorimeter, an oxygen consumption calorimetry technique [38], is often used to assess the fundamental flammability of a material, which includes heat release, mass loss, and smoke production rates. It has been the tool of choice to understand polymer nanocomposite flammability behavior, and some of the discussion in this chapter will refer to data generated by this technique. For how this technique correlates to other types of flammability tests, especially regulatory tests, the reader is directed to some papers on the topic [39–44], as this longer discussion is beyond the scope of this chapter. In regards to understanding polymer nanocomposite flammability, it is important to discuss

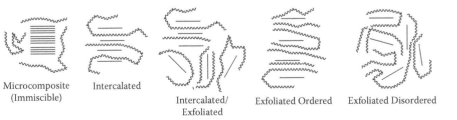

| Microcomposite (Immiscible) | Intercalated | Intercalated/ Exfoliated | Exfoliated Ordered | Exfoliated Disordered |

FIGURE 5.1
Layered nanoparticle polymer nanocomposite structures.

what the test has measured and what this measurement means in regard to the mechanism of flammability reduction.

When the cone calorimeter has studied polymer nanocomposite flammability, the measurements of heat release rate are typically recorded and used to describe the flammability reduction. The cone calorimeter was originally designed to serve as a fire safety engineering tool by assessing the heat release, especially peak heat release rate, of a burning material. The peak heat release rate (HRR) can be defined simply as the maximum of heat release rate at a point in time of the combustion of a material, but from a fire safety perspective, the peak HRR is the point in the burning of material that it may give off enough heat to ignite other nearby objects or greatly reduce the time to escape for people near the burning object [45,46]. For this reason, peak HRR is the measurement that most researchers of material flammability/flame retardants spend the most time studying due to its significance in fire safety engineering. Many of the other heat release and fire-related measurements from the cone calorimeter have been shown to be of equally important value [47,48], but peak HRR retains a great deal of importance; hence, it is emphasized when studying polymer nanocomposite flammability. For polymer-layered silicate nanocomposites, as well as nanocomposites made from double-layered hydroxides and graphite nanoparticles, significant reductions in peak HRR are almost always observed. A generic example of this is shown in Figure 5.2, where the large reduction in peak HRR compared to the base polymer can easily be seen. The general mechanism for this reduction in peak HRR, and likewise a change in

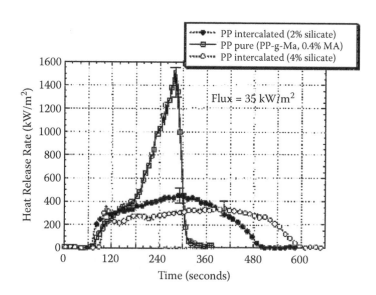

FIGURE 5.2
Heat-release rate curves for polypropylene and clay nanocomposites.

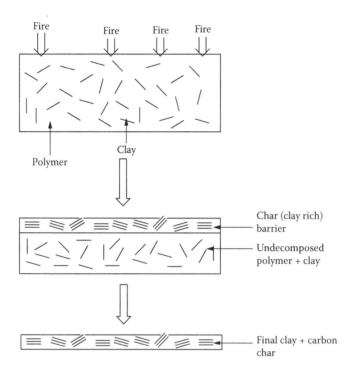

FIGURE 5.3
Idealized scheme showing how clay collapses down to form a protective barrier.

HRR curve shape seen in this figure due to the formation of a protective nanoparticle-rich barrier as the polymer pyrolyzes away. Whether this is caused by plates collapsing down to form the barrier [27,28,49–52] or nano-platelets migrating up toward the pyrolysis zone of the burning plastic [53] is not exactly clear, but most data support the collapse mechanism. Once the barrier sets up, it decreases the rate of mass loss/pyrolysis from the burning polymer, which in turn lowers heat release (less fuel burned = less heat release). The barrier also provides some thermal protection to the under-lying polymer, but eventually this too will be consumed. A general sche-matic of this mechanism is shown in Figure 5.3. In general, the heat release rate of a polymer plus a layered material nanocomposite closely follows the mass loss rate of the nanocomposite, indicating that polymer-layered material nanocomposites are primarily a condensed phase mechanism of flame retardancy. This mechanism is changed somewhat with the use of double-layered hydroxides that also release water from their structure upon thermal degradation, but even for these materials, the mechanism of flame retardancy is still primarily condensed phase with delays in mass loss rate being the primary feature [54,55]. For all of these nanocomposites, the final structure after the burning of the sample is an enhanced char structure, especially for polymers that typically form no char. For many materials,

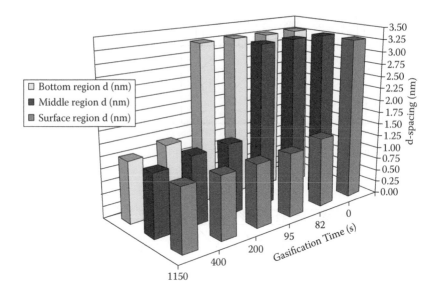

FIGURE 5.4
Plot of clay d-spacing versus heat exposure (gasification time) as a function of depth/analysis region for a polystyrene-clay nanocomposite [50].

the original shape of the object exposed to flame is maintained with the polymer nanocomposite, because the material does not drip and flow during burning, again unlike most commodity thermoplastics. The final chars typically show collapsed and concentrated layered particles that provided the protective barrier during the burning of the sample. A study published in 2006 studied this in more detail by evaluating changes in clay interlayer spacing of a polystyrene plus clay nanocomposite as a function of burning time in a modified cone calorimeter known as a gasification apparatus [50]. The samples are pyrolyzed under nitrogen conditions so that samples can be retrieved at different points of decomposition along the mass loss rate curve and analyzed. This study clearly found that the char formed at the surface was rich in clay content, and under the char layer, the clay would maintain its original clay d-spacing (as measured by X-ray diffraction [XRD]) until the heat penetrated deeper into the sample as time progressed (Figure 5.4). It is important to emphasize that this protective layer only slowed mass loss rate and penetration of heat into the sample, it did not stop it. More discussion on this point will be provided in the next section on the use of nanocomposites combined with other flame retardants.

5.2.2 Nanoplatelet Chemistry and Exfoliation versus Intercalation Effects on Heat Release

Keeping in mind that the mechanism of flammability reduction with a polymer nanocomposite is slowed mass loss rate, not the complete stop of mass loss, there are two issues to address about polymer nanocomposite flammability that require some specific discussion.

The first is nanoparticle chemistry, which can have some effect on heat release. As mentioned above, nanocomposites based upon double-layered hydroxides can also release water upon burning [54,55], but many layered silicates have some surface chemistry effects that activate only at elevated temperatures. Many layered silicates can serve as both hydrocarbon "cracking" catalysts as well as cycloaromatization catalysts at elevated temperatures, and this surface chemistry combined with prolonged residence time in the condensed phase can result in changes in polymer decomposition products. For many polymer clay nanocomposites, the chemical products of thermal decomposition are changed [56–58], and the products suggest both different condensed phase free-radical recombinations as well as catalyst-mediated reactions. This suggests that for clays at least, there may be a chemical pathway to the delayed mass loss and protective barrier formation that results in lowered heat release. From cone calorimeter results, the data suggest strictly a physical effect of flammability reduction, but from chemical analysis of the polymer thermal decomposition products, some chemical mechanisms may be at work as well [59].

The second issue to discuss is about whether exfoliation or intercalation is required for maximum reductions in flammability. The answer is most likely that both are needed and both work equally well. From surveying almost all of the literature on the subject with a wide range of polymer nanocomposite structures observed, both "intercalated" and "exfoliated" materials have shown equally good reductions in heat release. What appears to be more important is not the nanoscale structure (intercalated or exfoliated) of the layered material, but the overall macro- and microscale dispersion of the nanoparticles in the polymer matrix. If the overall dispersion of nanoparticles is poor, even if they are locally exfoliated [60], the flammability results are not as good as when the nanoparticles are evenly dispersed in the polymer matrix everywhere. Even dispersion of flame retardant throughout a polymer matrix is important for traditional flame retardants, and it is equally important for polymer nanocomposites. So even though the intercalated or exfoliated nature of a polymer-layered material nanocomposite may not be important in regard to flammability, it may be more important in regard to mechanical properties and for other enhanced mechanical and gas barrier properties found with nanocomposite materials.

5.2.3 Carbon Nanotube/Nanofiber Nanocomposite Flammability

Polymer nanocomposites can be synthesized with rod-shaped nanoparticles, with typical examples being carbon nanotubes (multiwall or single-wall) and vapor grown carbon nanofibers. These materials have also shown large reductions in HRR, and like that observed with polymer-layered silicate nanocomposites, the mechanism of flame retardancy is a condensed phase phenomena [61–65]. Specifically, the nanocomposite reduces mass loss rates which in turn reduce HRRs. Polymer nanotube and nanofiber nanocomposites achieve this reduced mass loss rate in a slightly different way than that observed with polymer-layered silicate nanocomposites. Rather than the nanotubes or nanofibers collapsing down to form a protective carbon-rich char barrier, they instead form a network structure as the polymer burns away, which in turn serves as a barrier to delay mass loss/fuel release [61]. The network structure of the carbon nanotubes/nanofibers is created during the synthesis of the polymer nanocomposite, but as the polymer burns away, the network remains intact and perhaps condenses and "tightens" up to form smaller and smaller holes in the network structure as the polymer pyrolyzes away.

To date, the flammability research on rod/tube nanocomposites has not been as extensive as that seen with layered materials, but new papers continue to be published every year. In general, polymer nanocomposites containing nanotubes and nanofibers can reduce flammability in similar amounts to that measured for polymer-layered silicate nanocomposites, but usually with less total nanoparticle loading. In some cases, flammability reductions were observed at loadings of single-wall nanotubes as low as 0.5 to 1 wt% [61–65]. This is a very good result especially considering the current high cost per kilogram of carbon nanotubes and even the cheaper vapor grown carbon nanofibers compared to other fillers. As an added bonus for the use of carbon nanotubes and nanofibers for flammability applications, these nanocomposites usually show some electrical conductivity as well. The same network structure that provides flammability reduction also serves as a semiconducting circuit through the polymer, meaning that nanocomposites made from these materials can not only be fire safe but may also be electrically conductive or electrically active multifunctional materials. These two features combined with enhanced mechanical properties makes this class of materials a very interesting and promising area for research.

As a final note about carbon nanotube and carbon nanofiber nanocomposites, the definitions of nanoparticle dispersion/nanocomposite structure used for layered-silicate materials do not apply for this class of materials. Carbon nanotubes and nanofibers cannot intercalate or exfoliate, but they do have primary particle agglomerates that must be broken apart either before entering the polymer matrix or during nanocomposite synthesis to obtain the flammability properties mentioned above. Carbon nanotubes can be agglomerated in micron-sized particles before being used, and if these large micron-sized particles are not broken up, there will be no significant reduction in polymer flammability because a microcomposite will have been prepared rather than a nanocomposite. Similar to the layered-silicate nanocomposites, carbon nanotube/nanofibers must be evenly dispersed at the micro- and macroscale throughout the polymer for the network structure to set up. Without this dispersion, there will be no network to provide flammability reduction or electrical conductivity in the final polymer nanocomposite. Finally, carbon nanotubes and carbon nanofibers have very different chemical structures and diameters. Carbon nanotubes tend to be much smaller and can be composed of a single wall of graphite rolled up to form a tube (single-wall carbon nanotube [SWNT]) or multiple walls (multi-wall carbon nanotube [MWNT]). Carbon nanofibers, on the other hand, are composed of a thin graphitic ribbon rolled up to form a continuous spiral such that it appears to be a tube from a distance but up close appears to be a stack of cups. More details on these chemical structures can be found in this book in other chapters, but it is clear that these differences in chemical structure result in differences in flammability performance, even if both form a network structure under burning.

5.2.4 Colloidal Nanocomposite Flammability

Out of the three general classes of polymer nanocomposites studied for flammability, this category is not as well studied as the other two. It covers a wide range of nanoparticles that may be irregular in shape but tend to be spherical and are nanoscale in all physical dimensions. One could argue that this category represents every nanoparticle tried for flammability that is not a layered material or a tube/fiber shape, and to some extent this is true. Most of the work done in this category of nanocomposite flammability research has been on studies with polyoctahedralsilasesquioxane (POSS) nanoparticles (Figure 5.5). POSS nanoparticles can be functionalized at the corners to yield a variety of structures that can work with a wide range of polymers. This versatility of structure has led to their use in many different polymers, and many of those resulting nanocomposites have been studied for flammability performance [66–70]. Flammability (heat release) reductions for POSS nanocomposites are not as great as those seen with nanotube or layered-material nanocomposites, but they are still greater than any flammability reduction from a traditional filled material. Further, higher loadings of POSS are required to obtain the same levels of heat release reduction seen for other

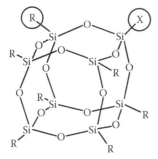

FIGURE 5.5
General POSS structure. R can be alkyl or aryl groups, or other chemistry as needed. X can be an R group as well, or can be Cl OH or other functional groups.

nanocomposites, which has limited their use due to the higher cost of POSS relative to some other nanofillers. The proposed mechanism of flammability release is likely a formation of a POSS-rich outer barrier that provides heat protection and reductions in mass loss rate, much like layered silicate nanocomposites [71–73], but there has not been a comprehensive study of the mechanism of flame retardancy of POSS nanocomposites at this time to prove this hypothesized mechanism. As with other nanocomposites, uniform nanoparticle dispersion throughout the polymer matrix is important to obtain reductions in flammability.

Related to POSS nanocomposite flammability work is some research on colloidal silicas (10 to 20 nm in size) dispersed into polymethylmethacrylate (PMMA) for flammability reduction [74]. For this system, the nanoparticles were found to disperse well into PMMA and, like with POSS, they began to form a silica-rich barrier that slowed mass loss as the polymer burned away. However, the HRR reduction was not as great as that seen with fumed silica at the same inorganic loading levels in PMMA [75,76]. Fumed silica creates a network structure while colloidal silica has to collapse down and have polymer burn away before its protective barrier can be formed. This suggests that for colloidal silicas, and perhaps even POSS, higher loading levels of nanofiller will be needed to obtain the same level of heat release reduction observed with layered silicates or carbon nanofibers. This may also be true for other small inorganic nanoparticles with the exception of nanoscale magnesium and aluminum hydroxides, which show great reductions in flammability compared to micron-sized magnesium and aluminum hydroxides. For these materials, having the same water-releasing flame retardant ($Mg(OH)_2$ or $Al_2O_3 \cdot 3H_2O$) in nanoscale form allows for lower total loading of flame retardant to be used while maintaining or exceeding flammability performance provided by the traditional filled system [77–81]. Although these systems have colloidal inorganic nanoparticles present, it is hard to say if these materials are truly nanocomposites, because the loading level of the nanoscale hydroxide is still quite high (>50%) in these systems, and it is likely that there are inorganic-rich domains microns in size composed of nanoscale-sized particles. Still, these nanoscale particles and other new results in colloidal nanoparticles yield new flammability results for this class of polymer nanocomposites, and more work is expected in the future.

5.2.5 General Nanocomposite Flammability Effects

As described in the previous section, polymer nanocomposites in general show reductions in HRR due to the reduction in mass loss rate when the polymer burns. What they also show is great increases in melt viscosity as the polymer burns. This is typically manifested during burning as "anti-drip" behavior. Polymers that normally melt and flow and drip when burned in a vertical orientation do not melt and flow when they are nanocomposites [29,30,82,83]. Instead, they hold in place and tend to maintain their shape.

From a flame spread and fire safety perspective, this physical effect of flame retardancy is quite useful, as it means that molten flaming polymer will not drip onto other objects and set them aflame as well. So, with reduced HRR and antidripping performance, one can say correctly that a polymer nanocomposite is a flame-retardant material, because it retards the growth and spread of flame. However, when polymer nanocomposites are set aflame, they are typically not self-extinguishing. Indeed, once ignited, they will burn very slowly until the entire sample is consumed, leaving behind the nanoparticle-rich char with some additional carbon trapped behind. So, by this definition, are they really flame retardant? This question has been critically asked in recent work [82,83], especially in light of the fact that when polymer nanocomposites are tested in regulatory fire tests, they do not pass them. The answer to this question is subjective and serves as a lead-in for the next section of this chapter. Polymer nanocomposites do indeed reduce heat release and provide antidrip behavior, a great improvement over the base polymers they are made from that can show very high heat release and flaming droplet behavior. What polymer nanocomposites really do is lower the baseline flammability of a polymeric material so that it requires less flame retardant to achieve a regulatory "pass" result of what would be defined as "flame retardant" in the codes, standards, and regulations that determine fire safety performance.

5.3 Polymer Nanocomposites Combined with Traditional Flame Retardants

With a lower base flammability and built-in antidrip effects, a polymer nanocomposite serves as a great place to start for making a flame-retardant material. When polymer nanocomposites have been combined with traditional flame-retardant additives, flammability performance is greatly enhanced, allowing these materials to pass a variety of regulatory flammability tests. Further, some of the flame-retardant additive can be removed and replaced with polymer nanocomposite on a more than 1:1 by weight replacement basis while still maintaining regulatory flammability performance. The effects of the polymer nanocomposite with traditional flame-retardant additives can be synergistic (greater than expected improvement in measured flammability) or strictly additive in the least of examples, and appear to be rather universal for most flame-retardant types. Polymer nanocomposite technology has worked with halogenated flame retardants [84–87], phosphorus-based flame retardants [88–95], intumescent systems [96–100], and mineral fillers [101–108] with great success. For the cases of halogenated and mineral fillers, commercial flame-retardant (FR) products have been introduced. Halogenated

FR additives have been combined with polyolefin nanocomposites to give a line of commercial UL-94 V rated materials under the Maxxam trade name from PolyOne as of 2006, but the nano-aspect of these products appears to have been deemphasized since their 2004 announcement, at least according to the PolyOne Web site [109]. Mineral (alumina trihydrate) fillers have been combined with poly(ethylene-co-vinyl acetate) (EVA) to yield commercial flame-retardant wire and cable jacket material in Europe from Kabelwerk Eupen (Beligum) [110].

While polymer nanocomposite technology has worked very well with many flame-retardant systems, there are some reports of antagonism between polymer nanocomposites and traditional flame-retardant additives. In some epoxy plus phosphorus flame-retardant systems, it was reported that the use of a layered silicate nanocomposite made flammability slightly worse rather than better [111–113]. The reason for this antagonism is not clear, but it may be that the nanocomposite delayed the release of a vapor-phase active phosphorus species when it was needed to put out the flame, or the nanocomposite interfered with the dispersion of the phosphorus flame retardant throughout the polymer matrix. Also, it may be that the clay was not as well dispersed in the polymer as the authors indicated in these reported examples, in which case, this antagonism could be due to flame retardant plus microcomposite antagonism. The exact reasons for the antagonism are unknown, but it was reported and is something the researcher of flame retardancy should be aware of when working on similar systems. In another example of antagonism, the use of a polymer nanocomposite made flammability worse in a polyamide-6 plus melamine cyanurate system [114]. In this case, the antidrip mechanism of the clay nanocomposite interfered with the prodripping mechanism of the melamine cyanurate. For some vertical burn tests, melamine cyanurate can be used to actually help depolymerize the polymer so that it melts (drips) away from the flame before it can ignite. This mechanism typically works only for polyamides where the cyanurate can attack the amide linkages to promote lower molten polymer viscosity upon exposure to flame, and so this mechanism of flame retardancy clashed with that of the polymer viscosity increase brought by the polymer nanocomposite. Even though this is the only reported case of nanocomposite antagonism with this type of flame-retardant additive, it is highly likely that polymer nanocomposites will be incompatible with any flame retardant that promotes melt flow and dripping to pass a flammability test. This is supported somewhat by results from a polyamide system with a sulfamate flame retardant that promoted dripping to yield passing vertical test ratings that failed when combined with a nanocomposite [115].

On a final note, the majority of the literature of polymer nanocomposites combined with other flame retardants has been with layered silicate (clay) nanocomposites and not with the other nanofillers described in this chapter. It is likely that flame-retardant synergism will be seen with carbon nanotube and nanofiber nanocomposites combined with traditional flame retardants, and it is further likely that this will also be found with colloidal nanoparticles, but the experiments need to be done and results reported before this can be stated definitively. One experiment that was reported in early 2008 used a multiwall carbon nanotube with mineral filler flame retardants in various wire and cable polymers [116,117]. This experiment was reported early in 2008 when this chapter was written and at the time was the only reported example of nanotubes being used with other flame retardants that could be found. It is highly likely that there will be more publications on this subject as more researchers study these nanotube-based nanocomposites in regard to flammability performance. Further, polymer nanocomposites are likely to continue to be commercialized as flame-retardant materials because the nanocomposite can bring a better balance of properties to the final material and, in some cases, can provide cost savings when the flame-retardant cost is reduced.

5.4 Making a Successful Flame-Retardant Polymer Nanocomposite

With the knowledge that polymer nanocomposites can provide flame retardancy in a wide range of polymers and with an equally wide range of flame-retardant additives, some consideration to how to prepare a successful flame-retardant nanocomposite is needed. Other parts of this book refer to all the synthetic details on making a polymer nanocomposite, so this section of this chapter will focus on some of the issues unique to flame-retardant polymer nanocomposites.

As mentioned earlier in the chapter, nanocomposites made with layered silicates are the most commonly studied class of nanocomposites in flammability applications. To date, the majority of these nanocomposites are made with organically treated montmorillonite clays that are available globally from a wide range of commercial suppliers. These organically treated clays (organoclays) are widely used as rheological modifiers for inks, paints, and drilling muds, and are typically treated with alkyl ammonium compounds. Although this keeps the cost of the organoclays quite low, it limits their use in some higher-temperature polymers because alkyl ammonium compounds are thermally unstable above 200°C. It has been shown that alkyl ammonium compounds on clays, much like the alkyl ammonium chemicals before they are put on clays, undergo a Hoffman degradation between 180 and 200°C depending upon alkyl ammonium structure [118–124] (Figure 5.6). When these alkyl ammonium compounds degrade, they cause the organophilic clay to revert to its hydrophilic (organophobic) nature, which means that the clays cannot form a polymer nanocomposite except with the most polar of

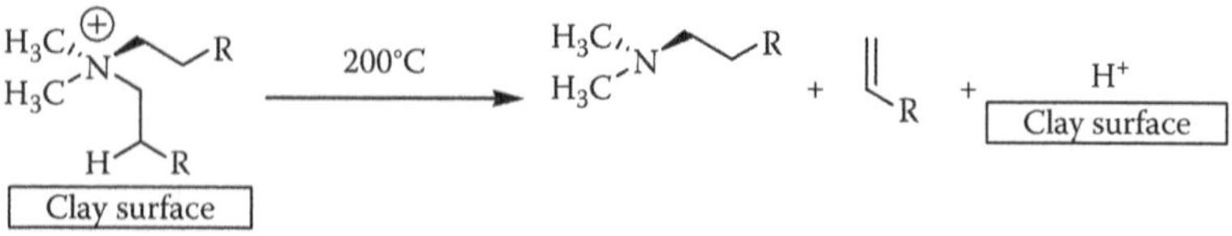

FIGURE 5.6
Idealized chemical scheme showing Hoffman degradation products from alkyl ammonium treated clays.

polymers. Once all nanocomposite structure is lost, the clay will produce only a microcomposite that yields no great reduction in heat release and does not provide antidrip effects during burning. So, for melt compounding operations with thermoplastics, taking the alkyl ammonium–treated organoclay above 200°C in an extruder will result in a microcomposite unless very short residence times are used in the extruder during the melt compounding step. Even then, the subsequent injection molding step of most thermoplastics will further degrade the alkyl ammonium treatment. The instability of the alkyl ammonium treatment is very likely the reason why flame-retardant polymer nanocomposites have only been commercialized in polyolefins that can be melt compounded well below 200°C. Therefore, the researcher wishing to incorporate layered silicates into a polymer with melt or processing temperatures above 200°C should consider alternate organic treatments that have higher thermal stability. Treatments shown to work include imidazolium [125–132], phosphonium [133,134], and some other exotic treatments [135–139] (stibonium, tropylium, etc.), but at this time none of these treatments are available on a commercial organoclay. This is not to say that these treatments are not commercially available. Imidazolium compounds are becoming more and more common as ionic liquids from several commercial suppliers, but at this time none of these compounds have been put onto commercially available clays. For now, imidazolium chemistry is in the category of "known to work," but it has not yet transitioned from the laboratory to the marketplace. The issues of thermal stability of organoclays are not relevant for carbon nanotubes, nanofibers, and colloidal nanoparticles, although the lack of an organic treatment on some of these nanoparticles can prevent nanocomposite synthesis in some polymers. Specifically, if the colloidal nanoparticles or carbon nanotubes/nanofibers are immiscible in the polymer matrix, they will either remain in their primary agglomerates or phase separate from the polymer after dispersion to return to their primary agglomerate state. In either case, the lack of an organic treatment for a nanoparticle can yield a microcomposite. This is a case-by-case scenario though, and again, one must look at nanoparticle and polymer chemistry at the same time to determine if organic treatment is needed, and if so, what the structure of the treatment should be.

Another issue to consider with polymer nanocomposites for flammability applications is evaluating flammability performance. The cone calorimeter s currently used the most to evaluate polymer nanocomposite flammability due to its ability to provide a quantitative assessment of flammability (heat release) and yield insight into the mechanistic behavior of polymer nanocomposites with other measurements from this instrument. Therefore, this instrument remains ideally suited for evaluating nanocomposite flammability behavior, but the cone calorimeter has limitations in correlating to regulatory tests [43,44], which can limit its usefulness to the industrial researcher looking to commercialize a product. The cone calorimeter measures flammability under well-ventilated, forced combustion conditions, which can be very different than some regulatory pass/fail flammability tests that may mimic limited ventilation, small ignition sources, or specific fire risk scenarios different than those measured by the cone calorimeter. Therefore, the polymer nanocomposite researcher working on flame-retardant materials should try to collect data from the cone calorimeter for scientific understanding of how the nanocomposite works and should prepare the materials for the final regulatory test if the researcher wants the nanocomposite to be commercially viable. The wide range of regulatory flammability tests is beyond the scope of this chapter, but some additional sources of fire test information and relevant fire safety engineering principles are as follows:

- *Fire Properties of Composite Materials,* Mouritz, A.P., Gibson, A.G. Eds., Series #143, in *Solid Mechanics and Its Applications,* Springer, Netherlands, 2006.
- *Fundamentals of Fire Phenomena,* Quintiere, J.G., John Wiley & Sons, England, 2006.
- *Fire Retardancy of Polymeric Materials,* Grand, A.F., Wilkie, C.A., Eds., Marcel-Dekker, New York, 2000.

5.5 Conclusions and Future Areas of Research

To conclude the chapter, polymer nanocomposites continue to be a very promising class of materials for future applications due to their multifunctional behavior. In regard to flammability performance, almost all polymer nanocomposites show some degree of heat release reduction and antidripping burning behavior. Therefore, this class of materials is inherently flame retardant, but to meet regulatory flammability tests, additional flame retardants are needed. On this point, polymer nanocomposite technology shows itself to be an almost universal synergist for all classes of flame retardants, allowing some of the traditional flame-retardant additive to be removed from a polymer and replaced with polymer nanocomposite while maintaining or exceeding flammability performance. For almost all of the classes of polymer nanocomposites described in this chapter, the primary mechanism of flammability/heat release reduction is that the nanocomposite slows the rate of mass loss/fuel release through the formation of protective barriers. These barriers vary in structure and mechanism of formation depending upon nanoparticle chemistry/structure, but with this understanding, new enhancements in flammability performance could be gained to either strengthen this protective barrier or cause it to completely prevent mass loss and fuel release.

5.5.1 Potential Issues/Research Areas for Flame-Retardant Polymer Nanocomposites

Flame-retardant material development is driven far more by regulations, codes, and standards than by customer demand or market forces. Material flammability performance is only one part of a long list of material properties that a product must have to be sold and used in commercial markets. The material performance characteristics are set by whatever fire code is relevant to the end-use fire risk scenario for that material. For example, the polymer used as a casing for a computer or television set will likely have its fire risk scenario as a small internal ignition caused by electrical failure/short circuit. So the polymers used in this type of application must show resistance to small ignition sources and the ability to self-extinguish and not propagate the flame further. This type of fire performance is readily obtainable today with a wide range of materials, but other regulations mandating recycling of the electronic products clouds the solutions to providing this level of fire safety. If the plastic that contains a flame retardant cannot easily be recycled, then in some countries it must go to landfill or incinerator. If the flame retardant in these landfill/incinerator-bound polymers gets out into the environment or causes toxic combustion products to be released upon incineration, then the polymeric material must not only meet fire performance standards, it must also meet new rules that can conflict with the existing chemistry used to meet the fire performance. There are now numerous laws in Europe such as the Reduction of Hazardous Substances (RoHS) and waste electronic and electrical equipment (WEEE) directives that are mandating better recycling of flame-retardant polymeric materials, and in some cases, the existing flame-retardant additives and materials are found to be in conflict with these rules [140–144]. This body of laws and study is too large to be presented in this chapter, but the point of mentioning it is that there is significant interest in seeing if polymer nanocomposites can replace some or all of these existing additives. However, there have not been any extensive studies on polymer nanocomposite recycling, and so this presents an opportunity for future polymer nanocomposite research. One cannot just stop at discovering and making the flame-retardant polymer nanocomposite that passes the end-use fire test; one must now also see how the product can be recycled or dealt with at its end-of-life use. Can a polymer nanocomposite be safely ground up and remelted, or should it go to incinerator? If burned, does it give off toxic by-products that have to be scrubbed out from the emission system? These questions and others related to it lead to the other unknown issue with polymer nanocomposites, the potential environmental, health, and safety issues of nanoparticle technology.

There are current concerns about nanoparticle safety related to human exposure and environmental disposition [145–147]. To some extent, nanoparticles are natural to our environment if we consider materials like montmorillonite clay which are used to make polymer–clay nanocomposites. What is different is that the clay in its natural form is micron sized or larger; chemical modification is required to get the clay plates to become individual nanoparticles. So the concern over nanoparticles relates to the individual nanoparticles when broken up, not in their primary particle form. Because these nanoparticles can be in their individual forms when dispersed in a polymer matrix, how that polymer nanocomposite is handled raises some questions. If the nanocomposite is broken, ground up, or drilled through, does this release the nanoparticles into the environment? Likewise, if the nanocomposite is burned, do all of the nanoparticles stay in the residual char or do they get volatized into the smoke and soot from the fire? At this time, these answers are completely unknown, and studies are needed to assess if nanoparticles can be released upon polymer nanocomposite regrinding and burning. Some of the data from flammability analysis that shows nanoparticle-rich char suggest that nanoparticles are not released during burning, but this needs to be confirmed. Further, many nanoparticles are fully encapsulated by polymer in a polymer nanocomposite, a situation that is required in successful nanocomposite manufacture. The nanoparticles would not be encapsulated by polymer if they were in a polymer microcomposite or if they were poorly dispersed. This suggests that grinding of polymer nanocomposites will likely not release nanoparticles into the environment, but it does not prove it either until data are collected. Finally, the fate of the nanoparticle after recycling through regrinding and remelting must be studied. As indicated in the previous section about the thermal instability of alkyl ammonium treatments on clay nanoparticles, prolonged heating above 200°C can degrade the interface between clay and polymer, resulting in loss of properties and destruction of nanocomposite structure. Therefore, regrinding and remelting of polymer nanocomposites may not be a practical recycling route for these materials, but it does need to be investigated and studied in detail to prove or disprove this hypothesis.

A final unknown with polymer nanocomposite technology is long-term aging and weathering of polymer nanocomposite technology. Along with end-of-life issues, balance of mechanical, flammability, and cost properties for a polymeric material, how the polymer maintains its properties over time in different environments is just as important as the other properties that the industrial researcher must put into a commercial product. For polymeric materials, compounds such as antioxidants, ultraviolet (UV) stabilizers, and thermal stabilizers are added to the polymer to enable it to handle a wide range of use conditions without breaking down or failing. For polymer nanocomposites, there have been some reports of accelerated UV-induced degradation [148–155], which may be due to nanoparticles catalyzing certain UV-activated reactions or may be due to the nanoparticles absorbing the antioxidants and UV stabilizers onto their surfaces which would prevent them from working in the polymer matrix. Long-term aging studies of these new materials are needed to better enable polymer nanocomposites to meet a wide range of end-use conditions and to ensure that these materials truly can enter into the marketplace as commercial products.

Even with the unknowns that need to be addressed for polymer nanocomposites, it is highly likely that this field of materials research will continue to grow. Already, new research on polymer nanocomposites containing more than one type of nanoparticle has been published in regard to flammability performance [156,157], and it is highly likely that more fundamental and applied flame-retardant polymer nanocomposite research will continue to be published at a growing pace in the coming years. Further, it is equally likely that the number of commercial products for flame retardancy applications based upon polymer nanocomposite technology will also grow over the coming decade, as more researchers take advantage of the synergistic flame-retardant effects that nanocomposites bring. As polymeric materials replace heavier materials such as ceramics and metals to save weight and improve fuel efficiency of engines and vehicles, more of these polymeric materials will require flame retardancy, not less. Therefore the demand for fire-safe, multifunctional materials is only likely to grow, which gives polymer nanocomposite technology good opportunities to grow as well. The demands on materials in general are only increasing, and not just in high-technology areas such as electronics and aerospace. Materials used in component and structure fabrication need to show multifunctionality to be truly innovative and in some cases just to meet the baseline performance for a new demanding application. Polymer nanocomposites in general are the most promising class of materials to provide multifunctional behavior, and so polymer nanocomposite use in general is expected to grow in high-technology areas first and later commodity applications. Certainly the unknowns of polymer nanocomposite recycling and nanoparticle health and safety need to be addressed, but the performance characteristics of these materials will drive researchers to come up with solutions to any possible problems that may arise from these materials. To conclude this chapter on polymer nanocomposite flammability, as it appears that almost all polymer nanocomposites tested to date show some level of flammability improvement (lowered heat release, antidrip behavior), it is highly likely that through the spread of polymer nanocomposite materials into more and more applications, fire safety for society will greatly improve without any deliberate efforts to improve it. This is not to say that by using polymer nanocomposites all of society's fire safety problems will be solved, but rather that the baseline flammability of man-made materials all around us will be greatly reduced.

Acknowledgments

The author wishes to thank the Omnova Foundation for partial funding of time to write this chapter. Additional funding for this work was provided through "Materials Development for Multifunctional Composite Survivability Systems" (Award # W911NF-07-2-0005) through the Army Research Laboratory, AMSRD-ARL-WM-MD.

References

1. "Synthesis of nylon 6-clay hybrid" Usuki, A., Kojima, Y., Kawasumi, M., Okada, A., Fukushima, Y., Kurauchi, T., Kamigaito, O. *J. Mater. Res.* 1993, *8*, 1179–1184.
2. "The chemistry of polymer-clay hybrids" Okada, A., Usuki, A. *Mat. Sci. Eng. C.* 1995, *3*, 109–115.
3. "Microstructural evolution of melt intercalated polymer-organically modified layered silicates nanocomposites" Vaia, R.A., Jandt, K.D., Kramer, E.J., Giannelis, E.P. *Chem. Mater.* 1996, *8*, 2628–2635.
4. "Polymer layered silicate nanocomposites" Giannelis, E.P. *Adv. Mater.* 1996, *8*, 29.
5. "Mechanism of clay tactoid exfoliation in epoxy-clay nanocomposites" Lan, T., Kaviratna, P.D., Pinnavaia, T.J. *Chem. Mater.* 1995, *7*, 2144–2150.
6. "Clay-reinforced epoxy nanocomposites" Lan, T., Pinnavaia, T.J. *Chem. Mater.* 1994, *6*, 2216–2219.
7. "Polymer/layered silicate nanocomposites: A review from preparation to processing" Ray, S.S., Okamoto, M. *Prog. Polym. Sci.* 2003, *28*, 1539–1641.
8. "Polymer nanocomposites containing carbon nanotubes" Moniruzzaman, M., Winey, K.I. *Macromolecules* 2006, *39*, 5194–5205.
9. "Synthetic, layered nanoparticles for polymeric nanocomposites (PNCs)" Utracki, L.A., Sepehr, M., Boccaleri, E. *Polym. Adv. Technol.* 2007, *18*, 1–37.
10. "Twenty years of polymer-clay nanocomposites" Okada, A., Usuki, A. *Macromol. Mater. Eng.* 2007, *291*, 1449–1476.
11. "Polymer nanocomposites with prescribed morphology: Going beyond nanoparticle-filled polymers" Vaia, R.A., Maguire, J.F. *Chem. Mater.* 2007, *19*, 2736–2751.
12. *Polymer Nanocomposites* Koo, J.H. McGraw-Hill, New York, 2006.
13. "Polymerization of adsorbed monolayers. I. Preparation of the clay-polymer complex" Blumstein, A. *J. Polym. Sci. Part A.* 1963, *3*, 2653–2664.

14. "Polymerization of adsorbed monolayers. II. Thermal degradation of the inserted polymer" Blumstein, A. *J. Polym. Sci. Part A.* 1963, 3, 2665–2672.
15. "Polymerization of adsorbed monolayers. III. Preliminary structure studies in dilute solution of the insertion polymers" Blumstein, A., Billmeyer, Jr., F.W. *J. Polym. Sci. Part A-2.* 1966, 4, 465–474.
16. "Association I: Two-dimensionally crosslinked poly(methyl methacrylate)" Blumstein, A., Blumstein, R. *J. Polym. Sci. Polym. Lett.* 1967, 5, 691–696.
17. "Branching in poly(methyl methacrylate) obtained by γ-ray irradiation" Blumstein, A. *J. Polym. Sci. Polym. Lett.* 1967, 5, 687–690.
18. "Tacticity of poly(methyl methacrylate) prepared by radical polymerization within a monolayer of methyl methacrylate adsorbed on montmorillonite" *J. Polym. Sci. Polym. Lett.* 1968, 6, 69–74.
19. "Polymerization of adsorbed monolayers. IV. The two-dimensional structure of insertion polymers" Blumstein, A., Blumstein, R., Vanderspurt, T.H. *J. Colloid Surf. Sci.* 1969, 31, 236–247.
20. "Polymerization of adsorbed monolayers. V. Tacticity of the insertion poly(methyl methacrylate)" Blumstein, A., Malhotra, S.I., Watterson, A.C. *J. Polym. Sci. Part A-2* 1970, 8, 1599–1615.
21. "Polymerization of monolayers. VI. Influence of the nature of the exchangeable ion on the tacticity of insertion (poly(methyl methacrylate)" *J. Polym. Sci. Part A-2* 1971, 9, 1681–1691.
22. "Polymerization of monolayers. VII. Influence of the exchangeable cation on the polymerization rate of methylmethacrylate monolayers adsorbed on montmorillonite" Malhotra, S.L., Parikh, K.K., Blumstein, A. *J. Colloid Surf. Sci.* 1972, 41, 318–327.
23. Bradbury, J.A., Rowlands, R., Tipping, J.W. US Patent #4447491, 1984.
24. Shain, A.L. US Patent #4582866, 1986.
25. Lee, J., Takekoshi, T., Giannelis, E. *Mater. Res. Soc. Symp. Proc.* 1997, 457, 512–518.
26. Gilman, J., Kashiwagi, T., Lomakin, S., Giannelis, E., Manias, E., Lichtenhan, J., Jones, P. in "Fire Retardancy of Polymers: The Use of Intumescence" The Royal Society of Chemistry, Cambridge, UK. 1998.
27. "Flammability and thermal stability studies of polymer layered-silicate (clay) nanocomposites" Gilman, J.W. *Applied Clay Science* 1999, 15, 31–49.
28. "Flammability properties of polymer-layered silicate nanocomposites. Polypropylene and polystyrene nanocomposites" Gilman, J.W., Jackson, C.L., Morgan, A.B., Harris, R., Manias, E., Giannelis, E.P., Wuthenow, M., Hilton, D., Phillips, S.H. *Chem. Mater.* 2000, 12, 1866–1873.
29. "Flame retarded polymer layered silicate nanocomposites: A review of commercial and open literature systems" Morgan, A.B. *Polym. Adv. Technol.* 2006, 17, 206–217.
30. *Flame Retardant Polymer Nanocomposites* Morgan, A.B., Wilkie, C.A. Eds. John Wiley & Sons, Hoboken, NJ. 2007.
31. "Polystyrene/graphite nanocomposites: Effect on thermal stability" Uhl, F.M., Wilkie, C.A. *Polym. Degrad. Stab.* 2002, 76, 111–122.
32. "Studies on the mechanism by which the formation of nanocomposites enhances thermal stability" Zhu, J., Uhl, F.M., Morgan, A.B., Wilkie, C.A. *Chem. Mater.* 2001, 13, 4649–4654.

33. "Flammability and thermal stability studies of styrene-butyl acrylate copolymer/graphite oxide nanocomposite" Zhang, R., Hu, Y., Xu, J., Fan, W., Chen, Z. *Polym. Degrad. Stab.* 2004, *85*, 583–588.

34. "Mechanistic aspects of nanoeffect on poly(acrylic ester)-GO composites: TGA-FTIR study on thermal degradation and flammability of polymer layered graphite oxide composites" Wang, J., Han, Z. *Fire and Polymers IV: Materials and Concepts for Hazard Prevention* ACS Symposium Series #922, Ed. Wilkie, C.A., Nelson, G.L. American Chemical Society, Washington, DC, 2005, pp. 172–184.

35. "Characterization of polymer-layered silicate (clay) nanocomposites by transmission electron microscopy and X-ray diffraction: A comparative study" Morgan, A.B., Gilman, J.W. *J. App. Polym. Sci.* 2003, *87*, 1329–1338.

36. "Evaluation of the structure and dispersion in polymer-layered silicate nanocomposites" Vermogen, A., Masenelli-Varlot, K., Seguela, R., Duchet-Rumeau, J., Boucard, S., Prele, P. *Macromolecules* 2005, *38*, 9661–9669.

37. "X-ray powder diffraction of polymer/layered silicate nanocomposites: Model and practice" Vaia, R.A., Liu, W. *J. Polym. Sci. B: Polym. Phys.* 2002, *40*, 1590–1600.

38. ASTM E1354 "Standard Heat Method for Heat and Visible Smoke Release Rates for Materials and Products Using an Oxygen Consumption Calorimeter" and "ISO/FDIS 5660-1 Reaction-to-fire tests—Heat release, smoke production and mass loss rate—Part 1: Heat release (cone calorimeter method)."

39. "Comprehensive fire behaviour assessment of polymeric materials based on cone calorimeter investigations" Schartel, B., Braun, U. *e-Polymers* 2003, No. 13. www.e-polymers.org/papers/schartel_010403.pdf.

40. "Mechanistic study of the combustion behavior of polymeric materials in bench-scale tests. I. Comparison between cone calorimeter and traditional tests" Costa, L., Camino, G., Bertelli, G., Borsini, G. *Fire and Materials* 1995, *19*, 133–142.

41. "Material fire properties and predictions for thermoplastics" Hopkins, D., Quintiere, J.G. *Fire Safety Journal* 1996, *26*, 241–268.

42. "The assessment of full-scale fire hazards from cone calorimeter data" Petrella, R.V. *J. Fire Sci.* 1994, *12*, 14–43.

43. "Heat release rate measurements of thin samples in the OSU apparatus and the cone calorimeter" Filipczak, R., Crowley, S., Lyon, R.E. *Fire Safety Journal* 2005, *40*, 628–645.

44. "Cone calorimeter analysis of UL-94 V-rated plastics" Morgan, A.B., Bundy, M. *Fire Mater.* 2007, *31*, 257–283.

45. "Specimen heat fluxes for bench-scale heat release rate testing" Babrauskas, V. *Fire and Materials* 1995, *19*, 243–252.

46. "Heat release rate: The single most important variable in fire hazard" Babrauskas, V., Peacock, R.D. *Fire Safety Journal* 1992, *18*, 255–272.

47. "Some comments on the use of cone calorimeter data" Schartel, B., Bartholmai, M., Knoll, U. *Polym. Degrad. Stab.* 2005, *88*, 540–547.

48. "Development of fire-retarded materials—Interpretation of cone calorimeter data" Schartel, B., Hull, T.R. *Fire and Materials* 2007, *31*, 327–354.

49. "An XPS study of the thermal degradation and flame retardant mechanism of polystyrene-clay nanocomposites" Wang, J., Du, J., Zhu, J., Wilkie, C.A. *Polym. Degrad. Stab.* 2002, *77*, 249–252.

50. "A study of the flammability reduction mechanism of polystyrene-layered silicate nanocomposite: Layered silicate reinforced carbonaceous char" Gilman, J.W., Harris, R.H., Shields, J.R., Kashiwagi, T., Morgan, A.B. *Polym. Adv. Technol.* 2006, *17*, 263–271.

51. "Flammability of polystyrene layered silicate (clay) nanocomposites: Carbonaceous char formation" Morgan, A.B., Harris, R.H., Kashiwagi, T., Chyall, L.J., Gilman, J.W. *Fire and Materials* 2002, *26*, 247–253.

52. "Kinetic analysis of the thermal degradation of polystyrene-montmorillonite nanocomposite" Bourbigot, S., Gilman, J.W., Wilkie, C.A. *Polym. Degrad. Stab.* 2004, *84*, 483–492.

53. "Maleated polypropylene OMMT nanocomposite: Annealing, structural changes, exfoliated and migration" Tang, Y., Lewin, M. *Polym. Degrad. Stab.* 2006, *92*, 53–60.

54. "Effect of hydroxides and hydroxycarbonate structure on fire retardant effectiveness and mechanical properties in ethylene-vinyl acetate copolymer" Camino, G., Maffezzoli, A., Braglia, M., De Lazzaro, M., Zammarano, M. *Polym. Degrad. Stab.* 2001, *74*, 457–464.

55. "Preparation and flame resistance properties of revolutionary self-extinguishing epoxy nanocomposites based on layered double hydroxides" Zammarano, M., Franceschi, M., Bellayer, S., Gilman, J.W., Meriani, S. *Polymer* 2005, *46*, 9314–9328.

56. "The thermal degradation of poly(methyl methacrylate) nanocomposites with montmorillonite, layered double hydroxides and carbon nanotubes" Costache, M.C., Wang, D., Heidecker, M.J., Manias, E., Wilkie, C.A. *Polym. Adv. Technol.* 2006, *17*, 272–280.

57. "Thermal degradation of ethylene-vinyl acetate copolymer nanocomposites" Costache, M.C., Jiang, D.D., Wilkie, C.A. *Polymer* 2005, *46*, 6947–6958.

58. "The effect of clay on the thermal degradation of polyamide 6 in polyamide 6/clay nanocomposites" Jang, B.N., Wilkie, C. A. *Polymer* 2005, *46*, 3264–3274.

59. "The influence of carbon nanotubes, organically modified montmorillonites and layered double hydroxides on the thermal degradation and fire retardancy of polyethylene, ethylene-vinyl acetate copolymer and polystyrene" Costache, M.C., Heidecker, M.J., Manias, E., Camino, G., Frache, A., Beyer, G., Gupta, R.K., Wilkie, C.A. *Polymer* 2007, *48*, 6532–6545.

60. "Fire properties of polystyrene-clay nanocomposites" Zhu, J., Morgan, A.B., Lamelas, F.J., Wilkie, C.A. *Chem. Mater.* 2001, *13*, 3774–3780.

61. "Effects of aspect ratio of MWNT on the flammability properties of polymer nanocomposites" Cipiriano, B.H., Kashiwagi, T., Raghavan, S.R., Yang, Y., Grulke, E.A., Yamamoto, K., Shields, J.R., Douglas, J.F. *Polymer* 2007, *48*, 6086–6096.

62. "Flammability properties of polymer nanocomposites with single-walled carbon nanotubes: Effects of nanotube dispersion and concentration" Kashiwagi, T., Du, F., Winey, K.I., Groth, K.M., Shields, J.R., Bellayer, S.P., Kim, H., Douglas, J.F. *Polymer* 2005, *46*, 471–481.

63. "Thermal degradation and flammability properties of poly(propylene)/carbon nanotube composites" Kashiwagi, T., Grulke, E., Hilding, J., Harris, R., Awad, W., Douglas, J. *Macromol. Rapid Commun.* 2002, *23*, 761–765.

64. "Thermal and flammability properties of polypropylene/carbon nanotube nanocomposites" Kashiwagi, T., Grulke, E., Hilding, J., Groth, K., Harris, R., Butler, K., Shields, J., Kharchenko, S., Douglas, J. *Polymer* 2004, *45*, 4227–4239.

65. "Nanoparticle networks reduce the flammability of polymer nanocomposites" Kashiwagi, T., Du, F., Douglas, J.F., Winey, K.I., Harris, R.H., Shields, J.R. *Nature Materials* 2005, *4*, 928–933.
66. "Polyurethane/clay and polyurethane/POSS nanocomposites as flame retarded coating for polyester and cotton fabrics" Devaux, E., Rochery, M., Bourbigot, S. *Fire and Materials* 2002, *26*, 149–154.
67. "Effects of surfactants on the thermal and fire properties of poly(methyl methacrylate)/clay nanocomposites" Jash, P., Wilkie, C.A. *Polym. Degrad. Stab.* 2005, *88*, 401–406.
68. "Metal functionalized POSS as fire retardants in polypropylene" Fina, A., Abbenhuis, H.C.L., Tabuani, D., Camino, G. *Polym. Degrad. Stab.* 2006, *91*, 2275–2281.
69. "Fire retardancy of vinyl ester nanocomposites: Synergy with phosphorus-based fire retardants" Chigwada, G., Jash, P., Jiang, D.D., Wilkie, C.A. *Polym. Degrad. Stab.* 2005, *89*, 85–100.
70. "Combustion and thermal properties of OctaTMA-POSS/PS composites" Liu, Lei, Hu, Yuan, Song, Lei, et al. *J. Mater. Sci.* 2007, *42*, 4325–4333.
71. "Synthesis and thermal properties of hybrid copolymers of syndiotactic polystyrene and polyhedral oligomeric silsesquioxane" Zheng, L., Kasi, R.M., Farris, R.J., Coughlin, E.B. *J. Polym. Sci. Part A, Polym. Chem.* 2002, 40, 885–891.
72. "Polyimide/POSS nanocomposites: Interfacial interaction, thermal properties and mechanical properties" Huang, J.C., He, C.B., Xiao, Y., Mya, K.Y., Dai, J., Siow, Y.P. *Polymer* 2003 44, 4491–4499.
73. "Silicon-based flame retardants" Kashiwagi T., Gilman, J.W. in *Fire Retardancy of Polymeric Materials*. Grand, A.F., Wilkie, C.A., eds. New York, Marcel Dekker, 2000, Chapt. 10: 353–389.
74. "Thermal and flammability properties of a silica-poly(methylmethacrylate) nanocomposite" Kashiwagi, T., Morgan, A.B., Antonucci, J.M., VanLandingham, M.R., Harris, R.H., Awad, W.H., Shields, J.R. *J. App. Polym. Sci.* 2003, *89*, 2072–2078.
75. "Flame retardant mechanism of silica gel/silica" Kashiwagi, T., Gilman, J.W., Butler, K.M., Harris, R.H., Shields, J.R. Asano, A. *Fire and Materials* 2000, *24*, 277–289.
76. "Flame-retardant mechanism of silica: Effects of resin molecular weight" Kashiwagi, T., Shields, J.R., Harris, R.H., Davis, R.D. *J. App. Polym. Sci.* 2003, *87*, 1541–1553.
77. "Effect of nano-Mg(OH)$_2$ on the mechanical and flame-retarding properties of polypropylene composites" Mishra, S., Sonawane, S.H., Singh, R.P., Bendale, A., Patil, K. *J. App. Polym. Sci.* 2004, *94*, 116–122.
78. "Effect of particle size on the properties of Mg(OH)$_2$-filled rubber composites" Zhang, Q., Tian, M., Wu, Y., Lin, G., Zhang, L. *J. App. Polym. Sci.* 2004, *94*, 2341–2346.
79. "Flame retardation of ethylene-vinyl acetate copolymer using nano magnesium hydroxide and nano hydrotalcite" Jiao, C.M., Wang, Z.Z., Ye, Z., Hu, Y., Fan, W.C. *J. Fire Sci.* 2006, *24*, 47–64.
80. "Investigation of interfacial modification for flame retardant ethylene vinyl acetate copolymer/alumina trihydrate nanocomposites" Zhang, X., Guo, F., Chen, J., Wang, G., Liu, H. *Polym. Degrad. Stab.* 2005, *87*, 411–418.

81. "Preparation and physical properties of flame retardant acrylic resin containing nano-sized aluminum hydroxide" Daimatsu, Kazuki, Sugimoto, Hideki, Kato, Yasunori, et al. *Polym. Degrad. Stab.* 2007, *92*, 1433–1438.

82. "Layered silicate polymer nanocomposites: New approach or illusion for fire retardancy? Investigations of the potentials and the tasks using a model system" Bartholmai, M., Schartel, B. *Polymers for Advanced Technologies* 2004, *15*, 355–364.

83. "Some comments on the main fire retardancy mechanisms in polymer nanocomposites" Schartel, B., Bartholmai, M., Knoll, U. *Polym. Adv. Technol.* 2006, *17*, 772–777.

84. "Fire retardant halogen-antimony-clay synergism in polypropylene layered silicate nanocomposites" Zanetti, M., Camino, G., Canavese, D., Morgan, A.B., Lamelas, F.J., Wilkie, C.A. *Chem. Mater.* 2002, *14*, 189–193.

85. "Preparation and combustion properties of flame retardant nylon-6/montmorillonite nanocomposite" Hu, Y., Wang, S., Ling, Z., Zhuang, Y., Chen, Z., Fan, W. *Macromol. Mater. Eng.* 2003, *288*, 272–276.

86. "Self-extinguishing polymer/organoclay nanocomposites" Si, M., Zaitsev, V., Goldman, M., Frenkel, A., Peiffer, D.G., Weil, E., Sokolov, J.C., Rafailovich, M.H. *Polym. Degrad. Stab.* 2007, *92*, 86–93.

87. "Cone calorimetric and thermogravimetric analysis evaluation of halogen-containing polymer nanocomposites" Wang, D., Echols, K., Wilkie, C.A. *Fire and Materials* 2005, *29*, 283–294.

88. "Synergy between conventional phosphorus fire retardants and organically-modified clays can lead to fire retardancy of styrenics" Chigwada, G., Wilkie, C.A. *Polym. Degrad. Stab.* 2003, *80*, 551–557.

89. "Cure kinetics of epoxy/anhydride nanocomposite systems with added reactive flame retardants" Torre, L., Lelli, G., Kenny, J.M. *J. App. Polym. Sci.* 2004, *94*, 1676–1689.

90. "Synthesis of phosphorus-based flame retardant systems and their use in an epoxy resin" Toldy, A., Toth, N., Anna, P., Marosi, G. *Polym. Degrad. Stab.* 2006, *91*, 585–592.

91. "Phosphonium-modified layered silicate epoxy resins nanocomposites and their combinations with ATH and organo-phosphorus fire retardants" Schartel, B., Knoll, U., Hartwig, A., Putz, D. *Polym. Adv. Technol.* 2006, *17*, 281–293.

92. "Intrinsically flame retardant epoxy resin—Fire performance and background—Part I" Toldy, A., Anna, P., Csontos, I., Szabo, A., Marois, Gy. *Polym. Degrad. Stab.* 2007, *92*, 2223–2230.

93. "Preparation and properties of halogen-free flame-retarded polyamide 6/organoclay nanocomposite" Song, L., Hu, Y., Lin, Z., Xuan, S., Wang, S., Chen, Z., Fan, W. *Polym. Degrad. Stab.* 2004, *86*, 535–540.

94. "Thermal oxidative degradation behaviours of flame-retardant copolyesters containing phosphorus linked pendent group/montmorillonite nanocomposites" Wang, D.-Y., Wang, Y.-Z., Wang, J.-S., Chen, D.-Q., Zhou, Q., Yang, B., Li, W.-Y. *Polym. Degrad. Stab.* 2005, *87*, 171–176.

95. "Fire retardancy of vinyl ester nanocomposites: Synergy with phosphorus-based fire retardants" Chigwada, G., Jash, P., Jiang, D.D., Wilkie, C.A. *Polym. Degrad. Stab.* 2005, *89*, 85–100.

96. "Intumescent flame retardant-montmorillonite synergism in polypropylene-layered silicate nanocomposites" Tang, Y., Hu, Y., Wang, S., Gui, Z., Chen, Z., Fan, W. *Polym. Int.* 2003, *52*, 1396–1400.

97. "Fire retardancy effect of migration in polypropylene nanocomposites induced by modified interlayer" Marosi, Gy., Marton, A., Szep, A., Csontos, I., Keszei, S., Zimonyi, E., Toth, A., Almeras, X., Le Bras, M. *Polym. Degrad. Stab.* 2003, *82*, 379–385.
98. "Fire safety assessment of halogen-free flame retardant polypropylene based on cone calorimeter" He, S., Hu, Y., Song, L., Tang, Y. *J. Fire Sci.* 2007, *25*, 109–118.
99. "The use of clay in an EVA-based intumescent formulation. Comparison with the intumescent formulation using polyamide-6 clay nanocomposite as carbonisation agent" Dabrowski, F., Le Bras, M., Cartier, L., Bourbigot, S. *J. Fire Sciences* 2001, *19*, 219–241.
100. "PA-6 clay nanocomposite hybrid as char forming agent in intumescent formulations" Bourbigot, S., Le Bras, M., Dabrowski, F., Gilman, J.W., Kashiwagi, T. *Fire and Materials* 2000, *24*, 201–208.
101. "The effectiveness of magnesium carbonate-based flame retardants for poly(ethylene-co-vinyl acetate) and poly(ethylene-co-ethyl acrylate)" Morgan, A.B., Cogen, J.M., Opperman, R.S., Harris, J.D. *Fire and Materials* 2007, *31*, 387–410.
102. "Flame retardant properties of EVA-nanocomposites and improvements by combination of nanofillers with aluminum trihydrate" Beyer, G. *Fire and Materials* 2001, *25*, 193–197.
103. "Intumescent mineral fire retardant systems in ethylene-vinyl acetate copolymer: Effect of silica particles on char cohesion" Laoutid, F., Ferry, L., Leroy, E., Lopez Cuesta, J.M. *Polym. Degrad. Stab.* 2006, *91*, 2140–2145.
104. "Thermal stability and flame retardancy of LDPE/EVA blends filled with synthetic hydromagnesite/aluminium hydroxide/montmorillonite and magnesium hydroxide/aluminium hydroxide/montmorillonite mixtures" Haurie, L., Fernandez, A.I., Velasco, J.I., Chimenos, J.M., Lopez-Cuesta, J.-M., Espiell, F. *Polym. Degrad. Stab.* 2007, *92*, 1082–1087.
105. "Influence of talc physical properties on the fire retarding behaviour of (ethylene-vinyl acetate copolymer/magnesium hydroxide/talc) composites" Clerc, L., Ferry, L., Leroy, E., Lopez-Cuesta, J.-M. *Polym. Degrad. Stab.* 2005, *88*, 504–511.
106. "Study of hydromagnesite and magnesium hydroxide based fire retardant systems for ethylene-vinyl acetate containing organo-modified montmorillonite" Laoutid, F., Gaudon, P., Taulemesse, J.-M., Lopez Cuesta, J.M., Velasco, J.I., Piechaczyk, A. *Polym. Degrad. Stab.* 2006, *91*, 3074–3082.
107. "Role of montmorillonite in flame retardancy of ethylene-vinyl acetate copolymer" Szep, A., Szabo, A., Toth, N., Anna, P., Marosi, G. *Polym. Degrad. Stab.* 2006, *91*, 593–599.
108. "An investigation into the decomposition and burning behaviour of ethylene-vinyl acetate copolymer nanocomposite materials" Hull, T.R., Price, D., Liu, Y., Wills, C.L., Brady, J. *Polym. Degrad. Stab.* 2003, *82*, 365–371.
109. www.polyone.com/prod/trade/trade_info.asp?ID={97982AED-1F22-46DC-BA4B-15CF4DCEC90D}&link=M.
110. "Flame retardancy of nanocomposites—From research to technical products" Beyer, G. *J. Fire Sci.* 2005, *23*, 75–87.
111. "Understanding the decomposition and fire performance processes in phosphorus and nanomodified high performance epoxy resins and composites" Liu, W., Varley, R.J., Simon, G.P. *Polymer* 2007, *48*, 2345–2354.

112. "Probing synergism, antagonism, and additive effects in poly(vinyl ester) (PVE) composites with fire retardants" Kandare, E., Chigwada, G., Wang, D., Wilkie, C.A., Hossenlopp, J.M. *Polym. Degrad. Stab.* 2006, *91*, 1209–1218.
113. "Effect of organo-phosphorus and nano-clay materials on the thermal and fire performance of epoxy resins" Hussain, M., Varley, R.J., Mathys, Z., Cheng, Y.B., Simon, G.P. *J. App. Polym. Sci.* 2004, *91*, 1233–1253.
114. "Preparation and combustion properties of flame retardant nylon-6/montmoril-lonite nanocomposite" Hu, Y., Wang, S., Ling, Z., Zhuang, Y., Chen, Z., Fan, W. *Macromol. Mater. Eng.* 2003, *288*, 272–276.
115. "Flammability of polyamide 6 using the sulfamate system and organo-lay-ered silicate" Lewin, M., Zhang, J., Pearce, E., Gilman, J. *Polymers for Advanced Technology* 2007, *18*, 737–745.
116. "Filler blend of carbon nanotubes and organoclays with improved char as a new flame retardant system for polymers and cable applications" Beyer, G. *Fire Mater.* 2005, *29*, 61–69.
117. "Flame retardancy of nanocomposites based on organoclays and carbon nan-otubes with aluminum trihydrate" Beyer, G. *Polym. Adv. Technol.* 2006, *17*, 218–225.
118. "Thermal degradation chemistry of alkyl quaternary ammonium montmoril-lonite" Xie, W., Gao, Z., Pan, W.-P., Hunter, D., Singh, A., Vaia, R. *Chem. Mater.* 2001, *13*, 2979–2990.
119. "Thermal decomposition of alkyl ammonium ions and its effect on surface polarity of organically treated nanoclay" Dharaiya, D., Jana, S.C. *Polymer* 2005, *46*, 10139–10147.
120. "Organoclay degradation in melt processed polyethylene nanocomposites" Shah, R.K., Paul, D.R. *Polymer* 2006, *47*, 4084.
121. "Catalytic charring-volatilization competition in organoclay nanocompos-ites" Bellucci, F., Camino, G., Frache, A., Sarra, A. *Polym. Degrad. Stab.* 2007, *92*, 425–436.
122. "Influence of compatibilizer degradation on formation and properties of PA6/organoclay nanocomposites" Monticelli, O., Musina, Z., Frache, A., Bullucci, F., Camino, G., Russo, S. *Polym. Degrad. Stab.* 2007, *92*, 370–378.
123. "Thermal degradation of commercially available organoclays studied by TGA-FTIR" Cervantes-Uc, J.M., Cauich-Rodriguies, J.W., Vasquez-Torres, H., Garfias-Mesias, L.F., Paul, D.R. *Thermochimica Acta* 2007, *457*, 92–102.
124. "Stability of organically modified montmorillonites and their polysty-rene nanocomposites after prolonged thermal treatment" Frankowski, D.J., Capracotta, M.D., Martin, J.D., Khan, S.A., Spontak, R.J. *Chem. Mater.* 2007, *19*, 2757–2767.
125. "Polymer/layered silicate nanocomposites from thermally stable trialkylim-idazolium-treated montmorillonite" Gilman, J.W., Awad, W.H., Davis, R.D., Shields, J., Harris, R.H. Jr., Davis, C., Morgan, A.B., Sutto, T.E., Callahan, J., Trulove, P.C., DeLong, H.C. *Chem. Mater.* 2002, *14*, 3776–3785.
126. "Polystyrene-clay nanocomposites prepared with polymerizable imidazolium surfactants" Bottino, F.A., Fabbri, E., Fragala, I.L., Malandrino, G., Orestano, A., Pilati, F., Pollicino, A. *Macromol. Rapid Commun.* 2003, *24*, 1079–1084.
127. "Melt-processable syndiotactic polystyrene/montmorillonite nanocomposites" Wang, Z.M., Chung, T.C., Gilman, J.W. Manias, E. *J. Polym. Sci. Part B.* 2003, *41*, 3173–3187.

128. "Thermal degradation studies of alkyl-imidazolium salts and their application in nanocomposites" Awad, W.H., Gilman, J.W., Nyden, M., Harris, R.H., Sutto, T.E., Callahan, J., Trulove, P.C., DeLong, H.C., Fox, D.M. *Thermochimica Acta* 2003, *409*, 3–11.
129. "Synthesis of imidazolium salts and their application in epoxy montmorillonite nanocomposites" Langat, J., Bellayer, S., Hudrlik, P., Hudrlik, A., Maupin, P.H., Gilman, J.W., Raghavan, D. *Polymer* 2006, *47*, 6698–6709.
130. "Recycled PET-organoclay nanocomposites with enhanced processing properties and thermal stability" Kracalik, M., Studenovsky, M., Mikesova, J., Kovarova, J., Sikora, A., Thomann, R., Friedrich, C. *J. App. Polym. Sci.* 2007, *106*, 2092–2100.
131. "Benzimidazolium surfactations for modifications of clays for use with styrenic polymers" Costache, M.C., Heidecker, M.J., Manias, E., Gupta, R.K., Wilkie, C.A. *Polym. Degrad. Stab.* 2007, *92*, 1753–1762.
132. "ABS/clay nanocomposites obtained by solution technique: Influence of clay organic modifiers" Modesti, M., Besco, S., Lorenzetti, A., Causin, V., Marega, C., Gilman, J.W., Fox, D.M., Trulove, P.C., De Long, H.C., Zammarano, M. *Polym. Degrad. Stab.* 2007, *92*, 2206–2213.
133. "Thermal stability of quaternary phosphonium modified montmorillonites" Xie, W., Xie, R., Pan, W.-P., Hunter, D., Koene, B., Tan, L.-S., Vaia, R. *Chem. Mater.* 2002, *14*, 4837–4845.
134. "Ionic liquid modification of layered silicates for enhanced thermal stability" Byrne, C., McNally, T. *Macromol. Rapid Commun.* 2007, *28*, 780–794.
135. "A carbocation substituted clay and its styrene nanocomposites" Zhang, J., Wilkie, C.A. *Polym. Degrad. Stab.* 2004, *83*, 301–307.
136. "Enhanced thermal properties of PS nanocomposites formed from montmorillonite treated with a surfactant/cyclodextrin inclusion complex" Yei, D.-R., Kuo, S.-W., Fu, H.-K., Chang, F.-C. *Polymer* 2005, *46*, 741–750.
137. "Polystyrene nanocomposites based on quinolinium and pyridinium surfactants" Chigwada, G., Wang, D., Wilkie, C.A. *Polym. Degrad. Stab.* 2006, *91*, 848–855.
138. "Ferrocene and ferrocenium modified clays and their styrene and EVA composites" Manzi-Nshuti, C., Wilkie, C.A. *Polym. Degrad. Stab.* 2007, *92*, 1803–1812.
139. "An optimum organic treatment of nanoclay for PMR-15 nanocomposites" Gintert, M.J., Jana, S.C., Miller, S.G. *Polymer* 2007, *48*, 7573–7581.
140. Waste electrical and electronic equipment directive (WEEE). http://europa.eu.int/eur-lex/pri/en/oj/dat/2003/l_037/l_03720030213en00240038.pdf
141. "Fire-LCA model: TV case study" Simonson, M., Blomqvist, P., Boldizar, A., Möller, K., Rosell, L., Tullin, C., Stripple, H., Sundqvist, J.O. SP Report 2000:13. Printed in 2000.
142. "Polybrominated diphenyl ethers in house dust and clothes dryer lint" Stapleton, H.M., Dodder, N.G., Offenberg, J.H., Schantz, M.M., Wise, S.A. *Environ. Sci. & Technol.* 2005, *39*, 925–931.
143. "The globalization of fire testing and its impact on polymers and flame retardants" Troitzsch, J.H. *Polym. Degrad. Stab.* 2005, *88*, 146–149.
144. "Waste electrical and electronic equipment plastics with brominated flame retardants—from legislation to separate treatment—thermal processes" Tange, L., Drohmann, D. *Polym. Degrad. Stab.* 2005, *88*, 35–40.

145. "Pulmonary toxicity of single-wall carbon nanotubes in mice 7 and 90 days after intratracheal instillation" Lam, C.W., James, J.T., McCluskey, R., Hunter, R.L. *Toxicol. Sci.* 2004, *77*, 126–134.
146. "Pulmonary effects of inhaled ultrafine particles" Oberdorster, G. *Int. Arch. Occup. Environ. Health* 2001, *74*, 1–8.
147. "Translocation of inhaled ultrafine particles to the brain" Oberdorster, G., Sharp, Z., Atudorei, V., Elder, A., Gelein, R., Kreyling, W., Cox, C. *Inhal Toxicol.* 2004, *16*, 437–445.
148. "Photo-oxidative degradation of polyethylene/montmorillonite nanocomposite" Qin, H., Zhao, C., Zhang, S., Chen, G., Yang, M. *Polym. Degrad. and Stab.* 2003, *81*, 497–500.
149. "Photo-oxidation of polymeric-inorganic nanocomposites: Chemical, thermal stability, and fire retardancy investigations" Tidjani, A., Wilkie, C.A. *Polym. Degrad. Stab.* 2001, *74*, 33–37.
150. "Photo-oxidation of polypropylene/montmorillonite nanocomposites. 1. Influence of nanoclay and compatibilizing agent" Morlat, S., Mailhot, B., Gonzalez, D., Gardette, J.-L. *Chem. Mat.* 2004, *16*, 377–383.
151. "Thermal stability and fire retardant performance of photo-oxidized nanocomposites of polypropylene-graft-maleic anhydride/clay" Diagne, M., Gueye, M., Vidal, L., Tidjani, A. *Polym. Degrad. Stab.* 2005, *89*, 418–426.
152. "The effect of photo-oxidation on thermal and fire retardancy of polypropylene nanocomposites" Diagne, M., Gueye, M., Dasilva, A., Vidal, L., Tidjani, A. *J. Mater. Sci.* 2006, *41*, 7005–7010.
153. "Environmental degradation of epoxy–organoclay nanocomposites due to UV exposure. Part I: Photo-degradation" Woo, Ricky S.C., Chen, Yanghai, Zhu, Honggang, et al. *Comp. Sci. Technol.* 2007, *67*, 3448–3456.
154. "Photooxidation of ethylene-propylene-diene/montmorillonite nanocomposites" Morlat-Therias, S., Mailhot, B., Gardette, J.-L., Da Silva, C., Haidar, B., Vidal, A. *Polym. Degrad. Stab.* 2005, *90*, 78–85.
155. "An overview on the degradability of polymer nanocomposites" Pandey, J.K., Reddy, K.R., Kumar, A.P., Singh, R.P. *Polym. Degrad. Stab.* 2005, *88*, 234–250.
156. "Use of oxide nanoparticles and organoclays to improve thermal stability and fire retardancy of poly(methyl methacrylate)" Laachachi, A., Leroy, E., Cochez, M., Ferriol, M., Cuesta, J.M.L. *Polym. Degrad. Stab.* 2005, *89*, 344–352.
157. "Flame retardancy of nanocomposites based on organoclays and carbon nanotubes with aluminum trihydrate" Beyer, G. *Polym. Adv. Technol.* 2006, *17*, 218–225.

Section II

Nano-Bio Composites

6

Animal-Based Fiber-Reinforced Biocomposites

Hoi-Yan Cheung
The Hong Kong Polytechnic University, Hong Kong

CONTENTS

6.1 Introduction

Bioengineering refers to the application of concepts and methods of the physical sciences and mathematics in an engineering approach toward solving problems in repair and reconstruction of lost, damaged, or deceased tissues. Any material that is used for this purpose can be regarded as a biomaterial. According to Williams [1], a biomaterial is a material used in implants or medical devices, intended to interact with biological systems. Those common types of medical devices include the substitute heart valves and artificial hearts, artificial hip and knee joints, dental implants, internal and external fracture fixators, and skin repair templates. One of the major features of composite materials is that they can be tailor made to meet different applications' requirements. The most common types of conventional composites are usually composed of epoxy, unsaturated polyester resin, polyurethanes, or phenolic reinforced by glass, carbon, or aramid fibers. These composite

139

TABLE 6.1

Key Factors for the Selection of Materials for Biomedical Applications [29]

Factors	Description		
	Chemical/ Biological Characteristics	Physical Characteristics	Mechanical/ Structural Characteristics
First-level material properties	Chemical composition (bulk and surface)	Density	Elastic modulus Shear modulus Poisson's ratio Yield strength Compressive strength
Second-level material properties	Adhesion	Surface topology Texture Roughness	Hardness Flexural modulus Flexural strength
Specific functional requirements (based on applications)	Biofunctionality Bioinert Bioactive Biostability Biodegradation behavior	Form and geometry Coefficient of thermal expansion Electrical conductivity Color, aesthetics Refractive index Opacity or translucency	Stiffness or rigidity Fracture toughness Fatigue strength Creep resistance Friction and wear resistance Adhesion strength Impact strength Proof stress Abrasion resistance
Processing and fabrication	Reproducibility, quality, sterilizability, packaging, secondary processability		
Characteristics of host: tissue, organ, species, age, sex, race, health condition, activity, systemic response			
Medical/surgical procedure, period of application/usage			
Cost			

structures lead to the problem of conventional removal after the end-of-life time, as the components are closely interconnected, relatively stable, and thus difficult to separate and recycle.

Within the past few years, there has been a dramatic increase in the use of natural fibers such as leaves from flax, jute, hemp, pineapple, and sisal for making a new type of environmentally friendly composite. Recent advances in natural fiber development, genetic engineering, and composite science offer significant opportunities for improved materials from renewable resources with enhanced support for global sustainability. A material that can be used for medical application must possess many specific character-istics. The most fundamental requirements are related to biocompatibility—that is, not having any adverse effects on the host tissues; therefore, those traditional composite structures with nonbiocompatible matrix or reinforce-ment are substituted by bioengineered composites. Table 6.1 summarizes several important factors that need to be considered in selecting a material for biomedical applications.

Biocomposites consist of biodegradable polymer as matrix material and usually biofibers as reinforcing elements that generally have low cost, low density, high toughness, acceptable specific strength properties, good thermal properties, ease of separation, enhanced energy recovery, and biodegradability. Biofibers are chosen as reinforcements, because they can reduce the chance of tool wear when processing, as well as reduce dermal and respiratory irritation. Conversely, these fibers are usually small in cross sections and cannot be directly used in engineering applications; they are embedded in matrix materials to form biocomposites. The matrix serves as a binder to bind the fibers together and transfer loads to the fibers. In order to develop and promote these natural fibers and their composites, it is necessary to understand their physicomechanical properties.

6.2 Silkworm Silk Fiber

Natural fibers are subdivided based on their origins, coming from plants, animals, or minerals. Generally, plant-based natural fibers are lignocelluloses in nature and are composed of cellulose, hemicellulose, and lignin, like flax, jute, sisal, and kenaf; whereas, animal-based natural fibers are of proteins, like wool, spider, and silkworm silk. The enhanced environmental stability of silk fibers in comparison to globular proteins is due to the extensive hydrogen bonding, the hydrophobic nature of much of the protein, and the significant crystallinity.

Silk proteins known as silk fibroins are stored in the glands of insects and spiders as an aqueous solution. During the spinning process, silkworm accelerates and decelerates its head in arcs to each change of direction, and the concentration of silk in the solution is gradually increased, and finally, elongation stress is applied to produce a partly crystalline, insoluble fibrous thread in which the bulk of the polymer chains in the crystalline regions are oriented parallel to the fiber axis. Faster spinning speed leads to stronger but more brittle fibers where slower speed leads to weaker and more extensible fibers. At even greater speed, silk toughness decreased, mainly due to the loss of extensibility [2].

Cocoons are natural polymeric composite shells made of a single continuous silk strand with length in the range of 1000 to 1500 m and conglutinated by sericin [3]. This protein layer resists oxidation, is antibacterial, ultraviolet (UV) resistant, and absorbs and releases moisture easily. This protein layer can be cross-linked, copolymerized, and blended with other macromolecular materials, especially artificial polymers, to produce materials with improved properties. On average, cocoon production is about 1 million tons worldwide, and this is equivalent to 400,000 tons of dry cocoon (see Figure 6.1).

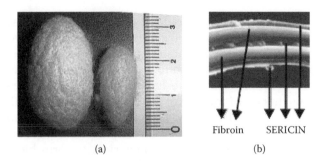

Fibroin   SERICIN

(a)   (b)

FIGURE 6.1
Raw cocoon silks (a) and side view of the silk fiber (b).

In the tissue engineering area, silks have been identified as a kind of bio-material, used in the healing process for bone, tendons, or ligament repairs. Slowly degrading biomaterials that can maintain tissue integrity following implantation, while continually transferring the load-bearing burden to the developing biological functional tissue, are desired. In such phenomena, the gradual transfer of the load-bearing burden to the developing or remodeling tissue should support the restoration and maintenance of tissue function over the life of the patient.

Silk fibers spun out from silkworm cocoons consist of fibroin in the inner layer and sericin in the outer layer, and all are protein based. From the outside to the inside layers of the cocoon, the volume fractions of sericin decrease while the relative content of fibroin increases. Also, it is known that silk fibroin consists of both hydrophilic and hydrophobic regions which are block-like polymeric systems. These fibers have a highly nonuniform cross-sectional geometry with respect to both shape and absolute dimensions. By changing the reeling conditions, silkworm silks can be stronger, stiffer, and more extensible, approaching the properties of spider dragline silks [4]. Each raw silk thread has a lengthwise striation, consisting of two separate but irregularly entwined fibroin filaments (brin) embedded in sericin. Silk sericin is a minor protein that envelops silk fibroin fibers and glues them together to form a cocoon shape. Fibroin and sericin in silk account for about 75 and 25 wt%, respectively. Silk fibers are biodegradable and highly crystalline with a well-aligned structure.

6.2.1 Mechanical Properties

Composition, structure, and material properties of silk fibers produced by spiders, silkworms, scorpions, mites, and flies may differ widely depending on the specific source and the uncontrollable reeling conditions of those insects. Spinning under controlled conditions will have more uniform cross-sectional area of silk fibers, reproducible molecular alignment, and fewer microstructural flaws. The size and weight of cocoons decrease with an increase in temperature, and cocoons can bear efficiently both external static forces and dynamic impact loadings [4]. A normal compact cocoon exhibits a high ability of elastic deformation with an elastic strain limit higher than 20% in both longitudinal and transverse directions. Anisotropic properties mainly due to the nonuniform distribution and orientations of silk segments and the inner layer of cocoon are low porosity (higher silk density) and smaller average diameter of silk; therefore, there is an increase in elastic modulus and strength from the outside to the inside layers. That is, the thinner the silk, the higher the elastic modulus and tensile strength and the maximum values at the innermost layer. On the other hand, at temperatures above the glass transition temperature, the cocoon and its layers become softer and softer and behave similar to a rubber-like material. Silk fibers have higher tensile strength than glass fiber or synthetic organic fibers, good elasticity, and excellent resilience [5]. They resist failure in compression, are stable at physiological temperatures, and the sericin coating is water-soluble proteinaceous glue. Table 6.2 shows a comparison of the mechanical properties of common silk types (silkworm and spider dragline) to several types of biomaterial fibers and tissues commonly used today.

Fibroin is a semicrystalline polymer of natural fibrous protein consisting mainly of two phases [6]: β-sheet crystals and noncrystalline, including microvoids and amorphous structure, by which the structure of sericin coating is amorphous, acting as an adhesive binder to maintain the fibroin core and the overall structural integrity of the cocoon. Degumming is a key process during which sericin is removed by thermochemical treatment of the cocoon. Although this surface modification can affect the tensile behavior and the mechanical properties of silk significantly, it is normally done to enhance interfacial adhesion between fiber and matrix.

In addition, according to Altman et al. [7], silks are insoluble in most solvents, including water, dilute acid, and alkali. Reactivity of silk fibers with chemical agents is positively correlated to the largeness of internal and external surface areas [8]. When fabricating silk-based composites, the amount of resin gained by fibers is strongly related to the degree of swelling of the noncrystalline regions—that is, the amorphous regions and the microvoids inside the fibers.

6.2.2 Applications

6.2.2.1 Wound Sutures

Silk fibers have been used in biomedical applications particularly as sutures by which the silk fibroin fibers are usually coated with waxes or silicone to enhance material properties and reduce fraying. But in fact, there are lots of confusing questions about the usage of these fibers, as there is an absence of detailed characterization of the fibers used, including the extent of extraction of the sericin coating, the chemical nature of wax-like coatings sometimes

TABLE 6.2

Tensile Properties of Plant- and Animal-Based Natural Fibers and Human Tissues

	UTS (Mpa)	Elongation at Break (%)	E (GPa)
Natural Fibers			
Flax	300–1500	1.3–10	24–80
Jute	200–800	1.16–8	10–55
Sisal	80–840	2–25	9–38
Kenaf	295–1191	3.5	2.86
Abaca	980		106 psi
Pineapple	170–1627	2.4	60–82
Banana	529–914	3	27–32
Coir	106–175	14.21–49	4–6
Oil palm (empty fruit)	130–248	9.7–14	3.58
Oil palm (fruit)	80	17	
Ramie	348–938	1.2–8	44–128
Hemp	310–900	1.6–6	30–70
Wool	120–174	25–35	2.3–3.4
Spider silk	875–972	17–18	11–13
Cotton	264–800	3–8	5–12.6
Human Tissues			
Hard tissue (tooth, bone, human compact bone, longitudinal direction)	130–160	1–3	17–20
Skin	7.6	78	
Tendon	53–150	9.4–12	1.5
Elastic cartilage	3	30	
Heart valves	0.45–2.6	10–15.3	
Aorta	0.07–1.1	77–81	

used, and many related processing factors. For example, the sericin glue-like proteins are the major cause of adverse problems with biocompatibility and hypersensitivity to silk. The variability of source materials has raised potential concerns with this class of fibrous protein. Yet, silk's knot strength, handling characteristics, and ability to lay low to the tissue surface make it a popular suture in cardiovascular applications, where bland tissue reactions are desirable for the coherence of the sutured structures [9].

6.2.2.2 Scaffold Tissue Engineering

A three-dimensional scaffold permits the *in vitro* cultivation of cell–polymer contructs that can be readily manipulated, shaped, and fixed to the defect site [10]. The matrix acts as the translator between the local environment (either *in vitro* or *in vivo*) and the developing tissue, aiding in the development of biologically viable functional tissue. However, during the 1960s through the early 1980s, the use of virgin silk negatively impacted the general acceptance of this biomaterial from the surgical practitioner perspective, for example, the reaction of silk to the host tissue and the inflammatory potential of silk. Recently, silk matrices are being rediscovered and reconsidered as potentially useful biomaterials for a range of applications in clinical repairs and *in vitro* as scaffolds for tissue engineering.

Silk, as a protein, is susceptible to proteolytic degradation *in vivo* and over a longer period of time *in vivo* will slowly be absorbed. Degradation rates mainly depend on health and physiological status of the patient, the mechanical environment of the implantation site, and types and dimensions of the silk fibers. The slow rate of degradation of silk *in vitro* and *in vivo* makes it useful in biodegradable scaffolds for slow tissue ingrowths, because the biodegradable scaffolds must be able to retain at the implantation site, including maintain their mechanical properties and support the growth of cells, until the regenerated tissue is capable of fulfilling the desired functions. The degradation rate should be matched with the rate of new tissue formation so as to compromise with the load-bearing capabilities of the tissue.

Additionally, scaffold structures including size and connectivity of pores determine the transport of nutrients, metabolites, and regulatory molecules to and from cells. The matrix must support cell attachment, spreading, growth, and differentiation. Meinel et al. [11] concentrated on cartilage tissue engineering with the use of silk protein scaffold, and the authors identified and reported that silk scaffolds are particularly suitable for tissue engineering of cartilage starting from human mesenchymal stem cells (hMSCs), which are derived from bone marrow, mainly due to their high porosity, slow degradation, and structural integrity. Figure 6.2 and Figure 6.3 show the basic principles of tissue engineering and three-dimensional scaffolding system.

Recent research with silk has focused on the development of a wire rope matrix for the development of autologous tissue engineered anterior cruciate ligaments (ACLs) using a patient's own adult stem cells [14]. Silk fibroin offers versatility in matrix scaffold design for a number of tissue engineering needs in which mechanical performance and biological interactions are major factors for success, including bone, ligaments, tendons, blood vessels, and cartilage. Silk fibroin can also be processed into foams, films, fibers, and meshes.

6.2.3 Silk-Based Biocomposites

Annamaria et al. [15] discovered in the studies that environment-friendly biodegradable polymers can be produced by blending silk sericin with other resins. Nomura et al. [16] identified that polyurethane foams incorporating sericin are said to have excellent moisture-absorbing and moisture-desorbing properties. Hatakeyama [17] has also reported producing sericin-containing polyurethane with excellent mechanical and thermal properties. Sericin blends well with water-soluble polymers, especially

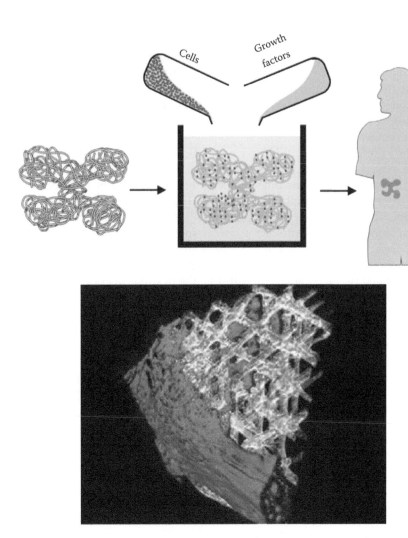

FIGURE 6.2
Background of tissue engineering (www.cs.cmu.edu/people/tissue/tutorial.html).

with polyvinyl alcohol (PVA). Ishikawa et al. [18] investigated the fine structure and the physical properties of blended films made of sericin and PVA. Moreover, a recent patent reported on a PVA/sericin cross-linked hydrogel membrane produced by using dimethyl urea as the cross-linking agent had high strength, high moisture content, and durability for usage as a functional film [19].

Silk fibroin film has good dissolved oxygen permeability in the wet state, but it is too brittle to be used on its own when in the dry state; whereas chitosan is a biocompatible and biodegradable material that can be easily shaped into films and fibers. Park et al. [20] and Kweon et al. [21] introduced an idea

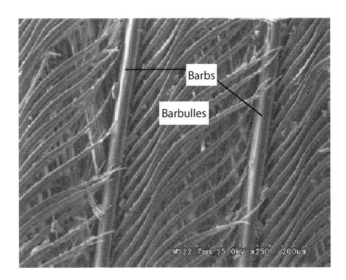

FIGURE 6.3
Three-dimensional tissue scaffold (www.eng.uab.edu/polymers/).

of silk fibroin/chitosan blends as potential biomedical composites, as the crystallinity and mechanical properties of silk fibroin are greatly enhanced with increasing chitosan content.

Another type of biocomposite is the silk fibroin/alginate blend sponges. For biotechnological and biomedical fields, silk fibroin's reproducibility, environmental and biological compatibility, and nontoxicity are benefits in many different clinical applications. As the collective properties, especially mechanical properties, of silk fibroin sponges in the dry state, are too weak to make it possible for them to be used as a wound dressing, they can be enhanced by blending silk fibroin films with other synthetic or natural polymers (e.g., polysaccharide–sodium alginate).

Furthermore, Katori and Kimura [22] and Lee et al. [23] examined the effect of silk/poly(butylenes succinate) (PBS) biocomposites. They found that the mechanical properties including tensile strength, fracture toughness, impact resistance, and the thermal stability of biocomposites would be greatly affected by their manufacturing processes. Moreover, good adhesion between the silk fibers and PBS matrix was found through observation and analysis by scanning electron microscope (SEM) imaging.

The mechanical properties of Bombyx mori, twisted Bombyx mori, and Tussah silk fibers were also investigated through tensile property tests. It was found that Tussah silk fiber exhibited better tensile strength and extensibility as compared with others. However, the stiffness of all samples was almost the same. This may be due to the distinction of the silkworm raising process, cocoon producing, and spinning conditions. Based on the Weibull analysis, it was shown that the Bombyx mori silk fiber has better reproducibility in terms of experimental measurement than the Tussah silk fiber. It may be due to the degumming treatment which has an effect on the microstructure of the fiber.

By using silk fiber as a reinforcement for biodegradable polymer, the mechanical properties do have a substantial change. Cheung et al. [24] demonstrated that the use of silk fiber to reinforce polylactic acid (PLA) can increase its elastic modulus and ductility to 40% and 53%, respectively, as compared with a pristine sample. It was also found that the biodegradability of silk/PLA biocomposites was altered with the content of the silk fiber in the composites. It reflects that the resorbability of the biocomposites used inside the human body can be controlled, and this is the key parameter of using this new type of material for bone plate development.

6.3 Chicken Feather Fiber (CFF)

Animal-based natural fibers, like CFF, have attracted much attention to different product design and engineering industries recently, and the use of CFF as reinforcement for polymer-based biodegradable materials has emerged gradually. The advantages of using these natural fibers over traditional reinforcing fibers in biocomposite materials include its low cost, low density, acceptable specific strength, recyclability, and biodegradability. Natural fibers generally have high specific mechanical properties.

Due to an increasing public awareness of environmental protection, the application of natural fibers in biocomposite materials has been increased rapidly in the past few years. CFFs, because of renewable and recyclable characteristics, have been appreciated as a new class of reinforcements for polymer-based biocomposites. However, a complete understanding of their mechanical properties, surface morphologies, environmental influences due to moisture and chemical attacks, bonding characteristics between silk fibroin and surrounding matrix, and manufacturing process is essential.

6.3.1 Chicken Feather

According to a survey conducted recently, a chicken processing plant produces about 4000 pounds of chicken feathers every hour. In most Western countries, these feathers are used as feather fiber feed; air filter elements that replace traditional wood pulp (retarding the cut-down rate for trees), and lightweight feather composites. Chicken feathers are approximately 91% protein (keratin), 1% lipids, and 8% water. The amino acid sequence of a chicken feather is similar to that of other feathers and has a great deal in common with reptilian keratins from claws [25]. The sequence is largely composed of cystine, glycine, proline, and serine, and contains almost no

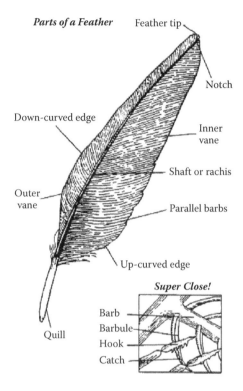

Parts of a Feather   Feather tip

Notch

Down-curved edge

Inner vane

Shaft or rachis

Outer vane

Parallel barbs

Up-curved edge

Super Close!

Barb
Barbule
Hook
Catch

Quill

FIGURE 6.4
Scanning electron microscope image of tertiary structure of a chicken feather (www.ornithology.com/lectures/Feathers.html).

histidine, lysine, or methionine. In fact, a CFF is made up of two parts: the fibers and the quills (Figure 6.4). The fibers are thin filamentous materials that merge from the middle core material called quills. In simple terms, the quill is the hard, central axis off which soft, interlocking fibers branch. Smaller feathers have a greater proportion of fiber, which has a higher aspect ratio than the quill. The presence of quill among fibers results in a more granular, lightweight, and bulky material. A typical quill has dimensions on the order of centimeters (length) by millimeters (diameter). Fiber diameters were found to be in the range of 5 to 50 μm. The density of CFF is lighter than the other synthetic and natural reinforcements; thus, CFF inclusion in a composite could potentially lower composite density, whereas the density of a typical composite with synthetic reinforcing increases as fiber content increases. Hence, lightweight composite materials can be produced by inclusion of CFF to plastics which even reduces the transportation cost. The barbs at the upper portion of the feather are firm, compact, and closely knit, while those at the lower portion are downy (i.e., soft, loose, and fluffy). The down feather provides insulation, and the flight feather provides an airfoil, protects the body from moisture, protects the skin from injury, and

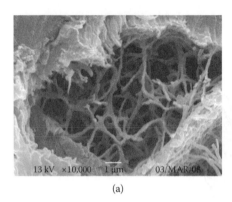
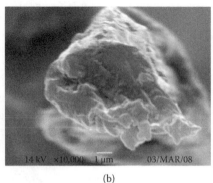

(a)   (b)

FIGURE 6.5
Flight of chicken feather fiber (a) and down chicken feather fiber (b).

provides colors and shapes for displays. Figure 6.5 shows the cross-sectional views of the flight and down feather fibers. It is obvious that flight feather fiber exists in a hollow form, while down fiber is in solid form. In terms of the purpose of fiber reinforcement, the use of down fiber appears much better than the use of flight fiber.

The moisture content of CFFs is an important factor that can highly influence their weight and mechanical properties. The moisture content of processed CFFs can vary depending upon processing and environmental conditions. The glass transition temperature (Tg) of the feather fibers and inner quills is approximately 235°C while that for the outer quills is 225°C. High Tg means that a tighter keratin structure is formed to which water is more strongly bonded. Fibers and inner quills do not begin to lose water below 100°C. The moisture evolution temperature of the CFF and quill occurs in the range of 100 to 110°C [26]. This suggests that it may be possible to have fully dry fibers and inner quills at 110°C.

The length and diameter (sometimes in the form of bundles) of CFF would highly affect their properties and permeability of resin into a resultant composite. Therefore, the control of resin temperature (thus, its viscosity), while at the same time to manage the sonication (ultrasonic vibration) time to facilitate the resin penetration rate into the fibres are essential. Short or longer fibres would highly affect the stress transferability as well as shear strength of the composites. The fibres, themselves also would be a barrier to the movement of polymer chains inside the composites and it may result in increasing their strength and thermal properties, but reduce their fracture toughness. These properties will be studied in detail, in this project.

Figure 6.6 and Table 6.3 show the SEM image of the down chicken feather fiber and its strength compared with other types of feathers. It was found that the development of chicken feather fiber biocomposites has been increasing in recent years, and the outcomes are expected to be able to alleviate the global waste problem.

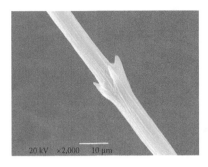

FIGURE 6.6
Scanning electron micrograph images of chicken feather fibers.

TABLE 6.3
Comparisons on Material Properties of Natural Fibers

Fiber	Fineness, denier	Length, cm	Strength, g/den	Elongation, %	Modulus, g/den	Moisture regain, %
Chicken barbs	76	1.5–4.5	1.44 ± 0.46	7.7 ± 0.85	35.6 ± 11.15	9.7
Turkey barbs, Pe	142	5.2	0.83	7.96	15.55	—
Turkey barbs, Pl	55.2	4.1	0.36	16.43	4.47	—
Wool	11	4.5–11.5	1.2–1.8	30–40	30–45	16.0

6.3.2 CFF/Polylactic Acid (PLA) Biocomposites

By mixing CFF with biopolymers, like PLA, a biodegradable composite can be formed and used for plastic products and implant applications. In preparation of the composites, chicken feather was immersed in alcohol for 24 h and then washed in a water-soluble organic solvent and dried at 60°C for 24 h [28]. CFFs with a diameter of about 5 μm and length of 10 to 30 mm were separated from the quill and then used. Figure 6.4 shows an SEM photograph of a CFF. Figure 6.7 shows the relations between CFF content and peak stress and modulus of elasticity, respectively. The modulus of elasticity of CFF/PLA composite increases with the CFF content and reaches the maximum modulus of 4.38 GPa (increment up to 35.6%) at the CFF content of 5 wt%. These reveal that the incorporation of CFF into the matrix is quite effective for reinforcement. The decrease of modulus for the composite with the CFF content above 5 wt% will be due to the insufficient filling of the matrix resin, designating 5 wt% CFF as the critical content.

It also can be found from the peak stress that the tensile strength of PLA after the addition of CFF is lower than that of pure PLA. This phenomenon, also reported by other researchers [1,11], is an indication of poor adhesion between the CFF and the matrix. Because the CFF surface is rough (Figure 6.8b) and because of the hydrophobic consistency of CFF and PLA, the adhesion between them is a problem. The stress could not be transferred from the matrix to the stronger fibers. Another possible explanation of this

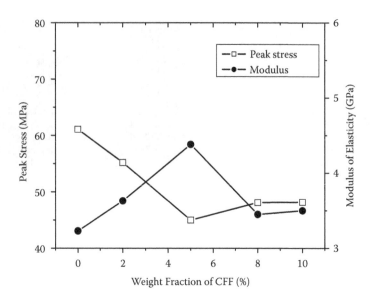

FIGURE 6.7
Relationship between tensile properties and chicken feather fiber content.

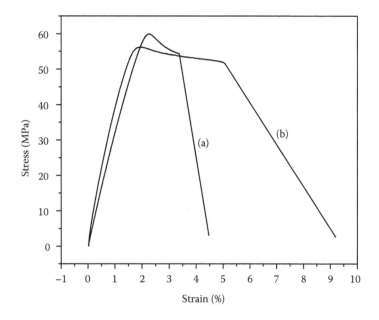

FIGURE 6.8
Stress–strain curves of (a) pure polylactic acid (PLA) sample; (b) 5 wt% chicken feather fiber/ PLA composite.

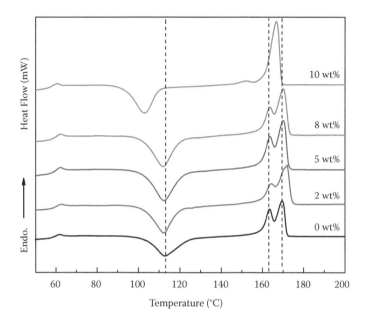

FIGURE 6.9
Differential scanning calorimetry curves of pure polylactic acid (PLA) and chicken feather fiber/PLA composites.

phenomenon could be that the CFFs were randomly oriented inside the composite; the failure of the composite might be initiated by the failure of the matrix, followed by fiber breakage. Figure 6.8 and Figure 6.9 show the stress–strain curves of the pure PLA and 5 wt% CFF/PLA composite. It is observed that a much longer plateau is located between a strain where the peak stress is reached and the strain is at break. It can be concluded that the proper content addition of CFF shows a positive effect on elongation to break for PLA, which was expected because of CFFs acting as bridges to prolong the fracture process of the CFF/PLA composite and because the failure of the composite was controlled by the bridging effect of CFF inside the composite. These conclusions could be proved by the fractured morphology of the microstructures observed by SEM. Thermal properties such as glass transition temperature (T_g), crystallization temperature (T_c), melting temperature (T_m), crystallization enthalpy (ΔH_c), and melting enthalpy (ΔH_m) obtained from the DSC studies are summarized in Table 6.4 and are plotted in Figure 6.10.

6.4 Summary

The mechanical and thermal properties of silk fiber/PLA and CFF/PLA biocomposites were investigated in-depth in the past few years. The

TABLE 6.4

Differential Scanning Calorimetry Results for Pure
Polylactic Acid (PLA) and Chicken Feather Fiber/PLA
Composites

CFF Content (wt%)	Tg (°C)	Tc (°C)	ΔHc (J/g)	Tm (°C)		ΔHm (J/g)
0	58.7	112.9	38.8	163.4	169.5	43.2
2	59.8	112.2	42.1	164.0	171.9	44.7
5	59.2	112.4	42.6	163.7	170.0	43.7
8	59.3	112.0	43.5	163.5	170.0	44.7
10	57.5	102.9	44.5	166.7		46.9

mechanical properties in terms of elastic modulus and ductility of these biocomposites increased substantially compared to the neat polymers. From the DMA results, incorporation of the fibers gave rise to a considerable increase of the storage modulus (stiffness) and to a decrease of the tan delta values. These results demonstrate the reinforcing effect of CFF on PLA matrix. The TGA thermograms reveal the thermal stability of the composites with respect to the pure PLA resin. In addition, the TMA results suggest that the biocomposite with small amount of animal fiber provided better thermal properties as compared with pristine polymer. The SEM investigations confirm that both fibers were well dispersed in the PLA matrix. Although plant- and animal-based fibers attracted much attention to product design and engineering and the bioengineering industry and have been subject to comprehensive research in the past few years, many works such as their interfacial bonding and stress transfer properties have not yet been solved. To widen the applications of these fibers in solving environmental problems, more studies have to be done in the future.

References

1. Williams D.F. Definitions in biomaterials. *Proceedings of a Consensus Conference of the European Society for Biomaterials,* Chester, England, 1986; 4, Elsevier, New York.
2. Shao Z. and Vollrath F. Surprising strength of silkworm silk. *Nature.* 2002; 418: 741.
3. Zhao H.P., Feng X.Q., Yu S.W., Cui W.Z. and Zou F.Z. Mechanical properties of silkworm cocoons. *Polymer.* 2005; 46: 9192–9201.
4. Atkins E. Silk's secrets. *Nature.* 2003; 424: 1010.
5. Perez-Rigueiro J., Viney C., Llorca J. and Elices M. Silkworm silk as an engineering material. *Journal of Applied Polymer Science.* 1998; 70: 2439–2447.

6. Jiang P., Liu H., Wang C., Wu L., Huang J. and Guo C. Tensile behavior and morphology of differently degummed silkworm (Bombyx mori) cocoon silk fibers.
Materials Letters. 2006; 60: 919–925.
7. Altman G.H., Diaz F., Jakuba C., Calabro T., Horan R.L., Chen J., Lu H., Richmond
J. and Kaplan D.L. Silk-based biomaterials. *Biomaterials.* 2003; 24: 401–416.
8. Kawahara Y. and Shioya M. Characterization of microvoids in Mulberry and
Tussah silk fibers using stannic acid treatment. *Journal of Applied Polymer Science.*
1999; 73: 363–367.
9. Postlethwait R.W. Tissue reaction to surgical sutures. In: Dumphy J.E., Van
Winkle W., eds. *Repair and Regeneration.* New York: McGraw-Hill, 1969. pp.
263–285.
10. Freed L.E., Grande D.A., Emmanual J., Marquis J.C., Lingbin Z. and Langer
R. Joint resurfacing using allograft chondrocytes and synthetic biodegradable
polymer scaffolds. *Journal of Biomedical Materials Research.* 1994; 28: 891–899.
11. Meinel L., Hofmann S., Karageorgiou V., Zichner L., Langer R., Kaplan D. and
Vunjak-Novakovic G. Engineering cartilage-like tissue using human mesenchymal stem cells and silk protein scaffolds. *Biotechnology and Bioengineering.* 2004; 88:
379–391.
12. www.cs.cmu.edu/people/tissue/tutorial.html.
13. www.eng.uab.edu/polymers.
14. Altman G.H., Horan R.L., Lu H., Moreau J., Martin I., Richmond J.C. and Kaplan
D.L. Silk matrix for tissue engineered anterior cruciate ligaments. *Biomaterials.*
2002; 23: 4131–4141.
15. Annamaria S., Maria R., Tullia M., Silvio S. and Orio C. The microbial degradation of silk: A laboratory investigation. *International Biodeterioration and
Biodegradation* 1998; 42: 203–211.
16. Nomura M., Iwasa Y. and Araya H. Moisture absorbing and desorbing polyurethane foam and its production. Japan Patent 07-292240A, 1995.
17. Hatakeyama H. Biodegradable sericin-containing polyurethane and its production. Japan Patent 08-012738A, 1996.
18. Ishikawa H., Nagura M. and Tsuchiya Y. Fine structure and physical properties
of blend film compose of silk sericin and poly(vinyl alcohol). *Sen'I Gakkaishi.*
1987; 43: 283–287.
19. Nakamura K. and Koga Y. Sericin-containing polymeric hydrous gel and method
for producing the same. Japan Patent 2001-106794A, 2001.
20. Park S.J., Lee K.Y., Ha W.S. and Park S.Y. Structural changes and their effect on
mechanical properties of silk fibroin/chitosan blends. *Journal of Applied Polymer
Science.* 1999; 74: 2571–2575.
21. Kweon H., Ha H.C., Um I.C. and Park Y.H. Physical properties of silk fibroin/
chitosan blend films. *Journal of Applied Polymer Science.* 2001; 80: 928–934.
22. Katori S. and Kimura T. Injection moulding of silk fiber reinforced biodegradable
composites. In Brebbia C.A. and de Wilde W.P. (ed), *High Performance Structures
and Composites.* WIT Press, Boston, 2002. Section 2: pp. 97–105.
23. Lee S.M., Cho D., Park W.H., Lee S.G., Han S.O. and Drzal L.T. Novel silk/poly
(butylenes succinate) biocomposites: The effect of short fiber content on their
mechanical and thermal properties. *Composites Science and Technology.* 2005; 65:
647–657.
24. Cheung H.Y., Lau K.T. and Tao X.M. A potential material for tissue engineering:
Silkworm silk/PLA biocomposite. *Comp. Pt. B: Engineering.* 2008; 36(6): 1026–1033.

25. Annual Report. Environment Hong Kong 2005. Environmental Protection Department, the Government of the Hong Kong Special Administrative Region. 2005.
26. Fraser, R.D.B. and Parry, D.A.D. The molecular structure of reptilian keratin. *International Journal of Biological Macromolecules.* 1996; 19: 207–211.
27. www.ornithology.com/lectures/Feathers.html.
28. Cheng S., Lau K.T., Liu T., Zhao Y.Q., Lam P.M., Ho M.P. and Yin Y.S. Preparation and mechanical properties of poly(lactic acid) composites containing chicken feather fibers. *Composites Part B.* 2008, in press.
29. Ramakrishna, S., Mayer, J., and Leong, K.W. Biomedical applications of polymer-composite materials: A review. *Composite Science and Technology.* 2001; 61(9): 1189–1224.

7

Biopolymeric Nanofibers for Tissue Engineering

Yogita Krishnamachari

University of Iowa, Iowa

CONTENTS

7.1 Nanomaterial Science in Tissue Engineering

Tissue engineering or regenerative medicine is a field of research that aims to replace or restore the anatomic structure and function of damaged or missing tissue. One of the most widely studied aspects of tissue engineering has been the design of polymeric scaffolds seeded with cells that have specific mechanical and biological properties similar to the native extracellular matrix (ECM). The interactions between cells and ECM control cellular activities such as migration, proliferation, differentiation, gene expression, and organogenesis [1]. Tissue engineering scaffolds aim to chemically, structurally, and functionally recreate microenvironments that closely resemble the *in vivo* environment. Tissue engineering scaffolds provide a three-dimensional framework for cells to attach, proliferate, and differentiate [1]. Along with performing the desired function, scaffolds should be biocompatible and not elicit any undesired immune response that results in rejection of the implant. Additionally, scaffolds should be porous to maximize cell attachment, proliferation, and angiogenesis. A number of scaffold designs have overcome several of the barriers to efficient regeneration but still have material-specific limitations [2]. The success of scaffolds used for tissue engineering is determined by the material properties of the scaffold. These include optimized mechanical, chemical, and geometrical properties of the scaffold for effective functioning at macroscopic and microscopic levels [3]. The scaffold must be able to provide mechanical stability to the tissue as it forms. At the microscopic level, cell growth and differentiation are dependent on how closely the scaffold can mimic the native ECM and provide the appropriate mechanical and signaling inputs to the cells [3]. Scaffolds composed of natural biopolymers made using conventional techniques have inadequate performance *in vivo* due to poor control of the scaffold geometry in terms of elasticity, compressibility, pore size and spatial distribution of pores, and lack of an internal fibrous network [3]. Along with these process-related shortcomings, natural materials have their own specific limitations like poor tensile strength and high probability of eliciting an immune response, resulting in inflammatory responses and rejection of the scaffold [4]. As a solution to these material-specific and process-specific limitations, material-based nanotechnology can be successfully exploited to effectively alter surface chemistry in terms of roughness, wettability, and mechanical strength [5]. Natural and synthetic materials, when designed at the nanometer size, have been shown to promote cell growth and differentiation [6]. This chapter introduces commonly used natural and synthetic materials for preparing nanoscale scaffolds for tissue engineering applications (Figure 7.1). We will also discuss how composites of these materials can provide the most optimal environment for tissue regeneration. The development of nanoscale scaffold materials is an area generating interest from technology leaders like Boston Scientific and Medtronic Inc. (Minneapolis, Minnesota). Living organisms are built of cells

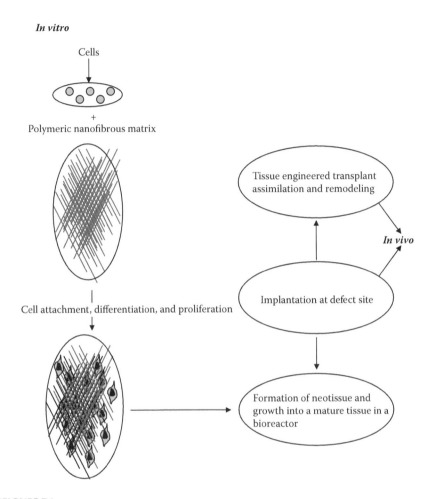

FIGURE 7.1
Tissue engineering process employing a nanostructured polymeric scaffold material.

that are typically around 10 to 50 μm in size. Cell parts and proteins exhibit smaller sizes as low as 5 nm. This has resulted in a strong interest in studying the biological processes and cellular machinery on a nanoscale level.

7.1.1 Nano-Biomaterials for Biomedical Applications

Creating nanofibrous scaffolds with material dimensions at the nanometer scale allows for the efficient replication of the physical structure of natural ECM [7]. Biomaterials fabricated as nanofibers can positively influence the physical and mechanical performance of the biomaterial scaffold. Scaffolds fabricated as nanofibers have been considered for use in the engineering of cartilage [8], bone [9], wound healing [10], and artificial blood vessel applications [11]. Some advantages of employing nanofibrous biomaterials for tissue engineering applications include physical mimicking of natural ECM, ease of surface functionalization, and improving scaffold mechanical properties [7]. Additionally, nanofibers can serve as a medium through which diffusion of soluble factors and cells can occur.

Nanostructured biomaterials have been developed with incorporation of gene delivery technology [12]. For example, scaffolds entrapping plasmid DNA have been fabricated using PLGA and PLA-polyethylene glycol (PEG). DNA release from the nanofibrous polymeric scaffold retained its activity with strong expression of the model protein β-galactosidase. Nanostructured biomaterials developed for wound-healing applications can stimulate activation of hemostasis, whereby bleeding at the site of injury is halted by a physiological process of coagulation. Hemostasis is promoted by the high surface area present on the nanostructured biomaterial [7]. Additionally, the high surface area/volume ratio of nanostructured biomaterials results in water absorption capacity that is 10- to 50-fold greater than conventional polymeric films. This is a critical feature in wound healing to ensure absorption of tissue and wound exudates. Polymers fabricated at the nanoscale also possess the ability to easily conform to the contour of the wound and tissue defect which is a critical parameter governing the success of the biomaterial for wound-healing and defect-filling applications [7].

7.1.2 Cellular Response to Nanomaterial Science in Tissue Engineering

A distinct feature of the native ECM is the ubiquitous presence of nanoscale components such as collagen. Mimicking these types of features requires scaffolds that are fabricated with nanoscale elements to have greater effect and control over cellular behavior. For example, osteoblast (mononucleate cells that are responsible for bone formation) and osteoclast (multinucleate type of bone cells responsible for bone resorption by removing mineralized matrix from bone tissue) activity can be increased by implantation of spherical nanoparticulate alumina which closely resembles the structure of natural hydroxyapatite crystals found in the human bone [13]. Carbon nanofibers also promote osteoblast proliferation significantly [14]. Nanofibers have a higher surface-to-volume ratio than most conventional micron-size materials, thus promoting cell adhesion and proliferation to a greater extent. The total amount of adsorbed proteins such as fibronectin, collagen, and laminin are significantly higher on nanoscale materials due to the large surface area exposed; hence, these proteins that mediate specific cellular functions show enhanced activity on nanophase materials [14,15]. These nanofibers exhibit diameters in the range of 50 to 500 nm that are several orders of magnitude smaller than that of the cells, thus closely resembling the native nanostructured ECM. One of the reasons attributed to the greater success of nanofibrous scaffold materials over traditional micron-scale particulate materials is the presence of a continuous structure [14]. Cells preferentially adhere and proliferate on nanofiber scaffolds and are found to cross-link nanofibers in the matrix to form a highly organized three-dimensional cellular network. Several techniques like self-assembly, phase separation, and electrospinning allow the production of polymer fibers with diameters ranging from 3 nm to 5 μm. One such technique that has gained increasing popularity for processing natural and synthetic composites for tissue engineering applications is electrospinning [16–18]. The main advantage of this technique is its cost-effectiveness which can be used to tailor the properties of the scaffold material and results in formation of a long, continuous fibrous network [4].

7.1.3 Electrospinning Theory and Process

The electrospinning technique developed by Reneker and Chun has been used to spin a variety of natural and synthetic polymers into nanofibers for applications in tissue engineering [16,19]. Electrospun fibers are being investigated as promising tissue engineering scaffolds because they mimic the nanoscale properties of the native ECM. The most attractive feature of electrospinning is the cost-effectiveness and noncomplex nature of the technique [4].

A typical electrospinning setup consists of a syringe pump, a high-voltage source, and a collector as shown in Figure 7.2. The application of an electric field using the high-voltage source causes a charge to be introduced in the polymer that leads to the development of repulsive forces in the solution. This electrostatic interaction overcomes the force of surface tension causing the initiation of a jet. As the solvent evaporates, the polymeric nanofibers

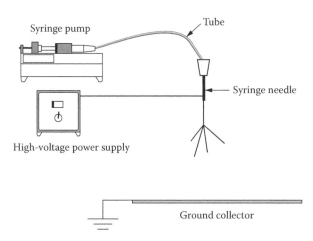

FIGURE 7.2
Electrospinning process to generate nanofibrous biomaterial. (Adapted from Zhang YZ, Ouyang HW, Lim CT, Ramakrishna S, Huang ZM. *Journal of Biomedical Materials Research Part B—Applied Biomaterials*, Jan 2005; 72B(1):156–165. Reprinted with permission from John Wiley and Sons Inc.)

FIGURE 7.3
Scanning electron micrograph of an electrospun nanofiber of a random polymer fiber mesh produced by electrospinning, Scale bar = 100 µm. (Adapted from Pham QP, Sharma U, Mikos AG. *Tissue Engineering*, May 2006; 12(5):1197–1211. Reprinted with permission of Mary Liebert Inc.)

formed can be collected by use of an appropriate collector [20–22]. Figure 7.3 represents a scanning electron microscopic (SEM) image of an electrospun nanofibrous polycaprolactone (PCL) polymeric matrix. Controlling process parameters like viscosity and concentration of polymeric solution, conductivity, dielectric constant and dipole moment, the strength and voltage of electric field, the flow rate, and the collector composition and geometry, produced polymeric nanofibers with desired porosity, geometry, and diameter to suit the specific needs of its final tissue engineering application [23,24].

Pore size, pore orientation, and fiber diameter strongly affect skin regeneration, and these properties can be effectively controlled by the electrospinning technique [21]. Cells seeded in the scaffold can move through the matrix, and these nanofibers then produce a flexible and dynamic architecture that enables cells to perform amoeboid movement [10]. The reduced resistance of cell movement allows the cells to grow, migrate, and proliferate into the nanofibrillar matrix [10].

7.2 Biodegradable and Bioresorbable Polymers

As defined by Vert [25], biodegradable polymers break down due to macromolecular degradation with dispersion *in vivo*. Bioresorbable polymers like polylactic acid (PLA) and polyglycolic acid (PGA) show bulk degradation and are eliminated through natural pathways or by simple filtration of their degradation products *in vivo* [26,27]. Biodegradable polymers can be natural or synthetic. Natural polymers include collagen [28], chitosan [29,30], fibrinogen [31], and silk [32]. Natural polymers induce positive cellular interactions with surrounding tissue and enhance biocompatibility. Synthetic biopolymers like PLA are formed with defined physicochemical properties through controllable polymerization techniques [4]. Hydrogels form the third type of biopolymers [33,34]. Injectable hydrogels can take the same shape as the site of implant, overcoming the need to prepare the scaffold in accordance with implantation site size and geometry [33,34].

In the following sections, we will introduce several types of synthetic and natural polymers. Each material has advantages and disadvantages. Composite scaffolds can be built by mixing two or more natural or synthetic biopolymers to combine the strengths of the individual components. This class of composite polymers is discussed in the last section of this chapter.

7.2.1 Natural Biopolymers

7.2.1.1 Collagen

Collagen, a fibrous protein that is the major component in connective tissue is an excellent example of a natural biopolymer [35]. Collagen has a highly organized structure that results in the ability to guide cell growth and differentiation at various stages of tissue development [10].

More than 25 distinct types of collagen have been identified. Types I and III are the most abundant forms of collagen in native tissue. All collagen is composed of three α polypeptide chains that are coiled into a left-handed helix [36]. These three chains are then woven around one another in a repeating motif fashion to give the final structure a rope-like appearance [4]. Purified collagen after extracting from natural sources is called reconstituted collagen [4]. Reconstituted collagen elicits only a small to negligible immune response [4]. At the ultrastructural level, this repeating motif exhibits a 67 nm bandwidth that imparts a characteristic binding pattern to the collagen fibrils [36]. The fibrils from collagens types I and III are the most abundant proteins in several mammalian bodies and are found throughout the interstitial spaces, providing the cells with appropriate cues for embryogenic development, organogenesis, cell growth, and wound repair [37].

The typical biochemical processes used to isolate and purify collagen may compromise the biological and structural properties. Barnes et al. [4] have shown that the production of collagen nanofibers by electrospinning yields scaffolds that retain collagen's biological and structural properties. The nanofibrous structure of collagen can mimic the ECM. The fibrillar structure of collagen is important for cell attachment and proliferation and is critical for

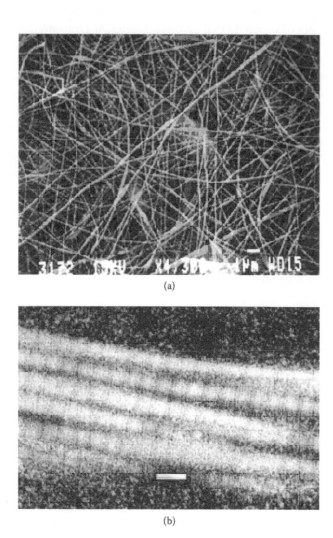

(a)

(b)

FIGURE 7.4
Scanning electron micrograph of electrospun collagen yielding fibers with diameter less than 100 nm (a) and transmission electron micrograph of the fibers exhibiting 67 nm bandwidth as native collagen (b). Scale bar = 100 nm. (Adapted from Matthews JA, Wnek GE, Simpson DG, Bowlin GL. *Biomacromolecules*, March–April 2002; 3(2):232–238. Reprinted with permission of American Chemical Society.)

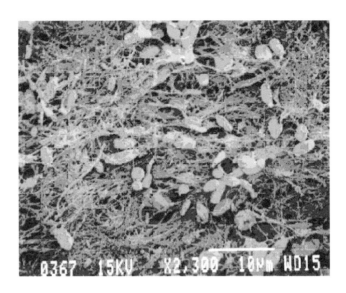

FIGURE 7.5
Scanning electron micrograph of nanofibrous collagen scaffold being infiltrated by smooth muscles cells at the end of 7 days. (Adapted from Matthews JA, Wnek GE, Simpson DG, Bowlin GL. *Biomacromolecules*, March–April 2002; 3(2):232–238[36]. Reprinted with permission of American Chemical Society.)

the development of engineered tissues that closely resemble the natural tissue both structurally and functionally [38]. Additionally, the high surface-to-volume ratio of these collagen nanofibers is believed to produce superior cell adhesion as compared to traditional scaffolds [36]. Collagen-based implants are degraded by mammalian collagenases. These are naturally occurring enzymes that attack the triple helical structure of collagen at specific locations [39].

7.2.1.1.1 Applications of Collagen Nanofibers in Tissue Engineering

Mathews et al. [36] showed that collagen nanofibers can be produced by preparing a solution of collagen in 1,1,1,3,3,3 hexafluoro-2-propanol (HFP) and collecting the fibers by electrospinning. Aortic smooth muscle cells suspended in a bioreactor showed that scaffolds can be densely populated with smooth muscle cells within 7 days (Figure 7.5) [36]. Cross-sectional analysis has shown that the nanofibrillar collagen promoted cell infiltration deep within the collagen matrix [36]. Fabrication of electrospun fibrous collagen matrices produces scaffolds that are superior to nonelectrospun collagen matrices in cellular compatibility and promoting cell growth.

Engineering a dermal substitute using collagen alone results in low mechanical strength. A composite (discussed in Section 7.3) [10] that can combine collagen with a synthetic polymer affords greater tensile strength, and it overcomes the specific limitation of collagen.

Gelatin, a derivative of collagen, has also been investigated for applications in tissue engineering [22,40]. Gelatin is obtained by extraction from collagen using acidic and alkaline processes and thermal denaturation [21]. Gelatin, like collagen, is nonimmunogenic. However, gelatin nanofibers have not been widely utilized due to concerns of loss of mechanical strength in an aqueous medium [4].

7.2.1.2 Chitosan

Chitosan is a deacetylated derivative of chitin, a high molecular weight natural biopolymer found abundantly in shells of marine crustaceans and cell walls of fungi [41]. Chitosan and collagen are the two most widely investigated natural polymers in tissue engineering [42]. However, the limited availability of collagen has hindered its widespread application on a commercial scale [42–44]. Chitosan is a linear polymer of natural origin composed of glucosamine and N-acetyl glucosamine units linked in a β (1-4) fashion [45]. The ratio of the glucosamine (NH_2) to the acetyl glucosamine ($CONH_2$) is referred to as the degree of deacetylation of the polymer. The higher this number, the greater is the number of primary amines in the structure. The molecular weight of chitosan ranges from 300 to over 1000 kD, and the polymer is readily soluble in dilute acids with pH less than 6 [42]. The cationic nature of chitosan results in the electrostatic interaction with anionic glycosaminoglycans (GAGs), proteoglycans, and other negatively charged molecules [43]. Chitosan has primarily found widespread applications in articular tissue engineering due to structural similarities with GAGs found in the articular cartilage [43]. This structural similarity affords chitosan a pivotal role in modulating chondrocyte morphology, differentiation, and function. Additionally, a large number of cytokines and growth factors are linked to cells interacting with GAG; hence, a chitosan–GAG complex can result in a concentration of growth factors in the scaffold released by the infiltrated and colonizing cells [43,44,46].

The degradation of chitosan follows an inverse relationship with the degree of deacetylation, whereas cell adhesion is directly correlated with the

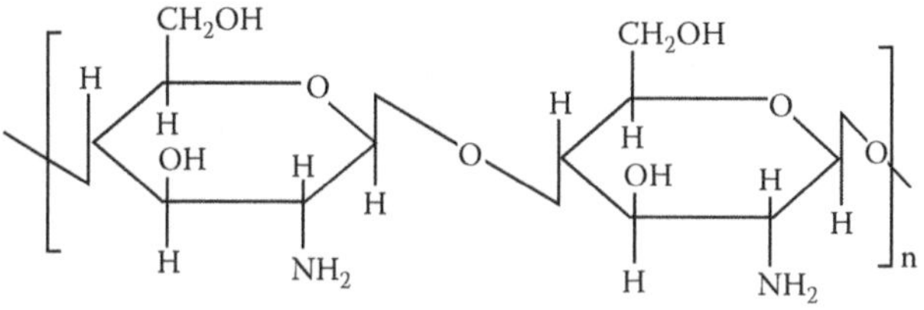

Chitosan

FIGURE 7.6
Structure of chitosan. (Adapted from Kumar M. *Reactive & Functional Polymers*, Nov 2000; 46(1):1–27. Reprinted with permission of Elsevier Inc.)

degree of deacetylation [43]. Within the body, chitosan is degraded by enzymatic hydrolysis. The enzyme, lysozyme, targets the acetylated residues [43]. Proteolytic enzymes also show some enzymatic activity with chitosan [43]. The degradation products are oligosaccharides of varying lengths. Highly deacetylated products last for several months following implantation and promote maximum cell adhesion and proliferation [47–50]. Several studies have shown that three-dimensional scaffolds of chitosan with adequate porosity and mechanical characteristics mimicking the native human tissue can be successfully fabricated [47,50].

Chitosan solutions possess high viscosity owing to which they cannot be readily electrospun to yield polymeric nanofibers, thus limiting its application in fabricating scaffolds. Polymer solution viscosity is an important parameter governing the ability to form nanofibers and its subsequent porosity [36,51,52]. The high viscosity of chitosan solutions is due to strong hydrogen bonding between the amino and hydroxyl group of the chitosan polymeric chains [51]. The high viscosity hinders the ability of the polymer chains to align in a way suitable to form a nanofiber matrix. In order to exploit the use of chitosan nanofibers in tissue engineering applications, there is a need to introduce an element that can disrupt the intermolecular interaction between the polymer chains. One approach to achieve this is to form a composite of chitosan with a biocompatible polyethylene oxide (PEO) synthetic polymer (discussed in Section 7.3) [51].

7.2.1.3 Silk Protein

Silk is a protein polymer spun into fibers by *Lepidoptera* larvae belonging to the family of silkworm spiders [53]. Although silk has been used in the textile industry for many decades, its application has recently attracted much attention as reinforcement in biomaterials for tissue engineering due to its unique mechanical property, biocompatibility, and biodegradability [54,55].

Silkworm silk is the most commonly used material for tissue engineering applications. Each fiber of silk is about 10 to 20 µm and consists of a core coating protein called sericin that holds the fibers together and serves as glue [32]. The core is composed of a heavy chain, a light chain, and a glycoprotein. The heavy chain and the light chains are interconnected by disulfide bonds [32]. The heavy chain, also called silk fibroin, is the main component determining the properties of silk for use in tissue engineering applications [56]. The strong mechanical properties of silk are due to the extensive hydrogen bonding, hydrophobic nature of the protein, and significant crystallinity associated with the protein [56]. It has a tensile strength of 610 to 690 MPa that is 10-fold greater than most other natural biopolymers [53]. The human bones and tendons exhibit a tensile strength of 160 to 200 MPa, thus making silk an ideal scaffold material to develop the bone [54,57].

The sericin component of silk has been identified to elicit immunological responses [58]. There are some limitations to using silk in tissue engineering applications. However, with the development of nanomaterial science and superior processing techniques, the sericin component can be easily removed, and such processed silk has been shown to be biocompatible and nonimmunogenic during *in vitro* macrophage assays [59]. Silk fibers have been shown to retain their tensile strength for up to one year *in vivo* with the degradation mainly occurring by bond cleavage as a result of the action of proteolytic enzymes [54]. The strength and toughness possessed by this natural biopolymer make it ideal for tissue engineering applications.

7.2.1.3.1 Applications of Silk Nanofibers in Tissue Engineering

Electrospun silk fibers are about 40 times smaller than native silk fibroin; hence, it is promising as a potential scaffold material with similar dimensions to the native ECM [32,58,60]. Jin et al. [32] investigated the use of nanoscale silk fibroin protein matrices for scaffolding. Nanofibrous polymer matrices of silk fabricated by electrospinning (Figure 7.7) were inoculated with 1 mL of bone marrow stromal cells (BMSCs) containing 2×10^6 cells. Following seeding, the silk matrices were cultured in an appropriate medium and observed up to 14 days. As seen in Figure 7.8, BMSC seeded on the nanoscale polymeric matrix reached confluency and resulted in complete matrix coverage and proliferation within 14 days of cultivation, whereas cells grown on a conventional micron-size polymeric matrix did not promote extensive growth (Figure 7.8b). These silk nanofibers were therefore able to support the attachment and growth of human bone marrow stoma cells for potential applications in regenerative stem cell therapy [32].

FIGURE 7.7
Scanning electron micrograph of electrospun nanofibers of silk for ligament and bone tissue engineering. (Adapted from Jin HJ, Chen JS, Karageorgiou V, Altman GH, Kaplan DL. *Biomaterials*, March 2004; 25(6):1039–1047. Reprinted with permission of Elsevier Inc.)

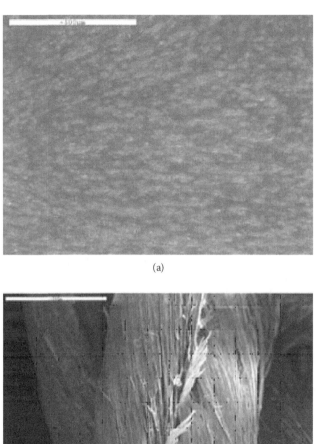

(a)

(b)

FIGURE 7.8
Scanning electron micrographs of bone marrow stromal cells (BMSCs) reaching confluence on electrospun silk fibroin matrix at the end of 14 days (a) and native silk fibroin with diameter 14 μm supporting the growth of very few cells (b). Scale bar = 500 μm. (Adapted from Jin HJ, Chen JS, Karageorgiou V, Altman GH, Kaplan DL. *Biomaterials*, March 2004; 25(6):1039–1047. Reprinted with permission of Elsevier Inc.)

Other natural molecules that have been successfully fabricated on a nano-scale for tissue engineering applications are DNA and fibrinogen [61–63].

7.2.2 Synthetic Biopolymers

Since the 1960s, synthetic biopolymers have been playing a pivotal role in tissue engineering. The advantages they offer over natural polymers include that they can be tailored to provide a wide range of properties, display reduced immunogenicity, and offer simplicity of processing and ease of sterilization [4]. The most common polymers under this category approved by the U.S. Food and Drug Administration (FDA) for use in tissue engineering applications are PLA, PGA, and polylactic-co-glycolic acid (PLGA). These polymers are degraded by simple hydrolysis, and in the absence of any major inflammation, there is little host-to-host variation in degradation rate [4].

7.2.2.1 Polylactic Acid (PLA) and Polyglycolic Acid (PGA)

PLA is a α-polyester with two enantiomeric forms. These are the L-lactide form and D-lactide form [4]. It occurs as a semicrystalline material with glass transition (Tg) temperatures between 60 and 65°C [4]. It exhibits high tensile in the order of 50 to 60 MPa which is significantly higher than those exhibited in biopolymer [64]. This is a feature that renders it useful for many tissue engineering applications [65,66]. PLA is comparatively hydrophobic due to its limited water absorption capacity and, hence, exhibits a slower degradation rate [35,65,66].

PGA is one of the simplest forms of the linear aliphatic polyesters. It is synthesized by the ring-opening polymerization of glycolide [4]. Because PGA is more hydrophilic than PLA, its degradation period is much shorter [4,24,67,68].

In general, polymers from the poly α-hydroxy acid family undergo bulk degradation. The molecular weight of these polymers starts to decrease upon prolonged contact with aqueous media. The molecular weight loss is triggered by a nonenzymatic hydrolysis reaction [69–71]. However, mass loss is not initiated until the molecular chains are reduced to a size that permits easy diffusion out of the polymeric matrix [27]. The mass loss occurs during the phase of accelerated polymer degradation, and the resorption kinetics continues until the physical integrity of the matrix is compromised. Molecular weight and subsequent mass loss may vary from a month to 18 months depending on a host of factors like chemical structure and composition, molecular weight distribution (polydispersity index), presence of any encapsulated growth factors, process parameters, nanofibrillar matrix design, crystalline versus amorphous morphology, and to a small extent the storage conditions and implant site [27,72].

The high mechanical integrity possessed by these classes of polymers means they are often used for bone and cartilage tissue engineering. The degradation and resorption kinetics need to be designed and controlled in such a way that the nanostructured scaffold retains its structural and functional integrity for at least 6 months (4 months for cell culture and generation of a premature tissue *in vitro* in a bioreactor and at least 2 months after implantation *in vivo)* [27,72]. The mechanical properties of the scaffold matrix should match those of the native host tissue as closely as possible at the time of implantation. It should possess sufficient tensile strength and stiffness for a considerable period until the *in vivo* tissue growth has replaced the slowly degrading polymeric scaffold matrix. Wang et al. [73] studied the degradation kinetics of a poly(D,L-lactide-co-glycolide) matrix under compressive loading conditions. The decrease in molecular weight is accompanied by a reduction in surface area due to hydrolysis. This continues until the matrix architecture no longer accommodates the mechanical loading and begins to lose its integrity [73]. This process can be designed to occur over long periods of time by manipulating various process and polymer characteristic variables. Engineering the scaffolds at the nanoscale can also enhance its ability to withstand the mechanical loading *in vitro* and *in vivo* [5].

7.2.2.1.1 Applications of Nanostructured PLA in Tissue Engineering

These biodegrdabale polymers, such as PLA and PLGA, can be readily reduced to nanoscale forms by a wide variety of methods including phase separation, electrospinning, and self-assembly [4].

F. Yang et al. [64] studied the application of nanoscale poly-l-lactic acid (PLLA) scaffolds for nerve tissue engineering. A phase separation technique using tetrahydrofuran (THF) as a solvent produced porous nanofibers in a controlled and reproducible manner. Scaffolds were seeded with a multipotent neuronal stem cell line (C17-2), which are neuron precursors leading to the development of cerebellum. The nanofibrous scaffold was designed with a structure similar to natural ECM to support neuron differentiation and neurite growth *in vivo.*

SEM images displayed in Figure 7.9 show a highly porous nanoscale structure. Upon culturing of the C17-2 cells in this nanofibrous matrix, it was observed that cells started attaching and growing from day one. Cells were randomly spread over the surface of the scaffold. However, only a few C17-2 cells were able to differentiate into neurons as displayed in Figure 7.10a. By the end of day 2, cell attachment had progressively increased over the matrix, and an increase in neurite (a projection from the body of a neuron into an axon or a dendrite, the growth of which is stimulated by nerve growth factor) length of the differentiated neurons was observed (Figure 7.10b).

In order to further increase the extent of cell attachment, a possible alternative would be fabrication of a nanocomposite of collagen and PLA [74]. The collagen component of the composite would permit increased attachment of neurons and thus favor neurite outgrowth. Additionally, the PLA component of the composite would aid the transport of attached neurons as a result of the higher tensile mechanical integrity and high elastic modulus.

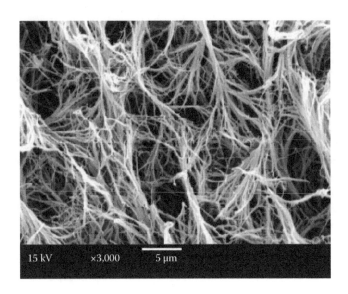

FIGURE 7.9
Scanning electron micrograph of nanostructured and porous fibrous matrix of polylactic acid (PLA) electrospun in 5% w/v solution of tetrahydrofuran (THF). (Adapted from Yang F, Murugan R, Ramakrishna S, Wang X, Ma YX, Wang S. *Biomaterials*, May 2004; 25(10):1891–1900. Reprinted with permission of Elsevier Inc.)

7.2.2.2 Miscellaneous Synthetic Polymers

7.2.2.2.1 Poly (E-Caprolactone) (PCL)

The ring-opening polymerization of *e*-caprolactone yields a semicrystalline polymer with a melting point of 58 to 63°C and a very low glass transition temperature of −60°C [27]. The repeating molecular structure of PCL homopolymer (Figure 7.11) consists of five nonpolar methylene groups and a single relatively polar ester group. The polymer degrades through hydrolytic scission with resistance to rapid hydrolysis [26]. This slow hydrolysis rate translates into a long degradation time in the order of 2 years [26]. To accelerate the degradation rate of PCL, copolymers of PCL with PLA and PGA have been synthesized [75]. The faster hydrolysis of PLA and PGA as compared to PCL results in formation of lactic acid and glycolic acid as degradation products, respectively. These acidic degradation products are thought to function as catalysts, aiding the breakdown of PCL fragments and hence accelerating the degradation rate [76].

7.2.2.2.2 Polyethylene Oxide (PEO)

PEO is often employed to prevent nonspecific adsorption of proteins and antibodies [77–79]. This assists in significantly reducing the likelihood of any adverse immune response. Additionally, the material possesses a high compressive modulus making it an interesting component for use in biocomposites [26].

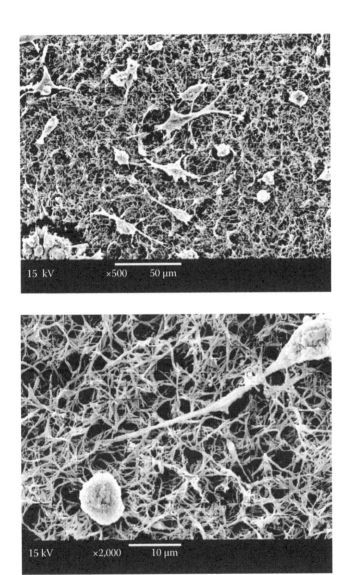

FIGURE 7.10
Scanning electron micrographs of C17-2 cells on the PLLA nanofibrous scaffold (5% wt/v) cultured for 1 day (a); scale bar = 50 μm and a differentiated cell with neurite penetration into the PLLA nanofibrous scaffolds (b); scale bar = 10 μm. (Adapted from Yang F, Murugan R, Ramakrishna S, Wang X, Ma YX, Wang S. *Biomaterials*, May 2004; 25(10):1891–1900. Reprinted with permission of Elsevier Inc.)

$$-\!\!\Big[\underline{O}\text{-CH}_2\text{-CH}_2\text{-CH}_2\text{-CH}_2\text{-CH}_2\text{-C}\!=\!\text{O}\Big]_n$$

FIGURE 7.11
Structure of polycaprolactone (PCL).

7.2.3 Hydrogels

Hydrogels form a subclass of natural and synthetic polymers. They are cross-linked polymers with hydrophilic characteristics that swell in an aqueous environment [33]. They can retain their shape even at high degrees of swelling. These hydrated forms resemble the native articular cartilage [26]. These are generally formulated as injectables, used to fill irregularly shaped defects, and thus involve the use of minimally invasive surgical procedures. They can also be used as carriers for bioactive molecules along with cells [33]. In cartilage engineering, hydrogels are used to engulf cells and growth factors in a polymer network, which immobilizes the cells and allows differentiation in chondrocytes [26]. The growth factors incorporated further help in cartilage tissue development by providing appropriate signaling cues [26]. Additionally, hydrogels exert controlled compressive forces on the encapsulated cells mimicking the physiological conditions.

Most of the natural polymers including collagen, chitosan, and fibrin can be formed into hydrogels. However, the hydrogels formed with natural polymers alone lack sufficient stiffness to function *in vivo*. For synthetic hydrogels, polyethylene glycol (PEG)–based polymers have been extensively studied. Because PEG lacks degradability, it is often linked with PLA or PGA [33].

7.2.3.1 Applications of Nanoparticulate-Based Hydrogels for Tissue Engineering

We investigated the use of a novel nanomaterial possessing the properties of solid porous matrices and *in situ* forming gels [34]. The nanoparticle-based porous scaffolds self-assembled at the site of injection [34]. Nanoparticles of PLA-PEG-biotin were prepared by a solvent evaporation technique and were self-assembled and cross-linked using avidin as a bridging molecule. Avidin is a tetrameric protein present in chicken egg white that can bind up to four molecules of biotin (vitamin H) with high specificity and affinity [34]. Human osteoblast sarcoma (HOS) cells were mixed with the self-assembling scaffold. Confocal microscopy images of the scaffold displayed viable single cells and clusters of cells evenly distributed throughout the matrix. Figure 7.12a shows that when the self-assembling porous scaffolds were injected into the surface of a choriallantoic membrane (CAM) of a day 10 chick embryo, widespread vascularization with invasion of blood vessels was observed after 7 days, suggesting the potential application of hydrogel-based scaffolds for angiogenesis. Injecting the cell-free scaffold forming material into wedge-shaped bone defects isolated from day 18 chick embryos resulted in rapid pervasive vascularization (Figure 7.12b) [34].

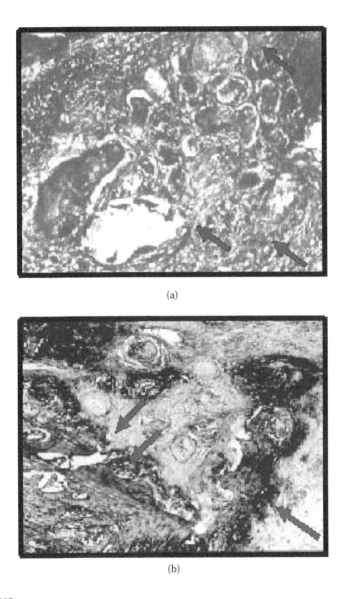

(a)

(b)

FIGURE 7.12
Paraffin section of the porous hydrogel scaffold on choriallantoic membrane (CAM) culture demonstrating CAM in-growth and matrix synthesis from CAM tissue (a). Paraffin section of scaffold within femur wedge defect displaying bone viability by alkaline phosphatase staining and integration of the scaffold with existing bone and regions of new tissue formation indicated by arrows (b). (Adapted from Salem AK, Rose F, Oreffo ROC, Yang XB, Davies MC, Mitchell JR, et al. *Advanced Materials*, Feb 2003; 15(3):210–213. Reprinted with permission of John Wiley and Sons Inc.)

TABLE 7.1

Mechanical Properties of Human Tissue

Tissue Type	Tensile Strength (MPa)	Young's Modulus (GPa)
Cortical bone	60–160	3–30
Cartilage	3.7–10.5	0.07–0.153
Ligament	13–46	0.065–0.541
Tendon	24–112	0.14–2.31

Source: Adapted from Yang SF, Leong KF, Du ZH, Chua CK. *Tissue Engineering* Dec 2001; 7(6):679–689. (Reprinted with permission from Mary Liebert Inc.)

7.3 Nano-Biopolymer Composites

Composite scaffolds consist of two or more materials. These materials together produce scaffolds that ideally draw from the strengths of the individual components properties [26]. The tensile strengths of various native tissues and the tensile strengths of the class of natural and synthetic polymers are summarized in Table 7.1 and Table 7.2. Most of the natural polymers like chitosan and collagen display significantly lower values of tensile strength and elasticity, even upon cross-linking, as compared to the synthetic class of biopolymers. Combining synthetic and natural polymer composites can overcome the limitations of natural or synthetic polymers alone [80]. In many cases, composites are designed by using a combination of natural and synthetic biopolymers in order to utilize the hallmarks of both of these classes of polymers. Some natural materials like silk protein fiber have been shown to have sufficient mechanical strength (600 to 650 MPa) [58] in comparison

TABLE 7.2

Mechanical Properties of Various Classes of Polymers

Polymer	Tensile Strength (MPa)	Young's Modulus (GPa)
Natural		
Collagen	0.9–7.4	0.002–0.04
B. Mori Silk (Purified)	610–690	15–17
Synthetic		
L-PLA	55.2–82.7	2.8–4.2
PGA	>68.9	>6.9
PLGA	41.4–55.2	1.4–2.8

Source: Adapted from Yang SF, Leong KF, Du ZH, Chua CK. *Tissue Engineering*, Dec 2001; 7(6):679–689. (Reprinted with permission from Mary Liebert Inc.)

to most other natural biopolymers like collagen and chitosan. As a result, silk has been utilized as a component of composites of two or more natural biopolymers.

7.3.1 Applications of Nano-Biopolymer Composites

7.3.1.1 Nano-HA/Collagen/PLA as Bone Scaffold Materials

Large bone fracture defects are difficult to treat. Autografts and allografts are some of the few available treatment options [81]. The main disadvantages of using these current treatment options are donor shortage, the risk of transmitting diseases, and immune responses resulting in rejection of the graft [82].

Nanostructured three-dimensional scaffold materials are promising options to mimic natural bone conformation and components and facilitate growth and vasculature resulting in new bone formation. For example, hydroxyapatite (HA)/collagen composites prepared by self-assembly processes can promote bone formation [28,83]. However, the HA/collagen composite lacks the mechanical strength needed for *in vivo* applications. Liao et al. [74] proposed a nanocomposite of a natural and synthetic biopolymer composed of HA/collagen and PLA to overcome the mechanical strength limitations (Figure 7.13). Natural bone mineral is a nonstoichiometric carbonated apatite with crystalline structure and nanometer-sized intricate hierarchical structures [74]. The axis of this mineral is oriented to the axis of

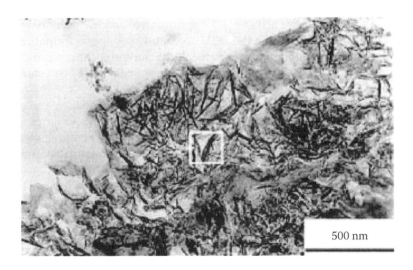

500 nm

FIGURE 7.13
Transmission electron micrograph of nano-hydroxyapatite (HA)/collagen/polylactic acid (PLA) composite possessing high mechanical strength for bone tissue engineering. (Adapted from Liao SS, Cui FZ, Zhang W, Feng QL. *Journal of Biomedical Materials Research Part B—Applied Biomaterials*, May 2004; 69B(2):158–165. Reprinted with permission of John Wiley and Sons Inc.)

collagen fibril which is also in the nanometer size range [74]. As a scaffold material, HA is osteoconductive but does not promote osteoinduction, and the rate of degradation is relatively slow [28]. To circumvent this disadvantage, it is often coupled with collagen to aid in osteoinduction [28,83]. The nano-HA/collagen/PLA composite mimics natural bone in addition to having a highly organized interconnected porous structure. The PLA gives the composite high compatibility and strength [74].

To prepare this composite, HA and collagen were assembled into mineralized fibrils that were about 6 nm in diameter, to mimic the structure of the natural structural protein of the bone [74]. These were then assembled into the PLA matrix to provide mechanical strength to the overall composite. PLA concentration was restricted up to a maximum of 12% by composite total weight to retain structural similarity with natural bone tissue. This composite scaffold was implanted into adult male New Zealand rabbits with segmental bone defects. New bone formation was indentified by the radiographic appearance of a calcified mass and bone trabeculae. The HA/collagen/PLA nanocomposite system produced new bone formation (Figure 7.14a), filling of segmental defect (Figure 7.14b), and new trabaculae formation (Figure 7.14c) *in vivo* over 12 weeks [74].

7.3.1.2 Nanofibers of PCL/Collagen Composite as a Dermal Substitute for Skin Regeneration

Biological activity is regulated by signals from the ECM and surrounding cells [10]. ECM molecules surround the cells to give them mechanical support and regulate cellular activity [10]. Fabrication of an ideal dermal wound-healing matrix should mimic the ECM to enable new tissue formation and simultaneously permit higher gas permeability and protect the wound from infection and dehydration [84]. Electrospun nanofibers have been shown to meet this requirement effectively.

A composite system consisting of natural biopolymer collagen and a synthetic bioresorbable polymer polycaprolactone (PCL) has shown potential for support and proliferation of dermal fibroblasts and keratinocytes [10]. Electrospun nanofibers immediately attach to wet wound surfaces without any fluid retention, and the rate of epithelialization is found to increase with the nanofibers [51,85]. The PCL in this nanocomposite system has been used as a resorbable material to provide biocompatibility with both soft and hard tissue and mechanical strength to function as a bone graft substitute [10].

The PCL/collagen nanofibrous composites were generated by an electrospinning technique to produce nanofibers in the size range of 200 to 250 nm [10]. In order to demonstrate the efficiency of the composite system over the use of scaffolds formed from each of the composite components separately, two controls were employed in the study. Figure 7.15 represents a graph of the growth rate of fibroblasts on the controls and the nanocomposite matrix [10]. The nanocomposite system showed a growth rate that was higher than

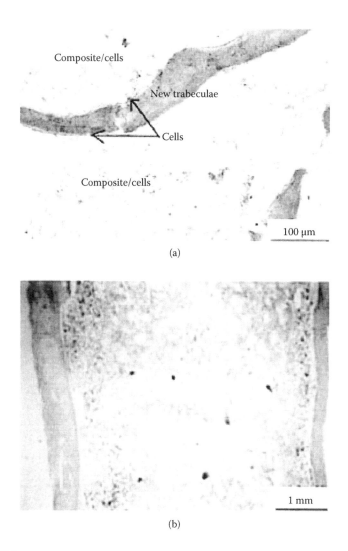

FIGURE 7.14
Nano-hydroxyapatite (HA)/collagen/polylactic acid composite implant at 8 weeks *in vivo* implant at 12 weeks (a), scale bar = 100 μm; implant after 12 weeks promoting filling of a large bone segmental defect (b), scale bar = 1 mm; and double cortical bone connected completely and bone marrow and new bone trabeculae formation at the end of 12 weeks (c), scale bar = 100 μm. (Adapted from Liao SS, Cui FZ, Zhang W, Feng QL. *Journal of Biomedical Materials Research Part B—Applied Biomaterials*, May 2004; 69B(2):158–165. Reprinted with permission of John Wiley and Sons Inc.)   *(Continued)*

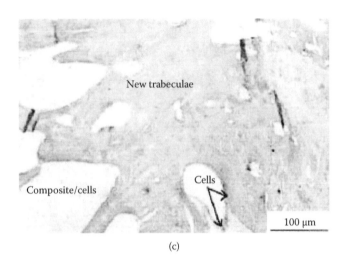

(c)

FIGURE 7.14
Continued.

that observed on the PCL matrix alone but slightly lower than the collagen matrix alone. Figure 7.16b shows that cell growth was present on PCL matrices but no confluence could be reached. The PCL fibers needed collagen for maximal proliferation and migration of cells inside the nanofibrous matrix (Figure 7.16a). Collagen alone has strong cell matrix interactions but lacks rigidity in the absence of PCL. Additionally, the high porosity present in the nanostructured composite matrix enables nutrient and metabolic waste exchange between the scaffold and the surrounding environment [10].

The HA/collagen/PLA and PCL/collagen composites clearly demonstrate the successful combination of a natural and synthetic polymer. Such a composite successfully combines the advantages of both natural and synthetic

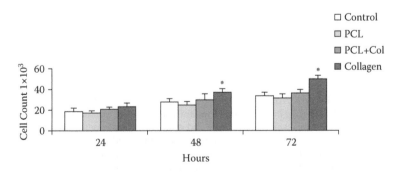

FIGURE 7.15
Growth rate of human dermoblasts on control groups and polycaprolactone/collagen nanomatrix. (Adapted from Venugopal J, Ramakrishna S. *Tissue Engineering*, May 2005; 11(5–6):847–854. Reprinted with permission of Mary Liebert Inc.)

(a)

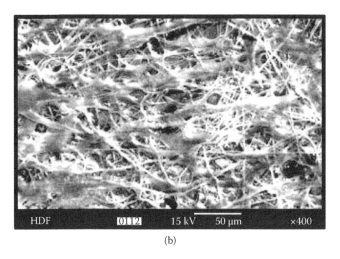

(b)

FIGURE 7.16
Polycaprolactone (PCL) matrix unable to promote cell proliferation to confluence (a), scale bar = 20 µm; and nanocomposite matrix of collagen/PCL promoting cell growth and proliferation to maximum (b), scale bar = 50 µm. (Adapted from Venugopal J, Ramakrishna S. *Tissue Engineering*, May 2005; 11(5–6):847–854. Reprinted with permission of Mary Liebert Inc.)

biopolymers, exhibiting the desired attributes of both in terms of mechanical integrity and surface chemistry.

7.3.1.3 Nanofiber Composite of Chitosan/PEO

Combining materials to prepare composites can also be carried out to enable the formation of a nanostructured scaffold that would not have been possible otherwise. An example of this is polyethylene oxide (PEO)/chitosan nanofibrous matrices [51]. As discussed earlier, chitosan cannot be readily spun into a nanofibrous system because of a high degree of intermolecular bonding. With increasing polymer concentration, the problem compounds further with the number of direct interacting chains of chitosan molecules increasing rapidly and reaching a critical threshold value. At this critical threshold value, the chitosan solution forms a three-dimensional networked high-viscosity gel that is not amenable to electrospinning [36,47,52]. Addition of a synthetic polymer like PEO reduces the viscosity of the chitosan solution by interacting through hydrogen bonding and renders the polymer spinnable even at high polymer concentrations [51].

Chitosan has been extensively studied for various wound dressing systems and tissue engineering applications [43]. It is shown to enhance bone formation both *in vitro* and *in vivo*. Figure 7.17a shows that a 2% polymer solution of chitosan was unable to yield a fibrous structure and above 2% polymer solution formed viscous unspinnable solutions. However, as seen in Figure 7.17b, when the chitosan was coupled with PEO in a ratio of 90:10 of chitosan to PEO and Triton-X-100 was added as a surfactant, the solution yielded a nanofibrous composite of PEO/chitosan [51].

Use of PEO alone as a scaffold for tissue engineering applications is not possible because PEO is not cell adherent, the polymer quickly dissolves in water and the fibrous structure of the matrix collapses in a few days without eliciting new tissue formation [33]. However, when PEO is coupled with chitosan and the latter forms the major component of the composite system, the natural polymer serves as a scaffold to support cell attachment, differentiation, and growth and to generate new tissue formation. Figure 7.18a shows that chitosan/PEO nanofibrous composite scaffolds generated a significantly higher attachment of chondrocytes when compared to a solvent cast conventional film of PEO/chitosan (Figure 7.18b) which does not have a high surface area and organized porous three-dimensional fibrillar networks [51].

7.3.1.4 Nanofiber Composite of Gelatin/PCL

Another interesting application of a composite system would be the use of one polymeric component that degraded faster than the other. This would result in increased porosity of the matrix and would lead to tissue growth within the microvoid spaces created within the matrix [21,26].

Zhang et al. [22] studied the application of one such system, where a natural biopolymer, gelatin, and a synthetic biodegradable polymer, PCL, were combined. The gelatin provides hydrophilic and cellular affinity properties and provides continuous release of the protein from the scaffold to create a favorable condition for cell attachment and proliferation. The gelatin added dissolves gradually, creating microvoid spaces and thus enhancing cell migration and greater infiltration into the scaffold as seen in Figure 7.19a. Addition of PCL provides good elongation, deformation, and superior mechanical

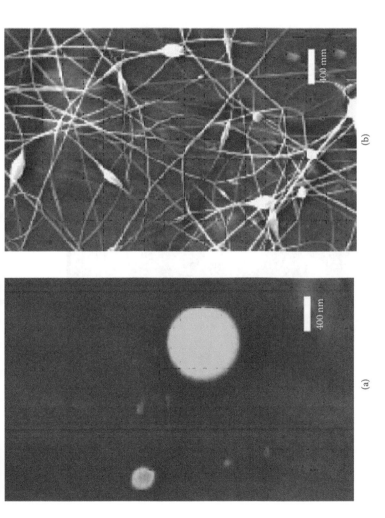

FIGURE 7.17

Scanning electron microscope images of electrospun chitosan in the absence of polyethylene oxide (PEO) (left) and a 10% PEO with 90% w/w chitosan resulting in a nanofibrous composite (right). (Adapted from Bhattarai N, Edmondson D, Veiseh O, Matsen FA, Zhang MQ. *Biomaterials*, Nov 2005; 26(31):6176–6184. Reprinted with permission of Elsevier Inc.)

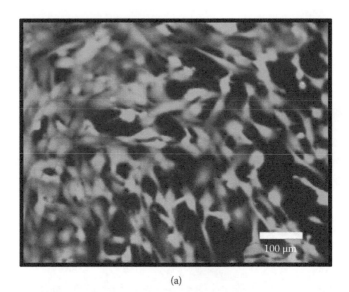

(a)

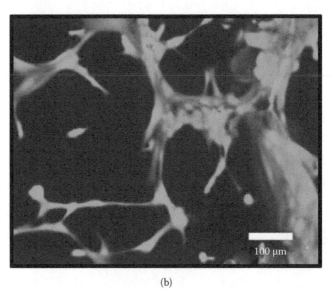

(b)

FIGURE 7.18
Chondrocyte proliferation on polyethylene oxide/chitosan nanoporous composite (a) and low cell differentiation on a traditional low porosity solvent cast composite film (b). (Adapted from Bhattarai N, Edmondson D, Veiseh O, Matsen FA, Zhang MQ. *Biomaterials*, Nov 2005; 26(31):6176–6184. Reprinted with permission of Elsevier Inc.)

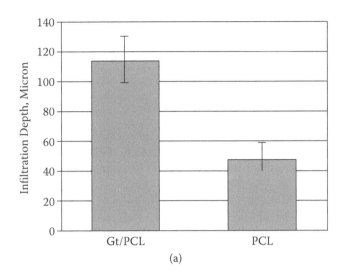

(a)

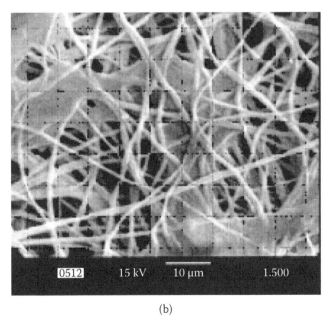

(b)

FIGURE 7.19
Extent of cell infiltration (depth μm) on polycaprolactone (PCL)/gelatin nanocomposite (blue bar) and nanofibrous scaffold using PCL alone (pink bar) (a) in and cell in-growth and inter-action demonstrated by multiple layers of cells over PCL/gelatin matrix (b). (Adapted from Zhang YZ, Ouyang HW, Lim CT, Ramakrishna S, Huang ZM. *Journal of Biomedical Materials Research Part B—Applied Biomaterials*, Jan 2005; 72B(1):156–165. Reprinted with permission from John Wiley and Sons Inc.)

properties to the composite, permitting cell penetration into deeper regions of the scaffold (Figure 7.19b) [22].

7.4 Summary

Nanomaterial science has created an exciting frontier in the area of tissue engineering. Current technology has enabled the mimicking of complex, hierarchical structures of the native tissue. Recent studies have shown that nanophase materials can significantly enhance cellular attachment, activity, response, and proliferation. Many of the barriers to using stand-alone materials such as collagen, chitosan, PLA, HA, and gelatin as scaffolds in tissue engineering can be overcome by forming composites of two or more different materials. The electrospinning technique can be used to fabricate a nanofibrous composite polymeric scaffold. These electrospun scaffolds yield materials with altered and favorable surface chemistry and greater control and reproducibility in designing the scaffold geometry to suit individual tissue engineering needs. Despite these advantages, material-specific concerns persist. *In vivo*, the release of acidic degradation products of bioresorbable polymers has resulted in inflammatory responses resulting from the inability of the tissues to eliminate these degradation by-products effectively. Hence, some crucial and key challenges that need to be addressed are identification of biomaterials that provide cell-specific interactions and elucidation of the molecular mechanisms of cell–biomaterial interactions. It is also imperative to study *in vivo* cellular and inflammatory responses in detail in order to evaluate the safety of the use of biomaterials on a long-term basis. The last decade has witnessed significant advances in the area of tissue engineering and the use of polymeric nanofibers as scaffold materials. These successful preliminary results have established a platform for further exciting research and outcomes in the field of tissue engineering.

References

1. Hubbell JA. Biomaterials in tissue engineering. *Bio-Technology*, June 1995; 13(6):565–576.
2. Langer R, Vacanti JP. Tissue engineering. *Science*, May 1993; 260(5110):920–926.
3. Drury JL, Mooney DJ. Hydrogels for tissue engineering: Scaffold design variables and applications. *Biomaterials*, Nov 2003; 24(24):4337–4351.
4. Barnes CP, Sell SA, Boland ED, Simpson DG, Bowlin GL. Nanofiber technology: Designing the next generation of tissue engineering scaffolds. *Advanced Drug Delivery Reviews* Dec 2007; 59(14):1413–1433.

5. Sato M, Webster TJ. Nanobiotechnology: Implications for the future of nanotechnology in orthopedic applications. *Expert Review of Medical Devices*, Sept 2004; 1(1):105–114.

6. Webster TJ, Siegel RW, Bizios R. Nanoceramic surface roughness enhances osteoblast and osteoclast functions for improved orthopaedic/dental implant efficacy. *Scripta Materialia*, May 2001; 44(8–9):1639–1642.

7. Zhang YZ, Lim CT, Ramakrishna S, Huang ZM. Recent development of polymer nanofibers for biomedical and biotechnological applications. *Journal of Materials Science–Materials in Medicine*, Oct 2005; 16(10):933–946.

8. Fertala A, Han WB, Ko FK. Mapping critical sites in collagen II for rational design of gene-engineered proteins for cell-supporting materials. *Journal of Biomedical Materials Research*, Oct 2001; 57(1):48–58.

9. Yoshimoto H, Shin YM, Terai H, Vacanti JP. A biodegradable nanofiber scaffold by electrospinning and its potential for bone tissue engineering. *Biomaterials*, May 2003; 24(12):2077–2082.

10. Venugopal J, Ramakrishna S. Biocompatible nanofiber matrices for the engineering of a dermal substitute for skin regeneration. *Tissue Engineering*, May 2005; 11(5–6):847–854.

11. Mo XM, Xu CY, Kotaki M, Ramakrishna S. Electrospun P(LLA-CL) nanofiber: A biomimetic extracellular matrix for smooth muscle cell and endothelial cell proliferation. *Biomaterials*, May 2004;25(10):1883–1890.

12. Luu YK, Kim K, Hsiao BS, Chu B, Hadjiargyrou M. Development of a nanostructured DNA delivery scaffold via electrospinning of PLGA and PLA-PEG block copolymers. *Journal of Controlled Release*, April 2003; 89(2):341–353.

13. Price RL, Gutwein LG, Kaledin L, Tepper F, Webster TJ. Osteoblast function on nanophase alumina materials: Influence of chemistry, phase, and topography. *Journal of Biomedical Materials Research Part A*, Dec 2003; 67A(4):1284–1293.

14. Ma ZW, Kotaki M, Inai R, Ramakrishna S. Potential of nanofiber matrix as tissue-engineering scaffolds. *Tissue Engineering*, Jan 2005; 11(1–2):101–109.

15. Xu CY, Inai R, Kotaki M, Ramakrishna S. Electrospun nanofiber fabrication as synthetic extracellular matrix and its potential for vascular tissue engineering. *Tissue Engineering*, July–Aug 2004; 10(7–8):1160–1168.

16. Frenot A, Chronakis IS. Polymer nanofibers assembled by electrospinning. *Current Opinion in Colloid and Interface Science*, March 2003; 8(1):64–75.

17. Huang ZM, Zhang YZ, Kotaki M, Ramakrishna S. A review on polymer nanofibers by electrospinning and their applications in nanocomposites. *Composites Science and Technology*, Nov 2003; 63(15):2223–2253.

18. Subbiah T, Bhat GS, Tock RW, Pararneswaran S, Ramkumar SS. Electrospinning of nanofibers. *Journal of Applied Polymer Science*, April 2005; 96(2):557–569.

19. Doshi J, Reneker DH. Electrospinning process and applications of electrospun fibers. *Journal of Electrostatics*, Aug 1995; 35(2–3):151–160.

20. Li WJ, Laurencin CT, Caterson EJ, Tuan RS, Ko FK. Electrospun nanofibrous structure: A novel scaffold for tissue engineering. *Journal of Biomedical Materials Research*, June 1960; (4):613–621.

21. Pham QP, Sharma U, Mikos AG. Electrospinning of polymeric nanofibers for tissue engineering applications: A review. *Tissue Engineering*, May 2006; 12(5):1197–1211.

22. Zhang YZ, Ouyang HW, Lim CT, Ramakrishna S, Huang ZM. Electrospinning of gelatin fibers and gelatin/PCL composite fibrous scaffolds. *Journal of Biomedical Materials Research Part B—Applied Biomaterials*, Jan 2005; 72B(1):156–165.

23. Lee JS, Choi KH, Do Ghim H, Kim SS, Chun DH, Kim HY et al. Role of molecular weight of atactic poly(vinyl alcohol) (PVA) in the structure and properties of PVA nanofabric prepared by electrospinning. *Journal of Applied Polymer Science,* Aug 2004; 93(4): 1638–1646.
24. Zong XH, Kim K, Fang DF, Ran SF, Hsiao BS, Chu B. Structure and process relationship of electrospun bioabsorbable nanofiber membranes. *Polymer,* July 2002; 43(16):4403–4412.
25. Vert M, Li SM, Spenlehauer G, Guerin P. Bioresorbability and biocompatibility of aliphatic polyesters. *Journal of Materials Science—Materials in Medicine,* Nov 1992; 3(6):432–446.
26. Cheung HY, Lau KT, Lu TP, Hui D. A critical review on polymer-based bioengineered materials for scaffold development. *Composites Part B—Engineering,* 2007; 38(3):291–300.
27. Hutmacher DW. Scaffolds in tissue engineering bone and cartilage. *Biomaterials,* Dec 2000; 21(24):2529–2543.
28. Du C, Cui FZ, Zhu XD, de Groot K. Three-dimensional nano-HAp/collagen matrix loading with osteogenic cells in organ culture. *Journal of Biomedical Materials Research,* March 1999; 44(4):407–415.
29. Grenga TE, Zins JE, Bauer TW. The rate of vascularization of coralline hydroxyapatite. *Plastic and Reconstructive Surgery,* Aug 1989; 84(2):245–249.
30. Ripamonti U, Duneas N. Tissue engineering of bone by osteoinductive biomaterials. *MRS Bulletin,* Nov 1996; 21(11):36–39.
31. Landis WJ, Song MJ, Leith A, McEwen L, McEwen BF. Mineral and organic matrix interaction in normally calcifying tendon visualized in 3 dimensions by high-voltage electron-microscopic tomography and graphic image-reconstruction. *Journal of Structural Biology,* Jan–Feb 1993; 110(1):39–54.
32. Jin HJ, Chen JS, Karageorgiou V, Altman GH, Kaplan DL. Human bone marrow stromal cell responses on electrospun silk fibroin mats. *Biomaterials,* March 2004; 25(6):1039–1047.
33. Lee KY, Mooney DJ. Hydrogels for tissue engineering. *Chemical Reviews,* July 2001; 101(7):1869–1879.
34. Salem AK, Rose F, Oreffo ROC, Yang XB, Davies MC, Mitchell JR, et al. Porous polymer and cell composites that self-assemble *in situ. Advanced Materials,* Feb 2003; 15(3):210–213.
35. Boland ED, Pawlowski KJ, Barnes CP, Simpson DG, Wnek GE, Bowlin GL. Electrospinning of bioresorbable polymers for tissue engineering scaffolds. In *Polymeric Nanofibers,* D.H. Reneker and H. Fong, Eds., 2006; 188–204.
36. Matthews JA, Wnek GE, Simpson DG, Bowlin GL. Electrospinning of collagen nanofibers. *Biomacromolecules,* March–April 2002; 3(2):232–238.
37. Grinnell F. Cell-collagen interactions—Overview. *Methods in Enzymology,* 1982; 82:499–503.
38. Shields KJ, Beckman MJ, Bowlin GL, Wayne JS. Mechanical properties and cellular proliferation of electrospun collagen type II. *Tissue Engineering,* Sept 2004; 10(9–10):1510–1517.
39. Lazarus GS, Daniels JR, Brown RS, Bladen HA, Fullmer HM. Degradation of collagen by a human granulocyte collagenolytic system. *Journal of Clinical Investigation,*1968; 47(12):2622–2629.

40. Ma ZW, He W, Yong T, Ramakrishna S. Grafting of gelatin on electrospun poly(caprolactone) nanofibers to improve endothelial cell spreading and proliferation and to control cell orientation. *Tissue Engineering*, July 2005; 11(7–8):1149–1158.
41. Kim IY, Seo SJ, Moon HS, Yoo MK, Park IY, Kim BC et al. Chitosan and its derivatives for tissue engineering applications. *Biotechnology Advances*, 2008; 26:1–21.
42. Madihally SV, Matthew HWT. Porous chitosan scaffolds for tissue engineering. *Biomaterials*, June 1999; 20(12):1133–1142.
43. Di Martino A, Sittinger M, Risbud MV. Chitosan: A versatile biopolymer for orthopaedic tissue-engineering. *Biomaterials*, Oct 2005; 26(30):5983–5990.
44. VandeVord PJ, Matthew HWT, DeSilva SP, Mayton L, Wu B, Wooley PH. Evaluation of the biocompatibility of a chitosan scaffold in mice. *Journal of Biomedical Materials Research*, March 2002; 59(3):585–590.
45. Kumar M. A review of chitin and chitosan applications. *Reactive and Functional Polymers*, Nov 2000; 46(1):1–27.
46. Hu QL, Li BQ, Wang M, Shen JC. Preparation and characterization of biodegradable chitosan/hydroxyapatite nanocomposite rods via *in situ* hybridization: A potential material as internal fixation of bone fracture. *Biomaterials*, Feb 2004; 25(5):779–785.
47. Chandy T, Sharma CP. Chitosan—as a biomaterial. *Biomaterials Artificial Cells and Artificial Organs*, 1990; 18(1):1–24.
48. Hirano S, Tsuchida H, Nagao N. N-Acetylation in chitosan and the rate of its enzymic-hydrolysis. *Biomaterials*, Oct 1989; 10(8):574–576.
49. Onishi H, Machida Y. Biodegradation and distribution of water-soluble chitosan in mice. *Biomaterials*, Jan 1999; 20(2):175–182.
50. Suh JKF, Matthew HWT. Application of chitosan-based polysaccharide biomaterials in cartilage tissue engineering: A review. *Biomaterials*, Dec 2000; 21(24):2589–2598.
51. Bhattarai N, Edmondson D, Veiseh O, Matsen FA, Zhang MQ. Electrospun chitosan-based nanofibers and their cellular compatibility. *Biomaterials*, Nov 2005; 26(31):6176–6184.
52. Jayaraman K, Kotaki M, Zhang YZ, Mo XM, Ramakrishna S. Recent advances in polymer nanofibers. *Journal of Nanoscience and Nanotechnology*, Jan–Feb 2004; 4(1–2):52–65.
53. Shao ZZ, Vollrath F. Materials: Surprising strength of silkworm silk. *Nature*, Aug 2002; 418(6899):741.
54. Altman GH, Diaz F, Jakuba C, Calabro T, Horan RL, Chen JS, et al. Silk-based biomaterials. *Biomaterials*, Feb 2003; 24(3):401–416.
55. Brooks G. Silk proteins in cosmetics. *Drug and Cosmetic Industry*, Oct 1989; 145(4):32.
56. Inoue S, Tanaka K, Arisaka F, Kimura S, Ohtomo K, Mizuno S. Silk fibroin of Bombyx mori is secreted, assembling a high molecular mass elementary unit consisting of H-chain, L-chain, and P25, with a 6:6:1 molar ratio. *Journal of Biological Chemistry*, Dec 2000; 275(51):40517–40528.
57. Kodrik D. Small protein-components of the cocoons in Galleria-Mellonella (Lepidoptera, Pyralidae) and Bombyx-Mori (Lepidoptera, Bombycidae). *Acta Entomologica Bohemoslovaca*, 1992; 89(4):269–273.

58. Jin HJ, Park J, Valluzzi R, Cebe P, Kaplan DL. Biomaterial films of Bombyx mori silk fibroin with poly(ethylene oxide). *Biomacromolecules*, May–June 2004; 5(3):711–717.

59. Panilaitis B, Altman GH, Chen JS, Jin HJ, Karageorgiou V, Kaplan DL. Macrophage responses to silk. *Biomaterials*, Aug 2003; 24(18):3079–3085.

60. Buchko CJ, Chen LC, Shen Y, Martin DC. Processing and microstructural characterization of porous biocompatible protein polymer thin films. *Polymer*, Dec 1999; 40(26):7397–7407.

61. Liu Y, Chen J, Misoska V, Wallace GG. Preparation of novel ultrafine fibers based on DNA and poly(ethylene oxide) by electrospinning from aqueous solutions. *Reactive and Functional Polymers*, May 2007; 67(5):461–467.

62. Takahashi T, Taniguchi M, Kawai T. Fabrication of DNA nanofibers on a planar surface by electrospinning. *Japanese Journal of Applied Physics Part 2—Letters and Express Letters*, 2005; 44(24–27):L860–L862.

63. Nagapudi K, Brinkman WT, Thomas BS, Park JO, Srinivasarao M, Wright E et al. Viscoelastic and mechanical behavior of recombinant protein elastomers. *Biomaterials*, Aug 2005; 26(23):4695–4706.

64. Yang F, Murugan R, Ramakrishna S, Wang X, Ma YX, Wang S. Fabrication of nano-structured porous PLLA scaffold intended for nerve tissue engineering. *Biomaterials*, May 2004; 25(10):1891–1900.

65. Katti DS, Robinson KW, Ko FK, Laurencin CT. Bioresorbable nanofiber-based systems for wound healing and drug delivery: Optimization of fabrication parameters. *Journal of Biomedical Materials Research Part B—Applied Biomaterials*, Aug 2004; 70B(2):286–296.

66. Zong XH, Bien H, Chung CY, Yin LH, Fang DF, Hsiao BS, et al. Electrospun fine-textured scaffolds for heart tissue constructs. *Biomaterials*, Sep 2005; 26(26):5330–5338.

67. Boland ED, Matthews JA, Pawlowski KJ, Simpson DG, Wnek GE, Bowlin GL. Electrospinning collagen and elastin: Preliminary vascular tissue engineering. *Frontiers in Bioscience*, May 2004; 9:1422–1432.

68. You Y, Lee SW, Youk JH, Min BM, Lee SJ, Park WH. *In vitro* degradation behaviour of non-porous ultra-fine poly(glycolic acid)/poly(L-lactic acid) fibres and porous ultra-fine poly(glycolic acid) fibres. *Polymer Degradation and Stability*, Dec 2005; 90(3):441–448.

69. Li SM, Garreau H, Vert M. Structure–property relationships in the case of the degradation of massive poly(alpha-hydroxy acids) in aqueous media. 3. Influence of the morphology of poly(L-lactic acid). *Journal of Materials Science—Materials in Medicine*, Nov 1990; 1(4):198–206.

70. Li SM, Garreau H, Vert M. Structure–property relationships in the case of the degradation of massive aliphatic poly-(alpha-hydroxy acids) in aqueous media. 1. Poly(Dl-lactic acid). *Journal of Materials Science—Materials in Medicine*, Oct 1990; 1(3):123–130.

71. Li SM, Garreau H, Vert M. Structure–property relationships in the case of the degradation of massive poly(alpha-hydroxy acids) in aqueous media. 2. Degradation of lactide-glycolide copolymers—Pla37.5ga25 and Pla75ga25. *Journal of Materials Science—Materials in Medicine*, Oct 1990; 1(3):131–139.

72. Agrawal CM, Athanasiou KA. Technique to control pH in vicinity of biodegrading PLA-PGA implants. *Journal of Biomedical Materials Research*, Summer 1997; 38(2):105–114.

73. Wang XD, Agrawal CM. Interfacial fracture toughness of tissue-biomaterial systems. *Journal of Biomedical Materials Research*, Spring 1997; 38(1):1–10.
74. Liao SS, Cui FZ, Zhang W, Feng QL. Hierarchically biomimetic bone scaffold materials: Nano-HA/collagen/PLA composite. *Journal of Biomedical Materials Research Part B—Applied Biomaterials*, May 2004; 69B(2):158–165.
75. Kwon IK, Kidoaki S, Matsuda T. Electrospun nano- to microfiber fabrics made of biodegradable copolyesters: Structural characteristics, mechanical properties and cell adhesion potential. *Biomaterials*, June 2005; 26(18):3929–3939.
76. Sung HJ, Meredith C, Johnson C, Galis ZS. The effect of scaffold degradation rate on three-dimensional cell growth and angiogenesis. *Biomaterials*, Nov 2004; 25(26):5735–5742.
77. Jeon SI, Andrade JD. Protein surface interactions in the presence of polyethylene oxide. 2. Effect of protein size. *Journal of Colloid and Interface Science*, March 1991; 142(1):159–166.
78. Jeon SI, Lee JH, Andrade JD, Degennes PG. Protein surface interactions in the presence of polyethylene oxide. 1. Simplified theory. *Journal of Colloid and Interface Science*, March 1991; 142(1):149–158.
79. Owens DE, Peppas NA. Opsonization, biodistribution, and pharmacokinetics of polymeric nanoparticles. *International Journal of Pharmaceutics*, Jan 2006; 307(1):93–102.
80. Yang SF, Leong KF, Du ZH, Chua CK. The design of scaffolds for use in tissue engineering. Part 1. Traditional factors. *Tissue Engineering*, Dec 2001; 7(6):679–689.
81. Damien CJ, Parsons JR. Bone-graft and bone-graft substitutes—A review of current technology and applications. *Journal of Applied Biomaterials*, Fall 1991; 2(3):187–208.
82. Hench LL, Wilson J. Surface-active biomaterials. *Science*, 1984; 226(4675): 630–636.
83. Du C, Cui FZ, Zhang W, Feng QL, Zhu XD, de Groot K. Formation of calcium phosphate/collagen composites through mineralization of collagen matrix. *Journal of Biomedical Materials Research*, June 2004; 50(4):518–527.
84. Ruszczak Z. Effect of collagen matrices on dermal wound healing. *Advanced Drug Delivery Reviews*, Nov 2003; 55(12):1595–1611.
85. Khil MS, Cha DI, Kim HY, Kim IS, Bhattarai N. Electrospun nanofibrous polyurethane membrane as wound dressing. *Journal of Biomedical Materials Research Part B—Applied Biomaterials*, Nov 2003; 67B(2):675–679.

8

Potential Use of Polyhydroxyalkanoate (PHA) for Biocomposite Development

Perrine Bordes, Eric Pollet, and Luc Avérous

Université de Strasbourg, France

CONTENTS

8.1 Introduction

8.1.1 Context

Most of today's synthetic polymers are produced from petrochemicals and are not biodegradable. It is widely accepted that the use of long-lasting polymers in products with a short life span, such as engineering applications, packaging, catering, surgery, and hygiene, is not recommended. Moreover, an increased concern exists today about the preservation of

193

ecological systems. Indeed, these persistent polymers ultimately generate significant sources of environmental pollution, damaging wildlife when they are dispersed in nature. For example, the disposal of plastic bags adversely affects sea-life [1]. Moreover, plastic waste management presents environmental issues as well. Energy production by incineration yields toxic emissions (e.g., dioxin), and material recycling is limited due to the difficulties in finding accurate and economically viable outlets. In addition, plastic recycling shows a negative ecobalance. This is often linked to the necessity to wash the plastic waste and to the energy consumption during the process phases (waste grinding and plastic processing). As plastics represent a large part of the waste collection at the local, regional, and national levels, institutions are now aware of the significant savings that would generate compostable or biodegradable materials. For these reasons, it is important to replace conventional plastics with degradable polymers for packaging applications.

The potential of biodegradable polymers has been recognized for a long time, as they could provide an interesting way to overcome the limitation of the petrochemical resources in the future. The fossil fuel and gas could be partially replaced by greener agricultural sources, which would also participate in the reduction of CO_2 emissions [1]. However, because of their high production costs, biodegradable polymers are not extensively used in industries to replace conventional plastics.

8.1.2 Biodegradability and "Compostability"

According to ASTM standard D-5488-94d, biodegradable means capable of undergoing decomposition into carbon dioxide, methane, water, inorganic compounds, or biomass. The predominant mechanism is the enzymatic action of microorganisms. It can be measured by standard tests over a specific period of time, with available disposal conditions. There are different media (liquid, inert, or compost medium) to analyze biodegradability. "Compostability" is material biodegradability using compost medium. Biodegradation is the degradation of an organic material caused by biological activity, such as microorganisms' enzymatic action. This leads to a significant change in the material chemical structure. The end-products are carbon dioxide, new biomass, and water (in the presence of oxygen: aerobic) or methane (oxygen absent: anaerobic), as defined in the European Standard EN 13432:2000. Depending on the type of standard to be followed (ASTM, EN), different composting conditions (humidity, temperature cycle) must be realized to determine the compostability level [2]. Thus, the comparison of the results obtained from different standards (ASTM, ISO, NF) seems to be difficult or impossible. It is required to take into account the amount of mineralization as well as the by-products of biodegradation [3]. The accumulation in the soil of contaminants from toxic residues of composts can cause plant growth inhibition. The key issue is to determine the environmental toxicity level for these by-products, which is known as ecotoxicity [4]. Some general rules enable the determination of the biodegradability evolution. For example, with the increase of hydrophobicity, the macromolecules' molar masses, and the crystallinity or the size of spherulites decreases the biodegradability [5].

8.1.3 Renewability and Sustainable Development

Renewability is linked to the concept of sustainable development. The United Nations World Commission on "Environment and Development in Our Future" defines sustainability as the development that meets the needs of the present time without compromising the ability of future generations to meet their own needs. According to Narayan [1], the manufactured products (e.g., packaging) must be designed and engineered from "conception to reincarnation," the "cradle-to-grave" approach. The use of annually renewable biomass must be understood in a complete carbon cycle. The carbon cycle is the complex process. Carbon is exchanged between the four main reservoirs of the planet. Those are lithosphere (e.g., limestone), biosphere (vegetal and animal), hydrosphere (e.g., bicarbonate dissolved in the oceans), and atmosphere (CO_2). Recent human activities (burning fossil fuel and massive deforestation) lead to an important imbalance in the carbon cycle with a huge and rapid release of CO_2 in the atmosphere which cannot be fully compensated for by the photosynthesis activity and dissolution in the oceans. It results in a large accumulation of carbon dioxide in the atmosphere. Greenhouse gases contribute to global warming. It is necessary to rebalance the carbon cycle by reducing the amount of CO_2 in the atmosphere. Part of the carbon cycle rebalancing concept is based on the development and the manufacture of products based on renewable and biodegradable resources. By collecting and composting biodegradable plastic waste, we can generate much-needed carbon-rich compost: humus materials. These valuable soil amendments can go back to the farmland and "reinitiate" the carbon cycle. Then, plant growth contributes in reducing CO_2 atmospheric accumulation through photosynthesis activity. Also, composting is an increasing key point to maintain the sustainability of the agricultural system by reducing the consumption of chemical fertilizers.

8.1.4 Biodegradable Polymer Classifications

Biodegradable polymers, also known as biopolymers, are a growing field [6–8]. A vast number of biodegradable polymers (e.g., cellulose, chitin, starch, polyhydroxyalkanoates, polylactide, polycaprolactone, collagen, and other polypeptides) have been synthesized or are formed in nature during the growth cycles of all organisms. Some microorganisms and enzymes capable of degrading such polymers have been identified [6,9]. Different classifications of the various biodegradable polymers have been proposed. As an example, Avérous [10] has proposed to classify the biodegradable polymers according to their synthesis process: (1) polymers from biomass such as the agro-polymers from agro-resources (e.g., starch or cellulose), (2) polymers obtained by microbial production as the polyhydroxyalkanoates (PHAs), and (3) polymers conventionally and chemically synthesized. In this case, the monomers are obtained from agro-resources (e.g., the polylactic acid [PLA]), (4) polymers and monomers are obtained conventionally, by chemical synthesis from petroleum-based resources. Only, the first three categories (1 through 3) are obtained from renewable resources. It is also possible to classify these biodegradable polymers into two main categories: the agropolymers (category 1) and the biodegradable polyesters or biopolyesters (categories 2 through 4).

To increase the properties and application ranges of these biopolymers, two different approaches need to be considered: (1) the chemical modification of the biopolymer and (2) its association with other compounds like another polymer (blends, multilayers), filler (biocomposites), or nanofiller (nanobiocomposites) [10]. The second approach (association with other polymers or fillers) is generally preferred because chemical modification is expensive and by-products may be toxic. The elaboration of nano-biocomposites will be described in detail below.

8.1.5 Nano-Biocomposites

In the scientific literature, the term "nano-biocomposites," also referred to as "nanobiocomposites" or "bionanocomposites," appeared in the beginning of the twenty-first century to describe a novel class of materials. The concept, however, was also taken on conventional nanocomposites and similarly defines nanostructured materials resulting from the addition of small amounts (less than 10 wt%) of nanosized fillers into biopolymer matrices. Studies on such materials started in the early 1990s due to growing interest in the development of "environmentally friendly" materials. Since then, projects and reports about nano-biocomposite structure and properties have continued to be published at their present sustained rate.

As with common nanocomposites [11], nano-biocomposites have been prepared using various nanofillers. These nanofillers can be classified according to their morphology, such as layered particles (e.g., clays); spherical particles (e.g., silica); or acicular particles (e.g., cellulose whiskers, carbon nanotubes). The properties of the resulting composites depend on the specific geometric dimensions of the particles, or their aspect ratio. Large enhancements of various composite properties can be achieved with strong polymer–nanofiller interactions and good dispersion.

These fillers are incorporated into biopolymer matrices to improve their properties while preserving their biodegradability or biocompatibility with living tissues. Biopolymers have raised more and more interest since decreasing fossil resources and increasing environmental concern have incited scientists and industrials to work on biodegradable and non-ecotoxic materials.

Ideally these biopolymers are produced from renewable resources and require low energy consumption. Nevertheless, most biopolymers are more expensive than synthetic polymers, and their properties are poor (low thermal stability, high moisture sensitivity) for final use. Nano-biocomposites thus appeared as a promising answer in a context of sustainable development policies.

Among the most studied biopolymer nanocomposites matrices are biopolyesters like polycaprolactone (PCL) [12–17], poly(lactic acid) (PLA) [18–26], and agropolymers [27–30] (e.g., plasticized starch also called "thermoplastic starch"). In the last decades, another class of biopolymers produced from microorganisms, the polyhydroxyalkanoates (PHAs), has gained major importance due to their wide range of properties. Unfortunately, some of the most common PHAs, like poly(hydroxybutyrate) (PHB) and poly(hydroxybutyrate-*co*-hydroxyvalerate) (PHBV), have drawbacks when compared to conventional polymers. Therefore, PHA-based nano-biocomposites have been recently investigated to develop environmentally friendly competitive materials that will ultimately be used to replace conventional synthetic and nondegradable polymers.

This chapter first gives an overview on the PHAs and highlights their main advantages as well as their drawbacks. Then, it will cover recent advances in PHA/nanoclays nano-biocomposites.

8.2 Polyhydroxyalkanoates

8.2.1 Synthesis

Poly(3-hydroxybutyrate) (PHB), the most common PHA, was first discovered in 1926 by Lemoigne [31,32]. It was not until the 1980s that PHAs were really exploited due to the continuous decrease in fossil fuel and the increase in public concern for environmental protection.

PHAs are naturally produced by microorganisms from various carbon substrates as a carbon or energy reserve. A wide variety of prokaryotic organisms [33,34] accumulate PHA from 30% to 80% of their cellular dry weight. PHB is produced under balanced growth conditions when the cells become limited for an essential nutrient but are exposed to an excess of carbon [33]. Depending on the carbon substrates and the metabolism of the microorganism, different monomers, and thus (co)polymers, could be obtained [33,35,36] (see Figure 8.1 and Table 8.1). The main polymer of the polyhydroxyalkanoates family is the poly(hydroxybutyrate) homopolymer (PHB), but different poly(hydroxybutyrate-*co*-hydroxyalkanoates) copolyesters exist such as poly(hydroxybutyrate-*co*-hydroxyvalerate)(PHBV), or poly(hydroxybutyrate-*co*-hydroxyhexanoate)(PHBHx), poly(hydroxybutyrate-*co*-hydroxyoctanoate)

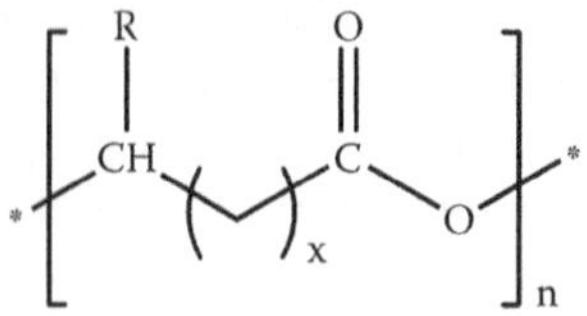

FIGURE 8.1
General chemical structure of polyhydroxyalkanoate (PHA).

TABLE 8.1
Most Common Polyhydroxyalkanoate (PHA) Homopolymers or Copolymers

n	R		PHA
1	H	Poly(3-hydroxypropionate)	P3HP
	CH_3	Poly(3-hydroxybutyrate)	P3HB (or PHB)
	CH_3, C_2H_5	Poly(3-hydroxybutyrate-*co*-3-hydroxyvalerate)	P(3HB-3HV) (or PHBV)
	CH_3, C_3H_7	Poly(3-hydroxybutyrate-*co*-3-hydroxyhexanoate)	P(3HB-3HHx) (or PHBHx)
	C_5H_{11}	Poly(3-hydroxyoctanoate)	P3HO (or PHO)
2	H	Poly(4-hydroxybutyrate)	P4HB
—	H	Poly(3-hydroxybutyrate-*co*-4-hydroxybutyrate)	P(3HB-4HB)

(PHBO) and poly(hydroxybutyrate-*co*-hydroxyoctadecanoate) (PHBOd). Moreover, depending on the side-chain length, two major classes of PHA can be considered. The first one referred to as short-chain-length PHA (scl-PHA) whose polymeric units are mainly constituted of five carbons or less, and the second one being the medium-chain-length PHA (mcl-PHA) with longer side-chain length (see Figure 8.1).

Thanks to the recent progress in biotechnology, it is possible not only for recombinant bacteria [34] but also for genetically modified plants [34,37,38] to produce such polymers. However, the recovery process (i.e., the extraction and purification steps) is critical to obtain pure PHA, and this process makes PHA expensive. Pure synthetic PHA can be produced by the ring opening polymerization (ROP) from butyrolactone and other lactones [39–44]. According to the synthesis route, different structures are obtained. For instance, we obtain isotactic with random stereosequences for the bacterial copolyesters and partially stereoregular block for the synthetic copolyesters. Recently, Monsanto Inc. developed a genetic modification to plants to make them produce small quantities of PHB [37,45,46].

8.2.2 Properties

PHB is a highly crystalline biopolyester (above 55%) with a glass transition temperature (Tg) just above 0°C (see Table 8.2). PHB is relatively stiff and brittle. Young's modulus reaches 3.5 GPa, and the elongation at break is less than

TABLE 8.2

Comparison of the Main Polyhydroxyalkanoate (PHA) Properties

Polymer	Tg (°C)	Tm (°C)	Crystallinity (%)	Young's Modulus (GPa)	Tensile Strength (MPa)	Elongation at Break (%)
PHB	0–10	170–180	>55	3.5	40	<5
P(3HB-3HV)			50–55			
3 mol% HV	8	170	—	2.9	38	—
9 mol% HV	6	162	—	1.9	37	20
20 mol% HV	–1	145	—	1.2	32	100
25 mol% HV	–6	137	—	0.7	30	—
P(3HB-4HB)						
3 mol% 4HB	—	166	—	—	28	45
10 mol% 4HB	—	159	46	—	24	240
64 mol% 4HB	—	50	—	30	17	590
90 mol% 4HB	—	50	—	100.0	65	1080
P4HB	–50	60	150	10	105	1000
P(3HB-3HHx)	–4	50	—	—	20	850
PHO	–35	60	30	—	6–10	300–450
PP	–10	170	50–70	1.7	38	400
PET	70	260	30–50	2.9	70	100
PLA	60	170–180	>40	2.1	—	9
PCL (CAPA®680)	–60	60	67	0.2	—	>500

Source: Data obtained from Avérous, L., *Journal of Macromolecular Science-Polymer Reviews* C44 (3), 231–274, 2004; Doi, Y., *Microbial Polyesters* John Wiley & Sons, Inc., New York, 1990; Kunioka, M. and Doi, Y., *Macromolecules* 23 (7), 1933–1936, 1990; and Khanna, S. and Srivastava, A. K., *Process Biochemistry* 40 (2), 607–619, 2005.

5%. The PHB melting point (Tm = 170 to 180°C) is rather high compared to the other biodegradable polyesters, and its degradation temperature is close to Tm. The PHB homopolymer thus shows a narrow window for processing conditions. The PHB thermal [47–54] and thermomechanical [55,56] stabilities have been well described in the literature. Thermal degradation occurs according to a one-step process—namely, a random chain scission reaction (see Figure 8.2). Under extrusion, increasing the shear level, the temperature, or the residential time [57] leads to fast decreases in the PHB viscosity and in the molecular weight due to chain cleavage.

Many routes were investigated to ease the PHB transformation [58], including plasticization with citrate ester. PHBV copolymers are more adapted for the melt process because an increase in HV content results in lower melting and glass transition temperatures [59]. Higher HV contents also lead to an increase in impact strength and to a decrease in crystallinity [60], tensile strength [61] (see Table 8.2), and water permeability [60]. Therefore, the material properties can be tailored by varying the HV content. PHBV properties

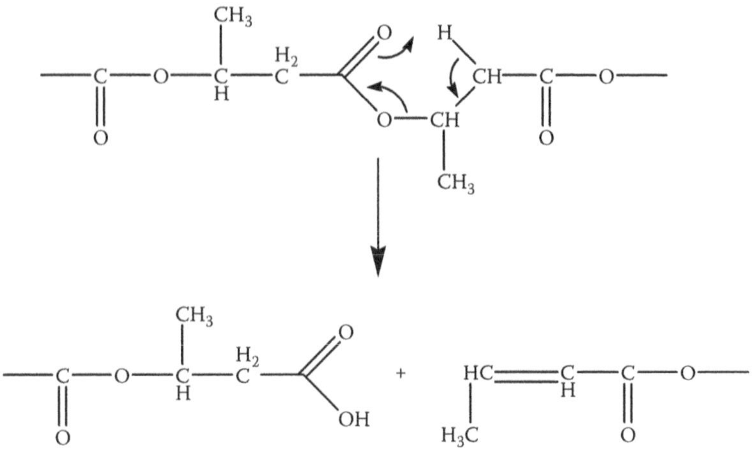

FIGURE 8.2
Poly(hydroxybutyrate) (PHB) random chain scission. (Adapted from Hablot, E., Bordes, P., Pollet, E., Avérous, L., *Polym. Degrad. Stab.*, 93, 413, 2008. With permission.)

can also be modified when plasticization occurs (e.g., with citrate ester, triacetin) [61,62].

Regarding biodegradable behavior, the enzymatic degradation rate is variable according to the crystallinity, the structure [5,63], and then, to the processing history [64]. Bacterial copolyesters biodegrade faster than homopolymers [65] and synthetic copolyesters [66].

8.2.3 Production and Applications

A large range of bacterial copolymer grades were industrially produced by Monsanto Inc. under the Biopol® trademark, with HV contents reaching 20%. The production was ceased at the end of 1999. Metabolix bought Biopol assets in 2001. Presently, Telles™, a joint venture between Metabolix and Archer Daniels Midlands Company (ADM), marketed the Mirel™ product as a new bio-based biodegradable plastic from corn sugar. ADM has begun to build the first plant in Clinton, Iowa, which will be able to produce 50,000 tons of resin per year. The start-up is scheduled for late 2008 (www.metabolix.com).

Different small companies produce bacterial PHA. For instance, PHB Industrial (Brazil) produces PHB and PHBV (HV = 12%) from sugar cane molasses [63]. This PHA production is planned to be 4000 tons/yr in 2008 and is then to be extended to 14,000 tons/yr [67]. In 2004, Procter & Gamble (Cincinnati, Ohio) and Kaneka Corporation (Osaka, Japan) announced a joint development agreement for the completion of R&D leading to the commercialization of Nodax, a large range of polyhydroxybutyrate-*co*-hydroxyalkanoates (PHBHx, PHBO, PHBOd) [68]. Although the industrial large-scale production was planned with a target price around 2€/kg, the Nodax development was stopped in 2006 [69].

PHA production aims at replacing synthetic nondegradable polymers for a wide range of applications [69], such as packaging, agriculture, and medicine [35,70]. Nevertheless, there are some limitations for using PHAs, such as brittleness and poor thermal stability, which consequently restrict development and uses. Therefore, nano-biocomposites appear as a possible answer to overcome these problems and to improve different properties, such as mechanical properties, thermal stability, and permeability.

8.3 PHA-Based Nanocomposites

As mentioned in the introduction, various nanofillers can be incorporated into polymers to prepare nano-biocomposites. Layered silicates or nanoclays are the most common nanofillers which have been used due to their high aspect ratio and large surface area.

8.3.1 PHA/Clay Nanocomposites

8.3.1.1 Introduction

Among clays, the 2:1 phyllosilicates (montmorillonite, saponite) are the most commonly used in nanocomposites. Their structure consists of layers made up of two tetrahedrally coordinated silicon atoms fused to an edge-shared octahedral sheet of either aluminium or magnesium hydroxide (see Table 8.3 and Figure 8.3). Each layered sheet is about 1 nm thick, and its length varies from tens of nanometers to more than one micron depending on the layered silicate. Layers stacking leads to a regular van der Waals gap between the platelets called the interlayer or the gallery. Isomorphic substitution may occur inside the sheet because Al^{3+} can be replaced by Mg^{2+} or Fe^{2+}, or Mg^{2+} by Li^+. The resulting global negative charges of the platelets are naturally

TABLE 8.3

Structural Characteristics of Principal 2:1 Layered Silicates

Phyllosilicates	Octahedra Occupancy	CEC (meq/100 g)	Aspect Ratio
Smectites			
Hectorite	Mg (3/3)	120	200–300
Montmorillonite	Al (2/3)	110	100–150
Saponite	Mg (3/3)	86.6	50–60

Source: Adapted from Sinha Ray, S. and Okamoto, M., *Prog. Polym. Sci.*, 28 (11), 1539–1641, 2003. (With permission.)

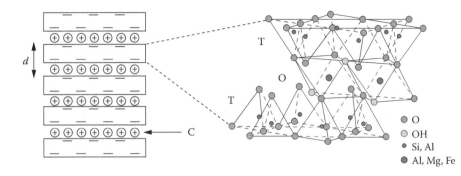

FIGURE 8.3
2:1 layered silicate structure (T, tetrahedral sheet; O, octahedral sheet; C, intercalated cations; d, interlayer distance). (Adapted from Lagaly, G., *Coagulation and flocculation—Theory and applications*, Dobias, B., Dekker, M., New York, 1993, pp. 427. With permission.)

counterbalanced by alkali and alkali earth cations (Na^+, Ca^{2+}) located in the galleries which increase the hydrophilic character of the clay. However, most of the polymers, particularly the biopolyesters, are considered as organophilic compounds. Thus, to obtain better affinity between the filler and the matrix, and eventually to improve final properties, the inorganic cations located inside the galleries (Na^+, Ca^{2+}) are generally exchanged by ammonium or phosphonium cations bearing at least one long alkyl chain and possibly other substituted groups. The resulting clays are called organo-modified layered silicates (OMLSs) and, in the case of montmorillonite (MMT), are abbreviated OMMT. This organo-modification, by improving the affinity between the matrix and the filler, can influence the material nanostructure and consequently the properties of the nanocomposites. Tables 8.3 and 8.4 present the characteristics of the main 2:1 layered silicates and commercial clays.

Another important key factor in the nano-(bio)composites preparation is the elaboration protocol. Three main methods are applied: (1) the solvent intercalation route that consists in swelling the layered silicates in a polymer solvent to promote the macromolecules diffusion in the clay interlayer spacing, (2) the *in situ* polymerization intercalation method for which the layered silicates are swollen in the monomer or monomer solution before polymerization, and (3) the melt intercalation process based on polymer processing in the molten state such as extrusion. Obviously, the last method is highly preferred in the context of sustainable development. It avoids the use of organic solvents that are not ecofriendly and then alter the life cycle analysis (LCA).

The addition of layered silicates nanofillers into the biopolymer matrix could lead to different structures of nano-biocomposites:

- A microcomposite when the clay layers are still stacked and the polymer is not intercalated within the (O)MMT's layers due to poor polymer–clay affinity, such a material presents phase separation.

TABLE 8.4

Some Commercial (Organo)Clays and Their Characteristics

Commercial Clays			
Supplier/Trade Name/ Designation	Clay Type	Organomodifier Type[a]	
Southern Clay Products (United States)			
Cloisite®Na	CNa	MMT	—
Cloisite®20A	C20A	MMT	$N^+(Me)_2(tallow)_2$
Cloisite®25A	C25A	MMT	$N^+(Me)_2(C_8)(tallow)$
Cloisite®30B	C30B	MMT	$N^+(Me)(EtOH)_2(tallow)$
Laviosa Chimica Mineraria (Italy)			
Dellite® 43B	D43B	MMT	$N^+(Me)_2(CH_2\text{-}\varphi)(tallow)$
CBC Co. (Japan)			
Somasif	MEE	SFM	$N^+(Me)(EtOH)_2(coco\ alkyl)$
	MAE	SFM	$N^+(Me)_2(tallow)_2$

[a] Tallow: mixture of long alkyl chains (~65% C_{18}; ~30% C_{16}; ~5% C_{14}).

- An intercalated nanocomposite for which the polymer is partially intercalated between the silicate layers; these latter are still stacked but the interlayer spacing has increased.
- An exfoliated nanocomposite showing individual and well-dispersed clay platelets into the matrix; in this case, the layered structure does not exist anymore.

The main interest of PHAs is to be produced from renewable resources (extracted from microorganisms). Thus, only the melt and the solvent intercalation routes are considered in the PHA-based nano-biocomposites elaboration studies.

The following part reports the main results obtained regarding the structural and properties aspects of PHB/clay and PHBV/clay materials.

8.3.1.2 PHA-Based Nanocomposite Structures and Properties

Initially, Maiti et al. [71] prepared PHB-based nanocomposites by melt extrusion. In their work, PHB was reinforced using organo-modified fluoromicas or OMMT containing 2 wt% and up to 4 wt% of clay, respectively. MEE and MAE fluoromicas (see Table 8.4) as well as OMMT modified with octadecylammonium ($MMT\text{-}NH_3^+(C_{18})$) were selected. X-ray diffraction (XRD) and transmission electron microscopy (TEM) revealed well-ordered intercalated nanocomposites with decreasing d-spacing when clay content increased (see Figure 8.4).

Dynamic mechanical analyses (DMAs) revealed a better reinforcing effect of fluoromica compared to OMMT. The storage modulus E′ increased with

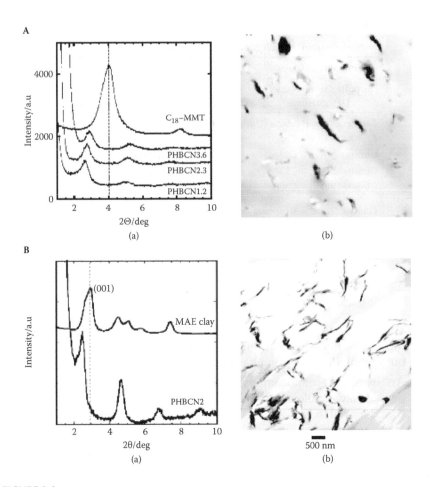

FIGURE 8.4
X-ray diffraction patterns (a) and bright field transmission electron micrograph images (b) of (A) PHB/C$_{18}$-MMT and (B) PHB/MAE-SFM nanocomposites. (From Maiti, P., Batt, C. A., and Giannelis, E. P., *Biomacromolecules* 8, 3393, 2007. With permission.)

clay content reaching an increment of 35% with 3.6 wt% of MMT-NH$_3^+$(C$_{18}$), +33%, and +40% with 2 wt% of MAE and MEE, respectively. Authors [71] explained this difference by a better dispersion and a higher aspect ratio in the case of fluoromicas compared to OMMT. Authors also stressed an enhanced PHB degradation that occurs in the presence of OMMT, due to ester linkages hydrolysis catalyzed by the Al Lewis acid sites located on the clay inorganic layers, resulting in the lower mechanical properties. Recently, Hablot et al. [72] reported that PHB degradation during melt processing can also be enhanced by decomposition products of clay organomodifiers.

The influence of organoclays on PHB properties was further studied and reported in a more recent article [73]. Regarding the thermal properties, it was considered that organoclays act as a better mass transport barrier because thermogravimetric curves are shifted to higher temperatures. It was also demonstrated that clay particles act as nucleating agents even if nanocomposites materials present lower crystallinity compared to neat polymer. However, this most recent publication [73] was mainly focused on biodegradation studies. Compost biodegradation studies were carried out at two distinct temperatures (20°C and 60°C). These studies were completed by XRD measurements and polarized optical microscope (POM) observation. The results led to the conclusion that crystallinity strongly affects the biodegradation, particularly at higher compost temperature. Indeed, authors showed that further crystallization occurs during biodegradation at 60°C leading to a higher amount of crystallinity and, thus, to a lower biodegradation rate at this compost temperature. Furthermore, this biodegradation rate was higher in the case of nanocomposites because of the higher amorphous content and the catalytic effect of clay. They eventually demonstrated that the control of materials processing and crystallization can lead to fine-tuning of the biodegradation rate [73].

A similar study on PHB-based nanocomposites was carried out by Lim et al. [74] using the solvent intercalation route. PHB/Cloisite®25A (C25A, see Table 8.4) nanocomposites were prepared, and whatever the clay content, intercalated structures were evidenced by XRD (see Figure 8.5). These

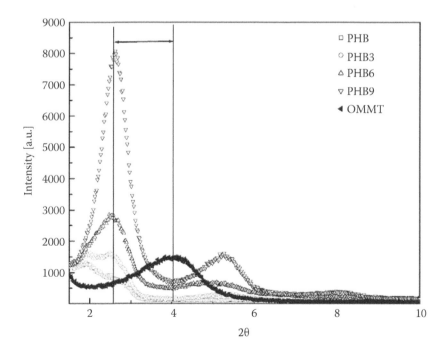

FIGURE 8.5
X-ray diffraction curves for various PHB/C25A nanocomposites with various C25A contents (3, 6, and 9 wt%). (From Lim, S. T., Hyun, Y. H., Lee, C. H., and Choi, H. J., *J. Mater. Sci. Lett.*, 22, 299, 2003. With permission.)

results were confirmed by FTIR analyses showing that two distinct different phases coexisted. These structural observations were completed by thermal stability investigation. Thermogravimetric analyses (TGA) results indicated an increase in the onset temperature and a decrease in the degradation rate with 3 wt% of C25A. This was attributed to the nanoscale dispersion of OMMT layers decreasing the diffusion of volatile decomposition products. At higher clay contents (>6 wt%), the onset of thermal degradation did not increase because of the organomodifiers' thermal sensitivity. But the nanocomposites' degradation rates decreased due to restricted thermal motion of the chains in the OMMT interlayer.

Scientists have been interested in the development of PHBV-based nanocomposites because PHBV presents better properties than PHB and better processability. In 2003, Choi et al. [75] described the microstructure, the thermal and mechanical properties of PHBV/Cloisite®30B (C30B, see Table 8.4) nanocomposites with lower clay content. These materials were prepared by melt intercalation using a Brabender mixer. XRD and TEM confirmed that intercalated nanostructures were obtained. Such structures were formed due to the strong hydrogen bond interactions between PHBV and the hydroxyl groups of the C30B organomodifier. Authors also demonstrated that the nanodispersed organoclay acted as a nucleating agent, increasing the temperature and rate of PHBV crystallization. Moreover, the DSC thermograms revealed that the crystallite size was reduced in the presence of nanodispersed layers because the PHBV melting temperature shifted to a lower temperature. Nanocomposites thermal stabilities were also studied. Thermogravimetric analyses revealed that the decomposition onset temperature increased with C30B content (+10°C with 3 wt% of filler). Authors explained these trends by the nanodispersion of the silicate layers into the matrix and thus concluded that the well-dispersed and layered structure accounted for an efficient barrier to the permeation of oxygen and combustion gas. Eventually, the mechanical properties showed that clays can also act as an effective reinforcing agent because the Young's modulus significantly increases from 480 to more than 790 MPa due to strong hydrogen bonding between PHBV and C30B.

Wang et al. [76] and Zhang and his group [77,78] also investigated the structure and the properties of PHBV/OMMT nanocomposites. They synthesized PHBV with 3 and 6.6 mol% of HV units as well as organo-modified MMT via cationic exchange in an aqueous solution with hexadecyl-trimethylammonium bromide (MMT-N^+(Me)$_3$(C$_{16}$)). Nanocomposites were prepared by solution intercalation method adding 1, 3, 5, or 10 wt% of the OMMT to a chloroform solution of PHBV. The resulting dispersions were exposed notably to an ultrasonication treatment. These conditions led to the formation of intercalated structures. The clay aggregation occurred when the clay content reached 10 wt%.

A detailed study of the PHBV/MMT-N^+(Me)$_3$(C$_{16}$) crystallization behavior was conducted. It was shown that OMMT acted as a nucleating agent in the PHBV matrix, which increased the nucleation and the overall crystallization rate, leading to more perfect PHBV crystals [77]. With an increasing amount of OMMT, the predominant crystallization mechanism of PHBV was shifted from the growth of crystals to the formation of crystalline nuclei. The nucleation effect of the organophilic clay decreased with the clay content increase. Wang et al. [76] postulated that the nanoscaled OMMT layers affect the crystallization in two opposite ways. A small part of OMMT can increase the number of crystalline nuclei, giving a more rapid crystallization rate. However, the interactions of OMMT layers with PHBV restrict the PHA chains' motion. Therefore, the crystallization rate increased, whereas the relative degree of crystallinity decreased with increasing amount of clay in the PHBV/OMMT nanocomposites. Furthermore, the PHBV processing behavior could be improved with OMMT-based nanocomposites because the processing temperature range enlarged by lowering the melting temperature with the increasing clay content. The tensile properties of the corresponding materials were improved by incorporation of 3 wt% of clay [78]. Above this clay content, aggregation of clay occurred, and tensile strength and strain at break decreased. From the modulus and the $T\alpha$ relaxation temperature, it was highlighted that the interface was maximized due to the restrictions of segmental motion near the organic–inorganic interface of nanocomposites. Thus, formation of intercalated nanocomposites was confirmed. Eventually, the biodegradability of these nanocomposites was investigated, and the results showed that biodegradation in soil suspension decreased with increasing amount of OMMT. This was related to greater interactions between PHBV and OMMT, lower water permeability, higher crystallinity, and the antimicrobial property of OMMT.

Eventually, Misra et al. developed a novel solvent-free method to prepare PHB functionalized by maleic anhydride (MA-g-PHB) [79]. The functionalization was successfully achieved by free radical grafting of maleic anhydride using a peroxide initiator by reactive extrusion processing. Then, they mixed MA-g-PHB with C30B to make the organomodifier hydroxyl functions react with the MA [80]. Although the d-spacing was comparable to PHB/C30B prepared by melt blending, the decrease in intensity of XRD signals and the TEM images showed that more delaminated platelets were obtained in the case of MA-g-PHB.

Therefore, most of the articles reported the preparation of PHA-based nanocomposite by solvent intercalation. Whatever the elaboration route, full exfoliation state was neither obtained nor clearly demonstrated. Exfoliated nanostructure does not seem so trivial to obtain, and explanations on this issue still require thorough studies.

Recently, further investigations were conducted to determine more precisely the nanostructure and the structure-properties relationships of PHA-based nano-biocomposites [81]. The results are presented in the next section.

8.3.1.3 Recent Advances

Because final materials properties are strongly dependent on the structure, the determination of clay organization into the matrix, using advanced characterization techniques such as XRD and TEM, is first reported. Further nanocomposites structural characterizations were carried out using a recent and innovative method based on solid-state NMR spectroscopy. Second, crystallization behavior as well as the thermal and mechanical properties were determined and correlated to the structure to establish the structure–properties relationships.

Different contents of Cloisite®30B (C30B, see Table 8.4; i.e., 1, 3, and 5 wt%) were added to PHA matrices by melt processing. The protocol setup allows, on one hand, high shear rate required for good clay delamination, and on the other hand, low viscous dissipation and short processing time to limit PHA degradation. For the sake of comparison, samples were also prepared using 3 wt% of Cloisite®Na (CNa), the nonmodified montmorillonite (see Table 8.4).

8.3.1.3.1 Structural Characterization

XRD patterns of PHB and PHBV with ~3 wt.% of nonmodified montmorillonite (PHB/CNa and PHBV/CNa) did not show any diffraction peak. On the contrary, CNa patterns exhibited a peak corresponding to $d_{001} = 12.8$ Å. Taking into account the highly hydrophilic character of CNa, the absence of diffraction peak in the composite materials was explained by the poor dispersion quality and the low clay content, rather than exfoliation. A similar observation was reported by Pluta et al. [82] on PLA/CNa microcomposites. On the contrary, diffraction peaks were detected for PHB/C30B and PHBV/C30B systems (see Figure 8.6 and Table 8.5). These peaks appeared at lower angles compared to C30B suggesting intercalated structures. The interlayer distances increased from 17.7 Å for C30B to 46.8, 40.4, and 39.9 Å for PHB/C30B containing 1.0, 2.2, and 5.0 wt.% of inorganic content, respectively. The same trends were obtained for PHBV/C30B nanocomposites, although the interlayer distances reached were lower compared to the PHB/C30B ones. Indeed, d_{001} equals 39.2, 38.5, and 36.6 Å for PHBV-based systems with 0.6, 2.6, and 4.4 wt.% of inorganic content, respectively. PHB/C30B materials display a large d_{001} increase, particularly for with 1 wt.% of clay content ($d_{001} = 46.8$ Å) and weak peaks intensity (see Figure 8.6). This suggests important polymer intercalation with probably exfoliated clay domains, or at least small tactoids. Furthermore, the peak intensity of the intercalated clay gradually increased with organoclay content indicating that the number and the size of tactoids slightly increase.

These results were complemented by a recent method using solid-state NMR which allows us to obtain quantitative information about the nano-biocomposite structure [83]. The NMR measurements lead to the determination of two coefficients, f and ε, respectively, the degree and the quality

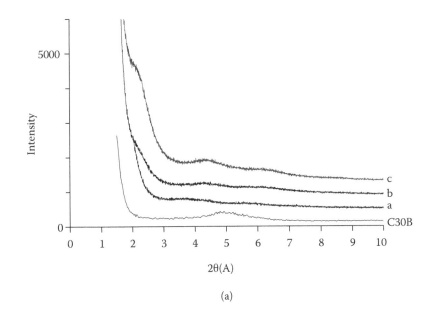

(a)

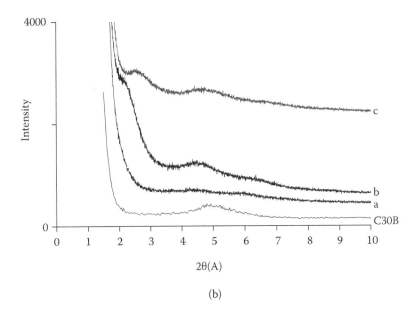

(b)

FIGURE 8.6
X-ray diffraction patterns of (A) PHB/C30B and (B) PHBV/C30B nanocomposites with (a)
1 wt.%, (b) 3 wt.%, and (c) 5 wt.% of clay inorganic content and the corresponding small-angle
X-ray scattering pattern of Cloisite®30B (C30B). (From Bordes, P., Pollet, E., Bourbigot, S., and
Averous, L., *Macromol. Chem. Phys.*, 209 (14), 1473, 2008. With permission.)

TABLE 8.5

Interlayer Distances (d_{001}) of PHA-Based Nanocomposites Obtained from X-Ray Diffraction

Polymer	None		PHB				PHBV		
Nanofiller	CNa	C30B	CNa	C30B			CNa	C30B	
wt.%[a] inorganic	—	—	2.9	1.0	2.2	5.0	2.3	0.6	2.6	4.4
d_{001} (Å)[b]	12.8	17.7	—	46.8	40.4	39.9	—	39.2	38.5	36.6

Source: From Bordes, P., Pollet, E., Bourbigot, S., and Avérous, L., *Macromol. Chem. Phys.*, 209, 1473, 2008. (With permission.)

[a] Content of inorganic matter determined from the thermogravimetric curves as the residue left at 600°C.

[b] For nanocomposites materials, d_{001} was determined from the d_{002} peak.

of clay dispersion (f = 1 and ε = 100% for a fully exfoliated and perfectly dispersed filler). These coefficients are calculated from experimental relaxation times T_1^H through a computational model that is already validated on other nanocomposite systems [83–88]. It was the first time that such a technique was reported on bacterial polyester-based systems. The results of solid-state NMR measurements (see Table 8.6) showed that the nonmodified montmorillonite (CNa) was still aggregated in the PHA matrix (f coefficients equal 0.1 in both PHB/CNa and PHBV/CNa materials). However, the large tactoids were reasonably dispersed (ε coefficients equal to 50% and 60% for PHB/CNa and PHBV/CNa, respectively). Regarding PHBV-based samples, relaxation times decrease with the incorporation of clay, particularly with the C30B increasing content. The f coefficient of PHBV/C30B with 1 wt.% of clay indicated the presence of very small tactoids (f = 0.7) with a reasonable nanodispersion (ε = 45%). However, at higher C30B content, the f and ε coefficients decreased, suggesting bigger tactoids and poorer nanodispersion (20 < ε < 30%). For the PHB-based systems, the data analysis was more complex because the relaxation times of PHB/C30B ($T_1^H{}_{PHB/C30B}$ = 2.46 and 2.43 s, for PHB/C30B with 1.0 and 2.2 wt.%, respectively) did not exhibit a reduction of T_1^H compared to the neat PHB reference ($T_1^H{}_{PHB}$ = 2.42 s). Further experiments were conducted to explain this observation. Therefore, the ε parameter was not taken into account, but it was still possible to reason about the degree of clay dispersion. Assuming that a real homogeneous nanodispersion was achieved, it appeared that the calculated f coefficients were lower than 0.5, indicating slightly delaminated tactoids.

TEM characterizations of PHB-based materials were consistent with the results obtained from solid-state NMR—that is, CNa was not intercalated showing well-stacked tactoids of hundreds of platelets (see Figure 8.7a), while small tactoids of about three to ten platelets were present in the PHB/C30B nanocomposites with a reasonable nanodispersion (see Figure 8.7b,c). This emphasized the higher compatibility of C30B with PHA, compared to CNa.

TABLE 8.6

Solid-State Nuclear Magnetic Resonance (NMR) Results for Poly(hydroxybutyrate) (PHB)- and Poly(hydroxybutyrate-*co*-hydroxyvalerate) (PHBV)-Based Materials and Their Structure Determined from X-Ray Diffraction, Transmission Electron Microscopy, and Solid-State NMR

	wt.% Clay	T_1^H (s)	f	ε (%)	Nanocomposite Structure
PHB	—	2.42	—	—	—
PHB/CNa	2.9	2.39	0.1	50	Microcomposite
PHB/C30B	1.0	2.46	0.27	—	Small and well-intercalated tactoids
	2.2	2.43	0.4	—	Small and well-intercalated tactoids
	5.0	2.34	0.3	—	Small and well-intercalated tactoids
PHBV	—	2.50	—	—	—
PHBV/CNa	2.3	2.35	0.1	60	Microcomposite
PHBV/C30B	0.6	2.32	0.7	42	Small intercalated tactoids with good nanodispersion
	2.6	2.19	0.45	27	Small intercalated tactoids with poor nanodispersion
	4.4	2.13	0.43	21	Small intercalated tactoids with poor nanodispersion

Source: From Bordes, P., Pollet, E., Bourbigot, S., and Avérous, L., *Macromol. Chem. Phys.*, 209, 1473, 2008. (With permission.)

The mobility and, thus, the chains' relaxation time can be affected by the crystallinity and the polymer molecular weight. Thus, these two parameters were measured to determine if their variations, depending on the nature and the clay content, could be significant. DSC characterizations showed quite constant crystallinity values regardless of the clay nature (see Table 8.7). It was concluded that crystallinity did not affect the relaxation time. On the contrary, the weight average molecular weights were considerably reduced with C30B (see Table 8.7). When C30B content is about 5 wt.%, the M_w decreased by 40% and 60% for the PHB- and PHBV-based systems, respectively, compared to the corresponding processed polymer. It is well known that PHA are highly temperature sensitive and degrade according to a random chain scission process, even just above their melting temperature [47–49] (see Figure 8.2). But in this case, it appeared that the C30B organomodifier (S-EtOH), and even the inorganic montmorillonite layer (i.e., the nonmodified montmorillonite CNa) enhance PHA degradation. Indeed, the M_w of PHA/CNa and PHA/S-EtOH decreased by 20% and 32% with CNa and S-EtOH, respectively, compared to the corresponding processed polymer. The enhanced PHA degradation observed with nonmodified layered silicates had already been reported. It was explained by the presence of Al Lewis acid sites in the inorganic layers [89,90] or by the presence of residual water, because CNa is hydrophilic. However, until recently, the organomodifier influence on PHA degradation

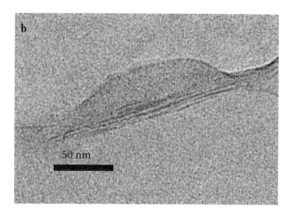

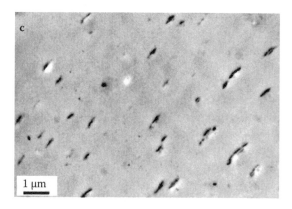

FIGURE 8.7
Transmission electron micrograph images of (a) PHB/CNa with 2.9 wt.% of inorganic content, and (b) and (c) PHB/C30B with 2.2 wt.% of inorganic content at high and low magnification, respectively. (From Bordes, P., Pollet, E., Bourbigot, S., and Avérous, L., *Macromol. Chem. Phys.*, 209, 1473, 2008. With permission.)

TABLE 8.7

Size Exclusion Chromatography (SEC) Results, Crystallinity (χ_l) of for Poly(hydroxybutyrate) (PHB)– and PHBV–Based Systems

System	PHB (T_p = 170°C)						PHBV (T_p = 160°C)					
	Neat PHA	S-EtOH	CNa		C30B		Neat PHA	S-EtOH	CNa		C30B	
wt.% clay	—	—	2.9	1.0	2.2	5.0	—	—	2.3	0.6	2.6	4.4
$M_w \times 10^{-3}$	320	217	258	297	209	194	542	364	418	437	324	220
PDI	2.1	2.1	2.0	2.0	2.0	2.2	2.2	2.1	2.2	2.1	2.3	2.4
χ_l (%)[a]	55	—	48	51	49	53	52	—	46	40	43	41

Source: From Bordes, P., Pollet, E., Bourbigot, S., and Avérous, L., *Macromol. Chem. Phys.*, 209, 1473, 2008. (With permission.)

[a] χ_l calculated from the ratio of experimental melting enthalpy measured on the first scan (not reported here) and $\Delta H_{m_{100\%}}$ crystalline $= 146\ J \cdot g^{-1}$ for PHB and PHBV.

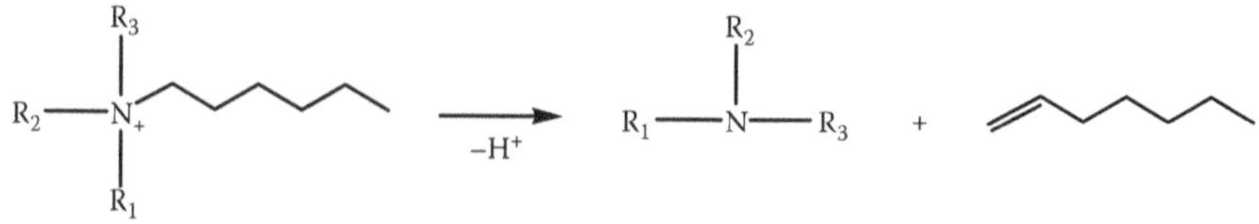

FIGURE 8.8
Hofmann elimination. (From Hablot, E., Bordes, P., Pollet, E., and Avérous, L., *Polym. Degrad. Stab.*, 93, 413, 2008. With permission.)

had not been investigated [72,91]. It was demonstrated that surfactants used to improve polymer–clay compatibility greatly boost the PHA degradation. Their decomposition releases a proton according to a Hoffmann elimination (see Figure 8.8) and then enhances the chain scission (see Figure 8.2) by an acidic catalysis reaction [72,91].

Typically, microcomposites are obtained with nonmodified montmorillonite, whereas a nanostructure was reached with small amounts (1 to 5 wt.%) of C30B due to higher polymer–clay affinity. Almost all the nanocomposites were well intercalated as demonstrated by XRD characterization, while the nanodispersion was poor for PHB/C30B systems and quite good for PHBV/ C30B samples. These particular structures were explained by the higher viscosity of PHBV-based systems which led to better clay delamination. The apparent discrepancy of solid-state NMR results about PHB/C30B samples might be attributed to further degradation phenomena. They occur during processing in the presence of clay and might also affect the relaxation time. Compared to previously published studies [71,73,75], a higher degree of intercalation/exfoliation was obtained here. Moreover, deeper morphological analyses were performed to assess the clay nanodispersion quality.

8.3.1.3.2 Properties

Considering these results, the crystallization behavior as well as the mechanical and thermal properties of the PHA/clay systems were determined and correlated to their structures.

8.3.1.3.2.1 Crystallization Behavior XRD spectra (not shown here) on PHB- and PHBV-based materials revealed no change in the crystalline structure of PHA, indicating that the addition of clay does not significantly modify the crystalline phase. On the contrary, DSC and POM measurements carried out on the same samples clearly demonstrated that clay enhances the crystallization (see Figures 8.9 and 8.10). This phenomenon was much more pronounced with the increasing polymer–clay affinity or filler content attesting the higher degree of C30B dispersion. The crystallization temperature and enthalpy increased, from about 70°C to more than 90°C and from 50 J · g^{-1} to more than 60 J · g^{-1}, for neat PHB and PHB/C30B, respectively. Also, the cold crystallization was no more observed for PHB/clay systems. Considering the

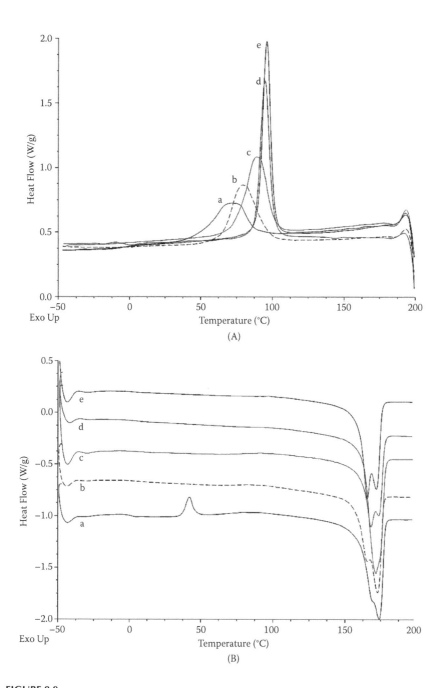

FIGURE 8.9
(A) Differential scanning calorimetry cooling and (B) second heating scans for (a) PHB, (b) PHB/CNa 2.9 wt.%, (c) PHB/C30B 1 wt.%, (d) PHB/C30B 2.2 wt.%, and (e) PHB/C30B 5 wt.%. (From Bordes, P., Pollet, E., Bourbigot, S., and Averous, L., *Macromol. Chem. Phys.*, 209 (14), 1473, 2008. With permission.)

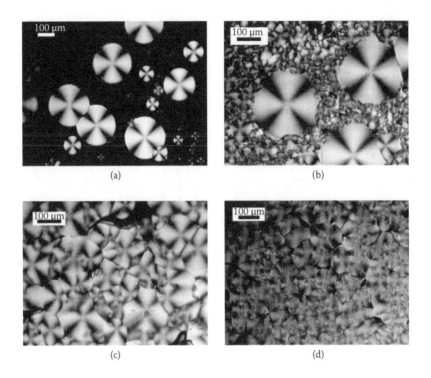

FIGURE 8.10
Polarized optical microscope photographs of (a) neat PHB, (b) PHB/C30B 1 wt.%, (c) PHB/C30B 2.2 wt.%, and (d) PHB/C30B 5 wt.%. (From Bordes, P., Pollet, E., Bourbigot, S., and Avérous, L., *Macromol. Chem. Phys.*, 209, 1473, 2008. With permission.)

crystallization kinetics, it was also shown that addition and dispersion of clay into the matrix make the crystallization occur faster, resulting in more homogeneous spherulite size (see Figure 8.10). This phenomenon is consistent with previous reported studies on the addition of nucleating agents to PHBV [92] or on nanocomposites [73,75,93–95].

Another noteworthy phenomenon that has been discussed is the double melting endotherms appearing during melting. This behavior was attributed to the well-known recrystallization phenomenon [96–98] rather than a consequence of the existence of bimodal size of crystallite or different types of crystallization as described by Wang et al. in PHBV/OMMT systems [76]. Moreover, it was pointed out the change in the ratio of the melting peak intensities with the clay content: T_{m1} peak intensity becomes preponderant in comparison with the one corresponding to the second melting peak while the clay content increases. This behavior, attesting for less recrystallization during melting, was attributed to the hindering effect of clay sheets on the recrystallization phenomenon. Such phenomenon of clay platelets hindering crystal growth while enhancing nucleation step has already been observed in other nano-biocomposite systems [94,95].

Because similar trends were observed in the case of PHBV-based materials, the interpretations drawn above could also be applied to PHBV.

8.3.1.3.2.2 Mechanical Properties   Regarding the tensile properties, higher increments were obtained in the case of PHA/C30B nanocomposites with low clay content due to better delaminated and nanodispersed clay as demonstrated by structural characterizations. The Young's modulus globally increases linearly with the C30B content reaching 14% and 17% improvements for PHB/C30B and PHBV/C30B with 5 wt.% of clay, respectively. Nevertheless, this enhancement was at the expense of the elongation at break. This parameter first increased with the addition of about 1 wt.% of clay (+21 and +85% for PHB- and PHBV-based nanocomposites, compared to the neat corresponding polymer). Then elongation at break decreased with increasing clay content. The material thus becomes more brittle at high clay contents. Considering the tensile strength, it was almost constant with C30B regardless of the clay content, reaching an increase of about +15% and +30% for PHB/C30B and PHBV/C30B, respectively. More generally, the enhancement of the tensile properties of PHA/C30B at low clay content (<3 wt.%) can be attributed to good compatibility between the PHA and the C30B clay that leads to intercalated structures with small tactoids dispersed into the polymer matrix. At higher clay content, the addition of clay is no more a benefit because the clay tends to form bigger tactoids with poorer nanodispersion. Compared to previously published results [75] regarding PHA/clay nanocomposites prepared by melt intercalation, a better balance was obtained between the stiffness and the strain at break. Indeed, Choi et al. [75] reached higher Young's modulus but lower elongation at break compared to neat PHBV, whatever the C30B content.

8.3.1.3.2.3 Thermal Stability   Eventually, thermal stability was investigated to determine the influence of the addition of clay into the PHA matrices. These results were compared to the corresponding processed polymer because it has been shown that these PHAs are prone to considerable degradation during processing (see Table 8.7 and Reference [95]). A considerable decrease in PHA/CNa thermal stability was observed because the characteristic degradation temperatures are significantly shifted to lower temperatures, particularly in the case of PHB-based systems. The onset and the maximal decomposition rate temperatures dropped off about 10°C and the offset degradation temperature decreased 26°C for PHB/CNa. This phenomenon was attributed to the polymer degradation in the presence of Al Lewis acid sites in the inorganic clay layers or due to the hydrophilic character of nonmodified montmorillonite as previously reported by Maiti et al. [71].

On the contrary, the thermal stabilities of PHA/C30B nanocomposites were similar or slightly higher than the corresponding neat PHA (without nanofiller). The onset degradation temperatures were, however, lower in the case of the addition of C30B due to the polymer degradation that was enhanced by the presence of both inorganic layers [89,90] and C30B organomodifier [72,91]. As a result, the decomposition temperature range enlarged with C30B suggesting that degradation is slowed, possibly by the very small and well-dispersed C30B tactoids into the PHA, compared to large aggregated tactoids of PHA/CNa samples. It was thus concluded that the well-dispersed nanofillers appear as effective barriers to heat and volatile compounds diffusion and thus counterbalance the degradation effect.

This most recent study brings new insight regarding the PHA/clay nano-biocomposites' structure by using an original method based on solid-state NMR as a complement to classical (XRD, TEM) characterization techniques. This study also emphasizes on the PHA degradation occurring in these systems, with the determination of the organomodifiers' negative influence on PHA thermal stability. This phenomenon was presented as a limit to the enhancement of PHA properties in the presence of nanoclays.

8.4 Summary

In the literature, nano-biocomposites based on polyhydroxyalkanoates, mainly PHB and PHBV, were prepared using different layered silicate clays and elaboration routes. The full exfoliation state was not previously reported even if a beginning of clay exfoliation was evidenced in few studies. Only intercalated or well-intercalated structures and microcomposites were obtained using, respectively, organomodified or unmodified layered silicates (clays).

Despite the fact that fully exfoliated structures were not obtained, the mechanical and thermal properties as well as the crystallization and biodegradation rates were improved. Through many characterization techniques, regarding structural aspects as well as materials properties, the role of nanoclays (type, content, and organization within the matrix) on the PHA properties was highlighted and understood. The structure–properties relationships for PHA/OMMT nano-biocomposites were established and are in agreement with the conclusions drawn in previously reported studies on synthetic polymer-based nanocomposites.

Further attention was paid to the PHAs' degradation in nanocomposites systems, because these polymers are very temperature sensitive. A possible effect of the clay organomodifier or the mineral clay layer itself was pointed out which can explain the limit of PHA improvements even with the addition of well-dispersed nanoclays. Considering these results, the poor thermal stability of PHAs and the role of (organomodified) clays on it represent particularly major obstacles to the elaboration of technically competitive materials.

Thus, scientists were interested in other PHA-based nanocomposites filled with layered double hydroxides (LDHs) [98,99], cellulose whiskers [100–102], and hydroxyapatite (HA) [103], this latter being particularly used for biomedical and tissue engineering applications.

LDH structures are similar to the layered silicate clays. Hsu et al. [98,99] reached exfoliated state using LDH organically modified by poly(ethylene glycol) phosphonates (PMLDH). The crystallization behavior of these PHB/PMLDH were comparable to PHB/OMMT nanocomposites [98,99].

In the case of PHA/cellulose whiskers materials, studies were conducted using a latex of poly(3-hydroxyoctanoate) (PHO) [104] as a matrix and a colloidal suspension of hydrolyzed cellulose whiskers as natural and biodegradable fillers. Due to the geometry and aspect ratio of the cellulose whiskers, the formation of a rigid filler network, called the percolation phenomenon, was observed, leading to higher mechanical PHO properties [100–102].

Eventually, regarding the nano-hydroxyapatite filler, like the polymer/clay nanocomposites, the good dispersion of inorganic fillers in the PHBV inevitably benefits the improvement of mechanical properties of the materials [103]. Furthermore, the study also pointed out the enhanced material bioactivity because this specific property is expected for the repair and replacement of bone.

"Green nanocomposites" based on PHAs appear as a next generation of environmentally friendly materials and broaden the range of PHAs applications by enhancing the polymer properties (ductility, melt viscosity, thermal stability). Thus, more appropriate new macromolecular architectures and nanoparticles-based systems should allow us, in the near future, to overcome the actual limits of these materials (high crystallinity, brittleness, poor thermal stability, etc.).

Acknowledgments

The authors thank Biocycle (Copersucar-Brazil), the PERF-LSPES (Ecole Nationale Supérieure de Chimie de Lille, France), and the GMI-IPCMS (Institut de Physique et Chimie des Matériaux de Strasbourg, France) for their technical support.

References

1. Narayan, R., Drivers for biodegradable/compostable plastics and role of composting in waste management and sustainable agriculture, *Orbit Journal* 1 (1), 1–9, 2001.

2. Steinbuchel, A., *Biopolymers, general aspects and special applications,* Wiley-VCH, Weinheim, Germany, 2003.

3. Avella, M., Bonadies, E., and Martuscelli, E., European current standardization for plastic packaging recoverable through composting and biodegradation, *Polymer Testing* 20 (5), 517–521, 2001.
4. Fritz, J., Link, U., and Braun, R., Environmental impacts of biobased/biodegradable packaging, *Starch* 53 (3–4), 105–109, 2001.
5. Karlsson, S. and Albertsson, A.-C., Biodegradable polymers and environmental interaction, *Polymer Engineering and Science* 38 (8), 1251–1253, 1998.
6. Kaplan, D.L., Mayer, J.M., Ball, D., McCassie, J., Allen, A.L., and Stenhouse, P., Fundamentals of biodegradable polymers, in *Biodegradable Polymers and Packaging*, Ching, C., Kaplan, D.L., and Thomas, E.L., Lancaster, PA, Technomic, 1993, pp. 1–42.
7. Van de Velde, K. and Kiekens, P., Biopolymers: Overview of several properties and consequences on their applications, *Polymer Testing* 21 (4), 433–442, 2002.
8. Rouilly, A. and Rigal, L., Agro-materials: A bibliographic review, *Journal of Macromolecular Science—Part C. Polymer Reviews* C42 (4), 441–479, 2002.
9. Chandra, R. and Rustgi, R., Biodegradable polymers, *Progress in Polymer Science* 23 (7), 1273–1335, 1998.
10. Avérous, L., Biodegradable multiphase systems based on plasticized starch: A review, *Journal of Macromolecular Science–Polymer Reviews* C44 (3), 231–274, 2004.
11. Sinha Ray, S. and Okamoto, M., Polymer/layered silicate nanocomposites: A review from preparation to processing, *Progress in Polymer Science (Oxford)* 28 (11), 1539–1641, 2003.
12. Messersmith, P.B. and Giannelis, E.P., Synthesis and barrier properties of polycaprolactone-layered silicate nanocomposites, *Journal of Polymer Science: Part A: Polymer Chemistry* 33 (7), 1047–1057, 1995.
13. Pantoustier, N., Alexandre, M., Degée, P., Calberg, C., Jerome, R., Henrist, C., Cloots, R., Rulmont, A., and Dubois, P., Poly(ε-caprolactone) layered silicate nanocomposites: Effect of clay surface modifiers on the melt intercalation process, *e-Polymers* no. 009, 2001.
14. Pantoustier, N., Lepoittevin, B., Alexandre, M., Kubies, D., Calberg, C., Jerome, R., and Dubois, P., Biodegradable polyester layered silicate nanocomposites based on poly(ε-caprolactone), *Polymer Engineering and Science* 42 (9), 1928–1937, 2002.
15. Lepoittevin, B., Pantoustier, N., Alexandre, M., Calberg, C., Jerome, R., and Dubois, P., Layered silicate/polyester nanohybrids by controlled ring-opening polymerization, *Macromolecular Symposia* 183, 95–102, 2002.
16. Pantoustier, N., Alexandre, M., Degée, P., Kubies, D., Jerome, R., Henrist, C., Rulmont, A., and Dubois, P., Intercalative polymerization of cyclic esters in layered silicates: Thermal vs. catalytic activation, *Composite Interfaces* 10 (4), 423–433, 2003.
17. Lepoittevin, B., Pantoustier, N., Devalckenaere, M., Alexandre, M., Calberg, C., Jerome, R., Henrist, C., Rulmont, A., and Dubois, P., Polymer/layered silicate nanocomposites by combined intercalative polymerization and melt intercalation: A masterbatch process, *Polymer* 44 (7), 2033–2040, 2003.
18. Ogata, N., Jimenez, G., Kawai, H., and Ogihara, T., Structure and thermal/mechanical properties of poly(L-lactide)-clay blend, *Journal of Polymer Science, Part B: Polymer Physics* 35 (2), 389–396, 1997.

19. Paul, M.-A., Alexandre, M., Degée, P., Calberg, C., Jerome, R., and Dubois, P., Exfoliated polylactide/clay nanocomposites by in-situ coordination-insertion polymerization, *Macromolecular Rapid Communications* 24 (9), 561–566, 2003.

20. Paul, M.-A., Alexandre, M., Degée, P., Henrist, C., Rulmont, A., and Dubois, P., New nanocomposite materials based on plasticized poly(L-lactide) and organo-modified montmorillonites: Thermal and morphological study, *Polymer* 44 (2), 443–450, 2003.

21. Pluta, M., Paul, M.-A., Alexandre, M., and Dubois, P., Plasticized polylactide/clay nanocomposites. I. The role of filler content and its surface organo-modification on the physico-chemical properties, *Journal of Polymer Science, Part B: Polymer Physics* 44 (2), 299–311, 2006.

22. Pluta, M., Paul, M.-A., Alexandre, M., and Dubois, P., Plasticized polylactide/clay nanocomposites. II. The effect of aging on structure and properties in relation to the filler content and the nature of its organo-modification, *Journal of Polymer Science, Part B: Polymer Physics* 44 (2), 312–325, 2006.

23. Sinha Ray, S., Yamada, K., Okamoto, M., and Ueda, K., Polylactide-layered silicate nanocomposite: A novel biodegradable material, *Nano Letters* 2 (10), 1093–1096, 2002.

24. Maiti, P., Yamada, K., Okamoto, M., Ueda, K., and Okamoto, K., New polylactide/layered silicate nanocomposites: Role of organoclays, *Chemistry of Materials* 14 (11), 4654–4661, 2002.

25. Sinha Ray, S., Yamada, K., Okamoto, M., Fujimoto, Y., Ogami, A., and Ueda, K., New polylactide/layered silicate nanocomposites. 5. Designing of materials with desired properties, *Polymer* 44 (21), 6633–6646, 2003.

26. Sinha Ray, S. and Okamoto, M., Biodegradable polylactide and its nanocomposites: Opening a new dimension for plastics and composites, *Macromolecular Rapid Communications* 24 (14), 815–840, 2003.

27. Park, H.-M., Li, X., Jin, C.-Z., Park, C.-Y., Cho, W.-J., and Ha, C.-S., Preparation and properties of biodegradable thermoplastic starch/clay hybrids, *Macromolecular Materials and Engineering* 287 (8), 553–558, 2002.

28. Park, H.-M., Lee, W.-K., Park, C.-Y., Cho, W.-J., and Ha, C.-S., Environmentally friendly polymer hybrids—Part I. Mechanical, thermal, and barrier properties of thermoplastic starch/clay nanocomposites, *Journal of Material Science* 38 (5), 909–915, 2003.

29. Dean, K., Yu, L., and Wu, D.Y., Preparation and characterization of melt-extruded thermoplastic starch/clay nanocomposites, *Composites Science and Technology* 67 (3-4), 413–421, 2007.

30. Chivrac, F., Pollet, E., and Avérous, L., New approach to elaborate exfoliated starch-based nanobiocomposites, *Biomacromolecules* 9 (3), 896–900, 2008.

31. Lemoigne, M., Products of dehydration and of polymerization of β-hydroxybutyric acid, *Bulletin de la Société de Chimie Biologique* 8, 770–782, 1926.

32. Lenz, R.W. and Marchessault, R.H., Bacterial polyesters: Biosynthesis, biodegradable plastics and biotechnology, *Biomacromolecules* 6 (1), 1–8, 2005.

33. Doi, Y., *Microbial polyesters* John Wiley & Sons, New York, 1990.

34. Madison, L.L. and Huisman, G.W., Metabolic engineering of poly(3-hydroxyalkanoates): From DNA to plastic, *Microbiology and Molecular Biology Reviews* 63 (1), 21–53, 1999.

35. Zinn, M., Witholt, B., and Egli, T., Occurrence, synthesis and medical application of bacterial polyhydroxyalkanoate, *Advanced Drug Delivery Reviews* 53 (1), 5–21, 2001.
36. Reddy, C.S.K., Ghai, R., Rashmi, and Kalia, V.C., Polyhydroxyalkanoates: An overview, *Bioresource Technology* 87 (2), 137–146, 2003.
37. Valentin, H.E., Broyles, D.L., Casagrande, L.A., Colburn, S.M., Creely, W.L., DeLaquil, P.A., Felton, H.M., Gonzalez, K.A., Houmiel, K.H., Lutke, K., Mahadeo, D.A., Mitsky, T.A., Padgette, S.R., Reiser, S.E., Slater, S., Stark, D.M., Stock, R.T., Stone, D.A., Taylor, N.B., Thorne, G.M., Tran, M., and Gruys, K.J., PHA production, from bacteria to plants, *International Journal of Biological Macromolecules* 25 (1–3), 303–306, 1999.
38. Poirier, Y., Polyhydroxyalkanoate synthesis in plants as a tool for biotechnology and basic studies of lipid metabolism, *Progress in Lipid Research* 41 (2), 131–155, 2002.
39. Hori, Y., Suzuki, M., Yamaguchi, A., and Nishishita, T., Ring-opening polymerization of optically active β-butyrolactone using distannoxane catalysts: Synthesis of high-molecular-weight poly(3-hydroxybutyrate), *Macromolecules* 26 (20), 5533, 1993.
40. Hori, Y., Takahashi, Y., Yamaguchi, A., and Nishishita, T., Ring-opening copolymerization of optically active β-butyrolactone with several lactones catalyzed by distannoxane complexes: Synthesis of new biodegradable polyesters, *Macromolecules* 26 (16), 4388, 1993.
41. Hori, Y. and Hagiwara, T., Ring-opening polymerisation of β-butyrolactone catalysed by distannoxane complexes: Study of the mechanism, *International Journal of Biological Macromolecules* 25 (1–3), 235–247, 1996.
42. Kobayashi, T., Yamaguchi, A., Hagiwara, T., and Hori, Y., Synthesis of poly(3-hydroxyalkanoate)s by ring-opening copolymerization of (R)-β-butyrolactone with other four-membered lactones using a distannoxane complex as a catalyst, *Polymer* 36 (24), 4707–4710, 1995.
43. Nobes, G.A.R., Kazlauskas, R.J., and Marchessault, R.H., Lipase-catalyzed ring-opening polymerization of lactones: A novel route to poly(hydroxyalkanoate)s, *Macromolecules* 29 (14), 4829, 1996.
44. Juzwa, M. and Jedlinski, Z., Novel synthesis of poly(3-hydroxybutyrate), *Macromolecules* 39 (13), 4627, 2006.
45. Asrar, J., Mitsky, T.A., and Shah, D.T., United States US6091002, 2000, Polyhydroxyalkanoates of narrow molecular weight distribution prepared in transgenic plants. 2000-07-18.
46. Asrar, J., Mitsky, T.A., and Shah, D.T., United States US6228623, 2001, Polyhydroxyalkanoates of narrow molecular weight distribution prepared in transgenic plants. 2001-05-08.
47. Grassie, N., Murray, E.J., and Holmes, P.A., Thermal degradation of poly(-(D)-β-hydroxybutyric acid): Part 3—The reaction mechanism, *Polymer Degradation and Stability* 6 (3), 127–134, 1984.
48. Grassie, N., Murray, E.J., and Holmes, P.A., Thermal degradation of poly(-(D)-β-hydroxybutyric acid): Part 1—Identification and quantitative analysis of products, *Polymer Degradation and Stability* 6 (1), 47–61, 1984.
49. Grassie, N., Murray, E.J., and Holmes, P.A., Thermal degradation of poly(-(D)-β-hydroxybutyric acid): Part 2—Changes in molecular weight, *Polymer Degradation and Stability* 6 (2), 95–103, 1984.

50. Kunioka, M. and Doi, Y., Thermal degradation of microbial copolyesters. Poly(3-hydroxybutyrate-co-3-hydroxyvalerate) and poly(3-hydroxybutyrate-co-4-hydroxybutyrate), *Macromolecules* 23 (7), 1933–1936, 1990.
51. Aoyagi, Y., Yamashita, K., and Doi, Y., Thermal degradation of poly[(R)-3-hydroxybutyrate], poly[ε-caprolactone], and poly[(S)-lactide], *Polymer Degradation and Stability* 76 (1), 53–59, 2002.
52. Li, S.-D., He, J.-D., Yu, P.H., and Cheung, M.K., Thermal degradation of poly(3-hydroxybutyrate) and poly(3-hydroxybutyrate-co-3-hydroxyvalerate) as studied by TG, TG-FTIR, and Py-GC/MS, *Journal of Applied Polymer Science* 89 (6), 1530–1536, 2003.
53. Carrasco, F., Dionisi, D., Martinelli, A., and Majone, M., Thermal stability of polyhydroxyalkanoates, *Journal of Applied Polymer Science* 100 (3), 2111–2121, 2006.
54. Abe, H., Thermal degradation of environmentally degradable poly(hydroxyalkanoic acid)s, *Macromolecular Bioscience* 6 (7), 469–486, 2006.
55. Renstad, R., Karlsson, S., and Albertsson, A.-C., Influence of processing parameters on the molecular weight and mechanical properties of poly(3-hydroxybutyrate-co-3-hydroxyvalerate), *Polymer Degradation and Stability* 57 (3), 331–338, 1997.
56. Melik, D.H. and Schechtman, L.A., Biopolyester melt behavior by torque rheometry, *Polymer Engineering & Science* 35 (22), 1795–1806, 1995.
57. Ramkumar, D.H.S. and Bhattacharya, M., Steady shear and dynamic properties of biodegradable polyesters, *Polymer Engineering and Science* 38 (9), 1426–1435, 1998.
58. Billingham, N.C., Henman, T.J., and Holmes, P.A., Degradation and stabilisation of polyesters of biological and synthetic origin, in *Developments in Polymer Degradation* Elsevier Applied Science, London, 1987, pp. 81–121.
59. Amass, W., Amass, A., and Tighe, B., A review of biodegradable polymers: Uses, current developments in the synthesis and characterization of biodegradable polyesters, blends of biodegradable polymers and recent advances in biodegradation studies, *Polymer International* 47 (2), 89–144, 1998.
60. Shogren, R., Water vapor permeability of biodegradable polymers, *Journal of Environmental Polymer Degradation* 5 (2), 91–95, 1997.
61. Kotnis, M.A., O'Brien, G.S., and Willett, J.L., Processing and mechanical properties of biodegradable poly(hydroxybutyrate-co-valerate)-starch compositions, *Journal of Environmental Polymer Degradation* 3 (2), 97–105, 1995.
62. Shogren, R.L., Poly(ethylene oxide)-coated granular starch-poly(hydroxybutyrate-co-hydroxyvalerate) composite materials, *Journal of Environmental Polymer Degradation* 3 (2), 75–80, 1995.
63. El-Hadi, A., Schnabel, R., Straube, E., Müller, G., and Henning, S., Correlation between degree of crystallinity, morphology, glass temperature, mechanical properties and biodegradation of poly(3-hydroxyalkanoate) PHAs and their blends, *Polymer Testing* 21 (6), 665–674, 2002.
64. Parikh, M., Gross, R.A., and McCarthy, S.P., The influence of injection molding conditions on biodegradable polymers, *Journal of Injection Molding Technology* 2 (1), 30, 1998.
65. Dos Santos Rosa, D., Calil, M.R., Fassina Guedes, C.d.G., and Rodrigues, T.C., Biodegradability of thermally aged PHB, PHB-V, and PCL in soil compostage, *Journal of Polymers and the Environment* 12 (4), 239–245, 2004.

66. Chiellini, E. and Solaro, R., Biodegradable polymeric materials, *Advanced Materials* 8 (4), 305–313, 1996.
67. Velho, L. and Velho, P., 2006. Innovation in resource-based technology clusters: Investigating the lateral migration thesis—The development of a sugar-based plastic in Brazil. Report of the Centre for Poverty, Employment and Growth. Published by the Human Science Research Council, Pretoria (South Africa), February 2006.
68. Noda, I., Green, P.R., Satkowski, M.M., and Schechtman, L.A., Preparation and properties of a novel class of polyhydroxyalkanoate copolymers, *Biomacromolecules* 6 (2), 580–586, 2005.
69. Philip, S., Keshavarz, T., and Roy, I., Polyhydroxyalkanoates: Biodegradable polymers with a range of applications, *Journal of Chemical Technology and Biotechnology* 82 (3), 233–247, 2007.
70. Williams, S.F., Martin, D.P., Horowitz, D.M., and Peoples, O.P., PHA applications: addressing the price performance issue I. Tissue engineering, *International Journal of Biological Macromolecules* 25 (1–3), 111–121, 1999.
71. Maiti, P., Batt, C.A., and Giannelis, E.P., Renewable plastics: Synthesis and properties of PHB nanocomposites, *Polymeric Materials Science and Engineering* 88, 58–59, 2003.
72. Hablot, E., Bordes, P., Pollet, E., and Avérous, L., Thermal and thermo-mechanical degradation of PHB-based multiphase systems, *Polymer Degradation and Stability* 93 (2), 413–421, 2008.
73. Maiti, P., Batt, C.A., and Giannelis, E.P., New biodegradable polyhydroxybutyrate/layered silicate nanocomposites, *Biomacromolecules* 8 (11), 3393–3400, 2007.
74. Lim, S.T., Hyun, Y.H., Lee, C.H., and Choi, H.J., Preparation and characterization of microbial biodegradable poly(3-hydroxybutyrate)/organoclay nanocomposite, *Journal of Materials Science Letters* 22 (4), 299–302, 2003.
75. Choi, W.M., Kim, T.W., Park, O.O., Chang, Y.K., and Lee, J.W., Preparation and characterization of poly(hydroxybutyrate-co-hydroxyvalerate)-organoclay nanocomposites, *Journal of Applied Polymer Science* 90 (2), 525–529, 2003.
76. Wang, S., Song, C., Chen, G., Guo, T., Liu, J., Zhang, B., and Takeuchi, S., Characteristics and biodegradation properties of poly(3-hydroxybutyrate-co-3-hydroxyvalerate)/organophilic montmorillonite (PHBV/OMMT) nanocomposite, *Polymer Degradation and Stability* 87 (1), 69–76, 2005.
77. Chen, G.X., Hao, G.J., Guo, T.Y., Song, M.D., and Zhang, B.H., Crystallization kinetics of poly(3-hydroxybutyrate-co-3-hydroxyvalerate)/clay nanocomposites, *Journal of Applied Polymer Science* 93 (2), 655–661, 2004.
78. Chen, G.X., Hao, G.J., Guo, T.Y., Song, M.D., and Zhang, B.H., Structure and mechanical properties of poly(3-hydroxybutyrate-co-3-hydroxyvalerate) (PHBV)/clay nanocomposites, *Journal of Materials Science Letters* 21 (20), 1587–1589, 2002.
79. Misra, M., Desai, S.M., Mohanty, A.K., and Drzal, L.T., Novel solvent-free method for functionalization of polyhydroxyalkanoates: Synthesis and characterizations, in *Proceedings of the 62nd Annual Technical Conference of the Society of Plastics Engineers*. Vol. 2, 2442–2446, 2004.
80. Drzal, L.T., Misra, M., and Mohanty, A.K., Sustainable biodegradable green nanocomposites from bacterial bioplastic for automotive applications, in *U.S. EPA STAR Progress Review Workshop—Nanotechnology and the Environment II*, Report number EPA/600/R-05/089; pp. 23–25, 2005.

81. Bordes, P., Pollet, E., Bourbigot, S., and Avérous, L., Structure and properties of PHA/clay nano-biocomposites prepared by melt intercalation, *Macromolecular Chemistry and Physics*, 209 (14), 1473–1484, 2008.

82. Pluta, M., Galeski, A., Alexandre, M., Paul, M.-A., and Dubois, P., Polylactide/montmorillonite nanocomposites and microcomposites prepared by melt blending: Structure and some physical properties, *Journal of Applied Polymer Science* 86 (6), 1497–1506, 2002.

83. Bourbigot, S., Vanderhart, D.L., Gilman, J.W., Awad, W.H., Davis, R.D., Morgan, A.B., and Wilkie, C.A., Investigation of nanodispersion in polystyrene-montmorillonite nanocomposites by solid-state NMR, *Journal of Polymer Science—Part B Polymer Physics* 41 (24), 3188, 2003.

84. Vanderhart, D.L., Asano, A., and Gilman, J.W., Solid-state NMR investigation of paramagnetic nylon-6 clay nanocomposites. 1. Crystallinity, morphology, and the direct influence of Fe3+ on nuclear spins, *Chemistry of Materials* 13 (10), 3781–3795, 2001.

85. Vanderhart, D.L., Asano, A., and Gilman, J.W., NMR measurements related to clay-dispersion quality and organic-modifier stability in nylon-6/clay nanocomposites, *Macromolecules* 34 (12), 3819–3822, 2001.

86. Vanderhart, D.L., Asano, A., and Gilman, J.W., Solid-state NMR investigation of paramagnetic nylon-6 clay nanocomposites. 2. Measurement of clay dispersion, crystal stratification, and stability of organic modifiers, *Chemistry of Materials* 13 (10), 3796–3809, 2001.

87. Bourbigot, S., Vanderhart, D.L., Gilman, J.W., Bellayer, S., Stretz, H., and Paul, D.R., Solid state NMR characterization and flammability of styrene-acrylonitrile copolymer montmorillonite nanocomposite, *Polymer* 45 (22), 7627–7638, 2004.

88. Bourbigot, S., Fontaine, G., Bellayer, S., and Delobel, R., Processing and nanodispersion: A quantitative approach for polilactide nanocomposite, *Polymer Testing* 27 (1), 2–10, 2008.

89. Xie, W., Gao, Z., Pan, W.-P., Hunter, D., Singh, A., and Vaia, R., Thermal degradation chemistry of alkyl quaternary ammonium montmorillonite, *Chemistry of Materials* 13 (9), 2979–2990, 2001.

90. Pandey, J.K., Kumar, A.P., Misra, M., Mohanty, A.K., Drzal, L.T., and Singh, R.P., Recent advances in biodegradable nanocomposites, *Journal of Nanoscience and Nanotechnology* 5 (4), 497–526, 2005.

91. Bordes, P., Hablot, E., Pollet, E., and Avérous, L., Effect of clay organomodifiers on polyhydroxyalkanoates degradation, *Polymer Degradation and Stability*, 94 (5), 789–796, 2009.

92. Liu, W.J., Yang, H.L., Wang, Z., Dong, L.S., and Liu, J.J., Effect of nucleating agents on the crystallization of poly(3-hydroxybutyrate-co-3-hydroxyvalerate), *Journal of Applied Polymer Science* 86 (9), 2145–2152, 2002.

93. Nam, J.Y., Sinha Ray, S., and Okamoto, M., Crystallization behavior and morphology of biodegradable polylactide/layered silicate nanocomposite, *Macromolecules* 36 (19), 7126–7131, 2003.

94. Di Maio, E., Iannace, S., Sorrentino, L., and Nicolais, L., Isothermal crystallization in PCL/clay nanocomposites investigated with thermal and rheometric methods, *Polymer* 45 (26), 8893–8900, 2004.

95. Chivrac, F., Pollet, E., and Averous, L., Nonisothermal crystallization behavior of poly(butylene adipate-co-terephthalate)/clay nano-biocomposites, *Journal of Polymer Science, Part B: Polymer Physics* 45 (13), 1503–1510, 2007.

96. Renstad, R., Karlsson, S., Albertsson, A.-C., Werner, P.-E., and Westdahl, M., Influence of processing parameters on the mass crystallinity of poly(3-hydroxy-butyrate-co-3-hydroxyvalerate), *Polymer International* 43 (3), 201–209, 1997.

97. Fujita, M., Sawayanagi, T., Tanaka, T., Iwata, T., Abe, H., Doi, Y., Ito, K., and Fujisawa, T., Synchrotron SAXS and WAXS studies on changes in structural and thermal properties of poly[(R)-3-hydroxybutyrate] single crystals during heating, *Macromolecular Rapid Communications* 26 (9), 678–683, 2005.

98. Hsu, S.-F., Wu, T.-M., and Liao, C.-S., Nonisothermal crystallization behavior and crystalline structure of poly(3-hydroxybutyrate)/layered double hydroxide nanocomposites, *Journal of Polymer Science, Part B: Polymer Physics* 45 (9), 995–1002, 2007.

99. Hsu, S.-F., Wu, T.-M., and Liao, C.-S., Isothermal crystallization kinetics of poly(3-hydroxybutyrate)/layered double hydroxide nanocomposites, *Journal of Polymer Science, Part B: Polymer Physics* 44 (23), 3337–3347, 2006.

100. Dufresne, A., Kellerhals, M.B., and Witholt, B., Transcrystallization in Mcl-PHAs/cellulose whiskers composites, *Macromolecules* 32 (22), 7396–7401, 1999.

101. Dubief, D., Samain, E., and Dufresne, A., Polysaccharide microcrystals reinforced amorphous poly(β-hydroxyoctanoate) nanocomposite materials, *Macromolecules* 32 (18), 5765–5771, 1999.

102. Dufresne, A., Dynamic mechanical analysis of the interphase in bacterial polyester/cellulose whiskers natural composites, *Composite Interfaces* 7 (1), 53–67, 2000.

103. Chen, D.Z., Tang, C.Y., Chan, K.C., Tsui, C.P., Yu, P.H.F., Leung, M.C.P., and Uskokovic, P.S., Dynamic mechanical properties and *in vitro* bioactivity of PHBHV/HA nanocomposite, *Composites Science and Technology* 67 (7–8), 1617–1626, 2007.

104. Dufresne, A. and Samain, E., Preparation and characterization of a poly(β-hydroxyoctanoate) latex produced by *Pseudomonas oleovorans, Macromolecules* 31 (19), 6426–6433, 1998.

105. Khanna, S. and Srivastava, A.K., Recent advances in microbial polyhydroxyal-kanoates, *Process Biochemistry* 40 (2), 607–619, 2005.

106. Lagaly, G., From clay mineral crystals to colloidal clay mineral dispersions, in *Coagulation and flocculation—Theory and applications*, Dobias, B. Dekker, New York, 1993, pp. 427.

9

The Reductionist Approach to the Molecular and Supramolecular Structures of Elastin

Antonio Mario Tamburro, Brigida Bochicchio, and Antonietta Pepe

Università della Basilicat, Italy

CONTENTS

9.1 Molecular Structures

Elastin is the protein responsible for the elasticity of tissues and organs such as lungs, skin, and arteries. Its insolubility is due to the presence of cross-links responsible for its resistance to fracture and rupture. Due to the extreme insolubility of elastin in common solvents, the use of classical spectroscopic techniques is precluded. For these reasons, past studies were confined to the use of soluble derivatives, such as α-elastin and κ-elastin, and to short synthetic peptides corresponding to repeating sequences. In some cases, peptides containing polymeric repeats of variable distribution of molecular weight were also studied [1,2].

On that basis, two research groups, first that of Urry in the United States and then that of Tamburro in Italy, started with the chemical synthesis of polypeptides including $(VPGVG)_n$ [3,4], $(VPGG)_n$ [5], $(VGVAPG)_n$ [6], and $(VPG)_n$ [7] sequences and on the other hand (VGG)n, (LGG)n [8], $(VGGVG)_n$ [9], $(VGGLG)_n$ [10], and $(LGGVG)_n$ [11] sequences.

The reason for studying these polymers was that they basically could, and indeed they were shown to be, very useful *models* for the entire protein structure. The rationale here is based on a "simple" assumption: the entire protein could be considered, "prima facie," as a repeating polypeptide. Obviously, one must be careful with this approach, because many features of the structure could be lost. However, by being cautious, the results may be, and indeed were, very effective.

Accordingly, two models for the elastin structure were developed [12,13], different in many aspects, but nevertheless sharing some others. The common point of view is the presence in both models of (mainly) type II β-turns. In the Urry model, essentially built up for the polymer poly(VPGVG), there is a regular array of type II β-turns having the sequence PG at the corners [14]. This helical structure, the β-spiral, has to be considered a dynamic structure undergoing librational motions in the segment GV, outside the β-turns [15].

On its turn, this motion causes an increase of entropy of the whole system, so explaining the elasticity of poly(VPGVG) and, by analogy, of elastin itself [16]. On applying a stretching force, a damping of librations will be produced and, therefore, a decrease of entropy, which will spontaneously increase by removal of the force. Although "librational" motions are undoubtedly present in some sequences of elastin, also according to our results, some crucial points argue against the β-spiral architecture.

First, the β-spiral has been put forward on the basis of the results obtained for poly(VPGVG), and very recently spectroscopic results showed that poly(VPGVG) does not adopt a β-spiral structure [17].

Molecular dynamics simulation pointed out that the β-spiral is a rather unstable, transient structure [18].

The work of Tamburro's group was based on glycine-rich repeating sequences of elastin such as XGG, GXGGZ [19,20]. Nonrecurring, isolated type II β-turns were proposed for GXGGZ repeating sequences (X, Z = V, L). These have XG or GG segments at the corner with 4 → 1 hydrogen bonds connecting the first and the fourth glycine or the X and Z residue. The turns are rather labile and, therefore, can interconvert each other, giving rise to dynamic β-turns sliding along the chain (Figure 9.1). The experimental support to this model came essentially from the analysis of NMR (nuclear magnetic resonance) data of pentapeptides, decapeptides, and polypentapeptides, containing the above-mentioned sequence GXGGZ [10,21]. In particular, the analysis of the temperature coefficients of the peptide amide resonances provided the most significant clue. The chemical shifts of the amide protons depend on the temperature, moving to higher fields as the temperature increases. The change of the chemical shift with temperature is reduced when the amide proton is involved in a hydrogen bond. As a consequence, the temperature coefficient is a measure of the stability of the hydrogen bonds.

Specifically, the occurrence of low temperature coefficients (<–4.0 ppb/K) for the amide protons is usually attributed to stable hydrogen bonds. In the case of β-turns, defined by four consecutive residues in the peptide sequence, the lowered temperature coefficient is due to the presence of a hydrogen bond between the amide proton of the fourth residue and the carbonyl group of the first residue of the turn. A value for the temperature coefficient in the

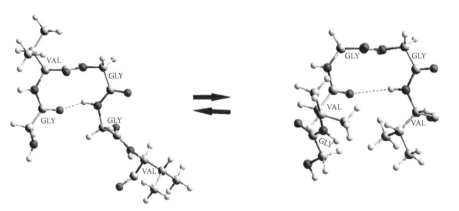

FIGURE 9.1
Molecular models describing the sliding β-turns in the elastin sequence –GVGGV–. The turn with –VG– at the corners (on the left) is in dynamic equilibrium "sliding" with the turn having –GG– at the corners (on the right).

range of 4.0 to 5.0 is usually attributed to a very weak H-bond or to an equilibrium between turn and other conformations devoid of H-bonds.

In the case of the $G^1X^2G^3G^4Z^5$ pentapeptide, temperature coefficients in the range of 4.5 to 4.8 ppb/K found for two sequential amide protons, G^4 and Z^5, were explained on the basis of an equilibrium involving two type II β-turns:

$$G^1 \, CO\cdots HN \, G^4 \rightleftarrows X^2 \, CO\cdots HN \, Z^5$$

inferring that the two β-turns were "sliding" along the polypeptide chain.

Finally, the fact that other residues, such as G^3, also exhibit moderately low temperature coefficients ($\Delta\delta/\Delta T = -4.6$) reinforces the idea of the presence of a complex, flexible pattern of labile hydrogen bonds interconverting their positions along the sequence and giving rise to transient equilibrium between folded and unfolded forms.

The sliding should give rise to an increase of the entropy of the chain, so contributing to the elasticity of the protein. The "sliding β-turn model" hypothesized for highly repetitive elastin polypeptides was further confirmed also for the hydrophobic domains of tropoelastin [22] (see below). Our results gave both a detailed, at atomic resolution, description of elastin structure and a self-consistent molecular mechanism of elastin elasticity [23]. An important point to be emphasized is that according to both NMR data and MD (molecular dynamics) simulations, the β-turns are characterized by an extensive sliding; that is, they are interconverting each other along the chain. This phenomenon is considered one of the possible sources of entropy in the relaxed state of elastin [24]. Recently, analogously to what was suggested for elastin and lamprin, an equilibrium including PPII, β-turns, and unordered conformation has been invoked as a main structural feature for elastic segments of titin [25].

9.2 The Reductionist Approach

The reductionist approach has been widely used in science. It consists, as is well known, of studying a complex system by isolating each variable and then by reconstructing it by assuming that the interaction factors are not dominating. Usually, the system is constituted by different molecules. In our case, the approach relates to a single macromolecule, which is "properly" dissected and then reassembled as a puzzle. In the case of tropoelastin, the reductionist approach is based on the observation that the tropoelastin gene exhibits a cassette-like organization, with a regular alternation of cross-linking domains followed by exons encoding hydrophobic polypeptide sequences, putatively responsible for elasticity (Figure 9.2) [26].

The hydrophobic domains are responsible for the elasticity of the protein and are rich in glycines, localized in the N-terminal and C-terminal regions of the protein, and in prolines, present mainly in the larger exons localized in the central part of the molecule. The cross-linking domains, alternating with the hydrophobic domains, consist of those rich in alanine (KA domains, central, and C-terminal region), and those containing proline (KP domains, confined principally to the N-terminal region of the molecule). Three other regions of the protein have unique compositions and functions: exon 1, which encodes a 26 amino acid signal peptide for extracellular transportation; exon

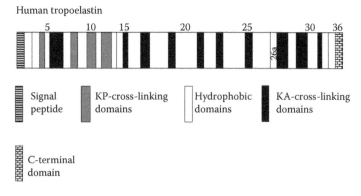

FIGURE 9.2
The domain structure encoded by the human tropoelastin gene. Human tropoelastin lacks exons 34 and 35, present in other mammalian genes. The exons 10, 11, 13, 14, 15, 20, 22, 23, 24, 25, 26, 27, 32, and 33 are alternatively spliced. (Adapted from Tamburro, A.M., Bochicchio, B., Pepe, A. 2003. *Biochemistry* 42(45): 13347–13362. Copyright 2003, American Chemical Society. With permission.)

26A, a hydrophilic exon that is rarely expressed whose function is unknown; and exon 36, which encodes a sequence responsible for interaction with other elastic fiber proteins [27].

The peculiar domain structure of tropoelastin prompted us to study the isolated domains encoded by the exons of tropoelastin, with the perspective of getting deep insights into the structural properties of the whole protein. The hypothesis is essentially based on the assumption that isolated domains of the protein—corresponding to the sequences coded by single exons—adopt a structure essentially similar to that they exhibit in the entire protein and have distinct structures independent of the neighboring domains. This approach is unfeasible for globular proteins, where different structural domains are mutually influenced. Furthermore, elastin has been demonstrated to be a fractal protein [28,29], which means that the protein is characterized by the property of statistical self-similarity. Accordingly, even very short sequences show molecular and supramolecular features similar to those of the whole protein [30,31]. This finding is most likely related to the abundance of repeating sequences of elastin, the repetition being found at different scales along the entire sequence of the molecule [32]. Previous results showed that other biological properties of the molecule, such as chemotaxis [33,34], vasodilatory activity [35], and upregulation of metalloproteinase expression [36] are also confined to the sequences coded by particular exons.

The reductionist approach has been applied by accomplishing the exon-by-exon full chemical synthesis of human tropoelastin and carrying out a complete conformational study on the synthetic polypeptides [22,37,38].

Different complementary spectroscopic techniques, circular dichroism (CD), Fourier transform infrared spectroscopy (FTIR), and Raman spectroscopies, were employed for the conformational studies of the polypeptides. Far-ultraviolet (UV) CD spectroscopy is a common technique in the conformational studies of peptides and proteins [39] and is based on the observation that different secondary structures show different and peculiar CD spectra. α-helices as well as β-sheets are easily identified by CD spectroscopy, as well as some secondary structures such as polyproline II (PP II). CD is the main spectroscopic technique that unambiguously reveals the presence of this conformation [40,41]. Also, some kinds of β-turns display well-defined CD spectra [42].

The conformational analysis of the different elastin domains was performed in a qualitative way, by comparing the curves with CD spectra of peptides of known secondary structures or of peptides adopting a mixture of known structures [43]. In particular, the analysis of the spectral features of the peptides as a function of solvents and temperature was employed to highlight the presence of different conformations in solution.

CD spectra analysis is more complex when different conformations are present or interconverting in highly flexible peptides. In that case, the CD spectra correspond to a linear combination of all the contributions due to different secondary structures. Also, the presence of chromophores other than amide peptide bonds can determine difficulties in the interpretation of the CD spectra of peptides. Nevertheless, this technique can reveal, by using very small amounts of sample, the preferred conformations adopted by the peptides.

Differently from CD spectroscopy, NMR requires higher amounts of sample for the analysis. In the case of NMR, great caution must be used in interpreting the data, when different conformations in rapid equilibrium are suggested [44]. As a matter of fact, the "slow" NMR time scale is not able to discriminate the contributions of the different conformations, and as a consequence, all the NMR parameters used for structural identification (i.e., chemical shift, coupling constants, temperature coefficients and nuclear Overhauser effect [NOEs]) are time- and ensemble-averaged values. However, NMR spectroscopy is able to give the average preferred secondary structure of a polypeptide sequence. For example, conformational studies on exon-5–coded domain indicate that the peptide has a high propensity to adopt short stretches of PPII structure, a conformation rather ubiquitous in the hydrophobic, putatively elastomeric sequences of elastin and recently suggested to play an important role both in the elasticity of elastin and in the self-assembly of extracellular matrix proteins such as elastin and lamprin [41,45,46]. On the other hand, β-turns were identified in 2,2,2-trifluoroethanol (TFE), a less polar solvent. Overall, the results obtained provided clear evidence for elastin-derived polypeptides large flexibility of the polypeptide chain that oscillates between rather extended conformations, such as PPII, and folded ones, such as β-turns [47]. Further support to this interpretation came from the results of MD simulations showing a quick interconversion between extended (mainly PPII) and folded (β-turns) conformations.

Summing up all CD and NMR data, we can infer that in aqueous solution, the dominant conformation is the PPII often in equilibrium with the "unordered" state. However, *in vivo*, the microenvironment that determines the conformation of the peptide chain is not known *a priori* and can be different from the bulk macroscopic solution conditions. Furthermore, the elastin's hydrophobicity and its highly cross-linked nature suggest a less polar internal environment than that of the surrounding solvent (i.e., aqueous solution).

For this reason, the spectra were acquired in both water and 2,2,2-trifluoroethanol (TFE) which is a less polar solvent than water. In TFE we found folded conformations, such as type I and type II β-turns, together with unordered conformations. Finally, we suggest that extended and folded structures are labile; that is, they are dynamically interchanging among themselves and also, as in the case of β-turns, sliding along the chain (Figure 9.1), as confirmed by MD simulations [20,48]. The rationale for proposing for elastin a general equilibrium

folded ⇌ extended

lies in the idea that the *in vivo* microenvironment conditions might be intermediate between water and TFE.

In regard to the regions containing lysines and polyalanine sequences (KA domains), the α-helix structure was found to dominate. On the contrary, the KP domains adopt either PPII or β-turns like the hydrophobic sequences [22,38].

In the solid state, the spectroscopies used to characterize the synthesized polypeptides were mainly Raman and FTIR. Although both are essentially based on the vibrational modes of polypeptides whose frequencies depend significantly on secondary structure, when used together they give additional and complementary information. The amide I band consists of amide carbonyl C=O stretching, with smaller contributions of C–N stretching and N–H bending. The amide II and III bands involve significant C–N stretching and N–H bending. EDP 3, 7, and 30 were also studied by Raman spectroscopy that evidenced in the solid state a specific sharp band at 1314 cm^{-1}, assigned to the PPII structure, and thus confirming the data obtained by CD spectroscopy in solution [49]. Furthermore, Raman spectra revealed for the EDPs studied the presence of α-helical conformations and β-strands although in higher amounts. The rationale for this finding is that in the solid state, PPII helices are easily converted into α-helices because "noncooperative PPII helices are obtained by minimization of chain packing density, whereas the α-helix is a compact structure that results from maximizing chain packing density due to the formation of cooperative attractive intra-chain hydrogen bonds and van der Waals contacts" [50]. Additionally, the conversion of PPII helix to β-strand is common to other proteins [51], this conformational transition being due to the increase of temperature or by the absence of water because of a difference in solvation between PPII and β-structures.

These considerations confirm the interpretation of the brittleness of elastin in the absence of water in terms of an increase of β-strands/β-structures as supported by earlier suggestions and findings [52,53]. Analogously, FTIR solid-state measurements made by Tamburro and coworkers on EDP 28 and 30 confirmed the presence of conformational equilibria between PPII and β-strands [23,54].

All the results, taken together, allowed the rising of a picture, shown in Figure 9.3, describing the reassembly of the whole molecule starting from the constituent domains, the assembly of the elastin "puzzle."

9.3 Supramolecular Organization of Elastin

The supramolecular studies on tropoelastin domains were based essentially on the examination of two aspects: (1) the capability of tropoelastin domains

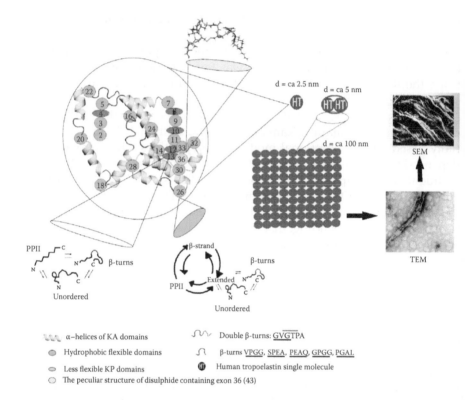

FIGURE 9.3
An idealized model of human tropoelastin molecular and supramolecular structures. The micrographs are referred to as *ligamentum nuchae* elastin. (Adapted from Tamburro, A.M., Pepe, A., Bochicchio, B. 2006. *Biochemistry*, 45, 9518. Copyright 2006, American Chemical Society. With permission)

to self-assemble, investigated by turbidimetry; and (2) the analysis of the morphology of the self-assembled peptides by microscopy techniques.

9.3.1 Self-Assembly Properties

On the basis of the "reductionist" approach cited above, we synthesized the sequences encoded by all exons of human tropoelastin [22] and analyzed their self-assembly properties. The polypeptide sequences studied are listed in Table 9.1.

The polypeptides belong to different regions of tropoelastin. So, EDP5 and EDP30 belong to the glycine-rich hydrophobic domains, responsible for elasticity as they contain more than 45% glycines. EDP19 belongs to the KA cross-linking domains characterized by a high content of alanine and lysine residues and by being responsible for the formation of cross-links. Finally, EDP 18, 20, 24, and 26 belong to the larger proline-rich hydrophobic domains also putatively responsible for the elasticity of the protein. Moreover, EDP24

TABLE 9.1

Sequences and Molecular Weights of the Elastin-Derived Peptides (EDP)

Peptide	Sequence	MW (Da)
EDP5	PGGLAGAGLGA	839.9
EDP18	GAAAGLVPGGPGFGPGVVGVPGAGVPGVGVPGAGIPVVPGAGIPGAAVP	4005.6
EDP19	GVVSPEAAAKAAAKAAKY	1702.9
EDP20	GARPGVGVGGIPTYGVGAGGFPGFGVGVGGIPGVAGVPSVGGVPGVGGVPGVGIS	4641.3
EDP24	GLVPGVGVAPGVGVAPGVGVAPGVGLAPGVGVAPGVGVAPGVGVAPGI	3932.6
EDP24s[a]	APAPVVAVVGAAAVVGGVGGGGGGPVVGGVGGPVGLPLPIPVPVVA	3932.6
EDP26	RAAAGLGAGIPGLGVGVGVPGLGVGAGVPGLGVGAGVPGFGA	3437.0
EDP26s[b]	GVGVGGAGGVAVGIGGAALRPAPLGPLGVLGGVAGGPAVGFG	3437.0
EDP28	GAAVPGVLGGLGALGGVGIPGGVV	1944.3
EDP30	GLVGAAGLGGLGVGGLGVPGVGGLG	1961.0

[a] Scrambled EDP24.
[b] Scrambled EDP26.

contains the VGVAPG motif, which is the most striking repeated unit present in the human tropoelastin, and EDP26 contains a repeated "nonapeptide" sequence: VGXGVPGZG (X=V, A; Z=L,F). Among the polypeptide sequences studied, only the proline-rich domains were able to coacervate under usual conditions (i.e., EDPs 18, 20, and 24). EDP26 coacervated only at high concentrations.

9.3.2 Coacervation of Elastin

The coacervation process consists of a temperature-induced phase transition, fully reversible, that gives rise to the formation of two phases: one rich in protein, called coacervate, and the other more rich in aqueous solvent. The process is typical of tropoelastin and of some soluble derivatives of elastin as α-elastin [55] and elastin-derived polypeptides [56].

The self-assembly of tropoelastin represents a fundamental step in the maturation process of elastin and is based on the ability of tropoelastin to coacervate with partial exclusion of water and occurs at 37°C in a pH range of 7 to 8 and in physiological salt conditions [57]. This process may hold an important role in correctly aligning the domains for the formation of intermolecular cross-links [58]. Coacervation is an intrinsic feature of tropoelastin, which strongly depends on the contributions from the individual hydrophobic domains and on the context of their position in the intact tropoelastin molecules. As a consequence, the identification of the role played by particular domains of tropoelastin in the self-assembly of the protein is of paramount importance.

Coacervation studies performed on isolated domains of tropoelastin showed that some polypeptides are able to coacervate in a manner similar, although not identical, to the entire protein [59]. In particular, proline-rich hydrophobic domains encoded by exon 18, 20, 24, 26 of the tropoelastin gene are able to self-assemble by coacervation. These polypeptides, made of about 40 to 54 amino acids, despite their reduced molecular weight related to the whole protein, show on heating the same reversible phenomenon of phase separation, indicating that the coacervation process is "intrinsic" to the hydrophobic domains.

Among these, elastin-derived polypeptide (EDP) 20, the largest peptide constituted by 54 amino acids, had the lowest critical temperature (Tc). The scrambled EDP24 and EDP26 having the same amino acid composition and, therefore, the same hydrophobic index, with respect to the parent peptides, were poorly soluble in the coacervation buffer and so were unable to reversibly coacervate. In order to verify the influence of a less polar solution condition in the coacervation process, the same experiments were performed in 7.5% trifluoroethanol (TFE) buffer solution. Table 9.2 reports the critical temperatures of the thermal transitions (Tc) of EDPs 18, 20, 24, and 26 either in the presence or in the absence of 7.5% TFE.

TABLE 9.2

Coacervation Temperatures at the Midpoint of
Transition (Tc) of Some Elastin Domains in the
Absence (Solution A, c.1 mM) and in the Presence
(Solution B, c. 1 mM) of 7.5% 2,2,2-trifluoroethanol
(TFE)

Peptide	Tc in A(°C)	Tc in B (°C)	ΔTc (°C)
EDP18	61	21	40
EDP20	44	29	15
EDP24	70	28	42
EDP26	53[a]	19	34

[a] c, 3 mM.

TFE significantly decreased their coacervation temperatures and increased the absorbance of samples. This suggests a facilitated phase transition of EDPs in the presence of TFE, as well as an increased dimension of the aggregates associated with coacervation. On the other hand, CD studies carried out on these solutions (data not shown) indicated the absence of conformational changes of peptide in the presence of TFE. In order to explain this behavior, Tamburro and collaborators have suggested [60] that in a mixture of water and TFE, TFE creates a layer onto the peptide surface, and thus displacing the water molecules. Therefore, the interactions water–peptide are weakened and substituted for by those peptide–peptide that favor the self-assembly.

On the contrary, EDP 30 and EDP28 did not coacervate and gave rise to an irreversible precipitation on increasing the temperature.

9.3.3 Studies by Microscopy Techniques

The ultrastructure of elastin has been extensively analyzed by different methodologies. Starting from the first descriptions, where elastin was depicted as an amorphous structure, more complex and, in some cases, variegated morphologies were revealed [61,62].

The ability to self-assemble of some isolated domains of tropoelastin was investigated by means of microscopy techniques, comparing the ultrastructure of the polypeptides with those observed for the whole tropoelastin molecule, in order to potentially identify the regions of tropoelastin responsible for its peculiar aggregation pattern [60].

The pioneering works of Gotte and coworkers [61,62] have shown that elastin and tropoelastin analyzed by transmission electron microscopy (TEM) revealed a common filamentous structure characterized by bundles of 3 to 5 nm diameter fibrils (Figure 9.3). Banded fibers were also obtained for tropoelastin [63]. The same supramolecular structures were observed for some elastin-related polypeptides [3].

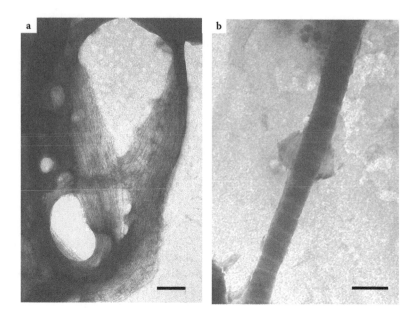

FIGURE 9.4
Transmission electron micrographs of exon 20 coded domain of human tropoelastin nega-
tively stained with uranyl acetate and incubated at 50°C for 24 h. (a) Bundles of 4 to 6 nm thick
filaments, bar: 250 nm; (b) banded fibers, bar: 100 nm. (Reprinted from Pepe, A., Guerra, D.,
Bochicchio, B., Quaglino, D., Gheduzzi, D., Pasquali Ronchetti, I., Tamburro, A.M. 2005. *Matrix
Biol.* 24(2): 96–109. With permission of Elsevier.)

Later, Tamburro and colleagues showed that exon 20 and exon 26 coded
domains formed supramolecular structures strongly resembling those
observed for elastin and tropoelastin [60].

In particular, EDP20, when observed by TEM after incubation at 50°C,
gives rise to flexible bundles of filaments, with diameter of 4 to 6 nm, and
to transverse banded fibers, too. Both structures, the filaments packed into
strictly interwoven bundles and the banded fibers, resemble those observed
for tropoelastin molecules (Figure 9.4) [60]. This strongly suggests that *in vivo*
these domains might play a pivotal role in the molecular assembly of tropoelastin molecules leading to the formation of elastic fibers.

Analogously, EDP26 has a great tendency to self-assemble in an ordered
manner as demonstrated by different microscopy techniques [64]. Figure 9.5a
shows the typical structural organization of EDP26 peptide after 24 h incubation at 35°C. At high magnification, these bundles were formed by densely
packed filaments, about 5 nm thick. On increasing the temperature to 50°C,
the number of bundles of filaments increased for incubation times up to 48 h
and to 10 days. Typical examples of these bundles are shown in Figure 9.5b.
Bundles were relatively long and displayed variable thickness within a given
sample. At high magnification, the bundles consisted of slightly undulating
5 nm thick filament (Figure 9.5b, arrows).

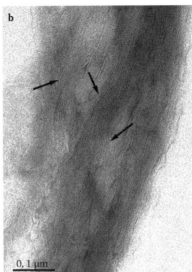

FIGURE 9.5
Transmission electron micrograph of EX26 incubated at 35°C for 24 h showing bundles consisting of densely packed well-aligned filaments. Figure 9.6b shows a similar filament aggregate obtained by incubation of EX26 at 50°C for 48 h. Arrows indicate that filaments appeared rather flexible. Bar: 0.1 μm. (Reprinted from Pepe, A., Flamia, R., Guerra, D., Quaglino, D., Bochicchio, B., Pasquali Ronchetti, I., Tamburro, A.M. 2008. *Matrix Biol.* 27(5): 441–450. With permission of Elsevier.)

At a micrometer (μm) scale, very long fibers as well as fractal aggregation patterns were observed. The aggregation properties are due to the peculiar sequence of EDP26, and not to its amino acid composition, as evidenced by the supramolecular analysis of a scrambled sequence of exon-26–coded domain of human tropoelastin, showing quite different aggregation patterns (data not shown).

These findings confirm that specific sequences can play a driving role in the aggregation process of tropoelastin molecule, and indicate exon-20– and exon-26–encoded domain among these sequences.

On the contrary, exon-24–coded domain, which is characterized by the most striking repeating sequence (VGVAPG), aggregates in rather rigid and relatively short filaments with low tendency to aggregate into ordered bundles (Figure 9.6). Parenthetically, the repeated sequence VGVAPG is important for remodeling the extracellular matrix, during wound healing, aging, and some pathological states [65,66]. In addition, it is well known that the released hexapeptide has strong chemotactic, cell-proliferation, and other bioactivities related to tissue repairing and healing and remodeling.

The polypeptide corresponding to the sequence encoded by exon 30 of human [37] and bovine [67] tropoelastin gene aggregated in very long and rather flexible filaments. This 25-residue long polypeptide aggregates in

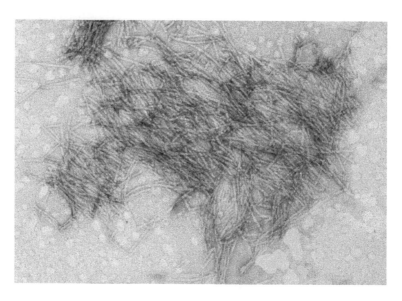

FIGURE 9.6
Transmission electron micrograph of exon 24 coded domain of human tropoelastin negatively stained with uranyl acetate and incubated at 50°C for 24 h. Bar: 500 nm. (Reprinted from Pepe, A., Guerra, D., Bochicchio, B., Quaglino, D., Gheduzzi, D., Pasquali Ronchetti, I., Tamburro, A.M. 2005. *Matrix Biol.* 24(2): 96–109. With permission of Elsevier.)

an oriented manner, forming precise entities with high tendency to grow in length and with low tendency to stick together. The filaments formed underwent a time-dependent reorganization into twisted-rope supramolecular structures. The helical filaments observed were strikingly similar to amyloid fibers (Figure 9.7). The amyloidogenic nature of this domain was further confirmed by spectroscopy techniques and by atomic force microscopy (AFM) [68]. Analogously, the sequence encoded by exon 28 of the human tropoelastin showed at a supramolecular level, as evidenced by AFM, a very big left-handed super-helix, 100 μm long, together with aggregates of different sizes, some of them being constituted by helically interwoven fibers [69].

The emerging picture clearly shows that some domains of human tropoelastin are able to form amyloid-like fibers. A suggestion has been put forward about the biological relevance of amyloid deposition of elastin fragments [68,70]. The presence of aggregates in elastic tissue under different pathological conditions is documented, and these are generically called "elastotic" material. In the case of acute interstitial lung disease, it has been shown that the elastin aggregates are amyloid-like in nature [71]. Being the deposition of elastotic material in aged, atherosclerotic blood vessels usually accompanied by an increased degradation of elastin by several proteases, we suggest that under certain pathological conditions, soluble peptides released from elastin by proteolytic degradation could

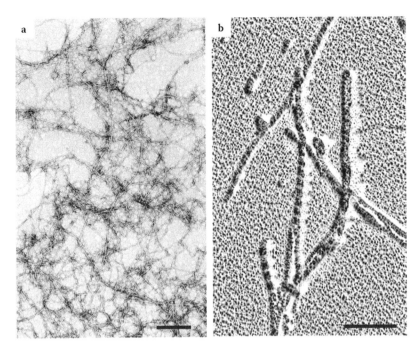

FIGURE 9.7
Transmission electron micrographs of exon 30 coded domain incubated at 50°C. (a) Network of amyloid-like filaments, negative staining, bar: 500 nm; and (b) twisted roped filaments, observed after incubation at 50°C for times longer than 48 h, rotary shadowing, bar: 500 nm. (Reprinted from Pepe, A., Guerra, D., Bochicchio, B., Quaglino, D., Gheduzzi, D., Pasquali Ronchetti, I., Tamburro, A.M. 2005. *Matrix Biol.* 24(2): 96–109. With permission of Elsevier.)

slowly aggregate because of the mutated microenvironment to form amyloid-like structures that explain the deposition of "elastotic material." The degradation of elastin by elastase is strongly enhanced by the presence of the unsaturated fatty acids [72] accumulated on elastic fibers in atheromathous plaques [73]. Therefore, lipids seem to have a double role (i.e., not only to favor the degradation of elastin by the proteases, but also to create a favorable environment for amyloid aggregation). Nowadays, after the proteolytic studies of Getie [74], we are aware that peptides comprising the general sequence "XGGZG" are delivered by enzymatic digestion from amyloid-forming polypeptide sequences (EX30: AGLGGLGVGGV, EX28: GGLGALGGVGIPGGVVGA), thus confirming the initial hypothesis.

On the whole, the results of the ultrastructural studies show that some domains, even if isolated, exhibit supramolecular features similar to those exhibited by the entire protein. Accordingly, these findings not only let us carefully identify which domains are able to behave as the entire molecule, but also to ascertain different functions for each of them, some of them being biologically or pathologically relevant [60,68,69].

9.4 Summary

The important conclusion to be drawn from the results described above is that in the case of a particular fibrous, not globular, protein as elastin, the reductionist approach works even at an intramolecular level. What is particularly astonishing is to find both the molecular and the supramolecular features of the entire protein are well reproduced even by single domains of elastin. The possible reasons for the described behavior could be the following:

- The specific cassette-like structure of tropoelastin gene (Figure 9.2). This renders autonomous, both functionally and structurally, the polypeptide sequences coded by the single exons. Therefore, the structure of a single domain is quite significant in terms of the whole structure of the protein, so making feasible the assembly of the structured "puzzle."

- The fractal aspects of elastin structure. Nowadays, the fractality of elastin has been experimentally demonstrated. Furthermore, it can be theoretically predicted on the basis of the extensive sequence repetition. Accordingly, one can make the synthesis of some domains, and even of some simple model polypeptides, quite representative of the physico-chemical properties of elastin. Parenthetically, this may give rise to the production of elastin-inspired polymeric biomaterials [11,75].

As a concluding remark, we note that the reductionist approach may be used in principle also with other elastomeric proteins. As a matter of fact, lamprin [45], abductin [76,77], and resilin [46], which are all characterized by extensive sequence repetition and then are presumably self-similar in terms of fractal geometry, have been preliminarily studied by using a reductionist approach. What is already clear is that the conformational space explored by those proteins is the same as that of elastin. Accordingly, a common mechanism of elasticity has been suggested [77], possibly for most elastomeric proteins, based on that widely previously investigated [24,37,38].

References

1. Foster, J.A., Bruenger, E., Gray, W.R., Sandberg, L.B. 1973. Isolation and amino acid sequences of tropoelastin peptides. *J. Biol. Chem.* 248(8): 2876–2879.
2. Sandberg, L.B., Leslie, J.G., Leach, C.T., Alvarez, V.L., Torres, A.R., Smith, D.W. 1985. Elastin covalent structure as determined by solid phase amino acid sequencing. *Pathol. Biol. (Paris)* 33(4): 266–274.

3. Urry, D.W., Long, M.M., Cox, B.A., Ohnishi, T., Mitchell, L.W., Jacobs, M. 1974. The synthetic polypentapeptide of elastin coacervates and forms filamentous aggregates. *Biochim. Biophys. Acta* 371(2): 597–602.

4. Urry, D.W. 1988. Entropic elastic processes in protein mechanisms. II. Simple (passive) and coupled (active) development of elastic forces. *J. Protein Chem.* 7: 81–114.

5. Long, M.M., Rapaka, R.S., Volpin, D., Pasquali-Ronchetti, I., Urry, D.W. 1980. Spectroscopic and electron micrographic studies on the repeat tetrapeptide of tropoelastin: (Val-Pro-Gly-Gly)n. *Arch. Biochem. Biophys.* 201(2): 445–452.

6. Urry, D.W., Onishi, T., Long, M.M., Mitchell, L.W. 1975. Studies on the conformation and interactions of elastin: Nuclear magnetic resonance of the polyhexapeptide. *Int. J. Pept. Protein Res.* 7(5): 367–378.

7. Daga Gordini, D., Guantieri, V., Tamburro, A.M. 1989. Electron microscopic evidence for elastin-like supramolecular organization in synthetic polytripeptides. *Connect. Tissue Res.* 19(1): 27–34.

8. Castiglione-Morelli, A., Scopa, A., Tamburro, A.M., Guantieri, V. 1990. Spectroscopic studies on elastin-like synthetic polypeptides. *Int. J. Biol. Macromol.* 12(6): 363–368.

9. Flamia, R., Zhdan, P.A., Martino, M., Castle, J.E., Tamburro, A.M. 2004. AFM study of the elastin-like biopolymer poly(ValGlyGlyValGly). *Biomacromolecules* 5(4): 1511–1518.

10. Tamburro, A.M., Guantieri, V., Gordini, D.D. 1992. Synthesis and structural studies of a pentapeptide sequence of elastin. Poly(Val-Gly-Gly-Leu-Gly). *J. Biomol. Struct. Dyn.* 10(3): 441–454.

11. Martino, M., Perri, T., Tamburro, A.M. 2002. Biopolymers and biomaterials based on elastomeric proteins. *Macromol. Biosci.* 2: 319–328.

12. Venkatachalam, C.M., Urry, D.W. 1981. Development of a linear helical conformation from its cyclic correlate. β-Spiral model of the elastin poly(pentapeptide) (VPGVG)n. *Macromolecules* 14: 1225–1229.

13. Tamburro, A.M., Pepe, A., Bochicchio, B., Quaglino, D., Ronchetti, I.P. 2005. Supramolecular amyloid-like assembly of the polypeptide sequence coded by exon 30 of human tropoelastin. *J. Biol. Chem.* 280(4): 2682–2690.

14. Urry, D.W. 1983. What is elastin; what is not. *Ultrastruct. Pathol.* 4(2–3): 227–251.

15. Urry, D.W., Henze, R., Redington, P., Long, M.M., Prasad, K.U. 1985. Temperature dependence of dielectric relaxations in alpha-elastin coacervate: Evidence for a peptide librational mode. *Biochem. Biophys. Res. Commun.* 128(2): 1000–1006.

16. Urry, D.W. 1991. Thermally driven self-assembly, molecular structuring and entropic mechanisms in elastomeric polypeptides. *Molecular Conformation and Biological Interactions.* P. Balaram and S. Ramaseshan. Bangalore, India, Indian Academy of Science: 555–583.

17. Gross, P.C., Possart, W., Zeppezauer, M. 2003. An alternative structure model for the polypentapeptide in elastin. *Z. Naturforsch* 58: 873–878.

18. Li, B., Daggett, V. 2002. Molecular basis for the extensibility of elastin. *J. Muscle Res. Cell Motil.* 23(5–6): 561–573.

19. Tamburro, A.M., Guantieri, V., Pandolfo, L., Scopa, A. 1990. Synthetic fragments and analogues of elastin. II. Conformational studies. *Biopolymers* 29(4–5): 855–870.

20. Lelj, F., Tamburro, A.M., Villani, V., Grimaldi, P., Guantieri, V. 1992. Molecular dynamics study of the conformational behavior of a representative elastin building block: Boc-Gly-Val-Gly-Gly-Leu-OMe. *Biopolymers* 32(2): 161–172.
21. Tamburro, A.M. 1990. Order-disorder in the structure of elastin: A synthetic approach. *Elastin: Chemical and Biological Aspects.* A.M. Tamburro and J.M. Davidson. Potenza, Galatina Congedo Editore: 127–145.
22. Tamburro, A.M., Bochicchio, B., Pepe, A. 2003. Dissection of human tropoelastin: Exon-by-exon chemical synthesis and related conformational studies. *Biochemistry* 42(45): 13347–13362.
23. Tamburro, A.M., Bochicchio, B., Pepe, A. 2005. The dissection of human tropoelastin: From the molecular structure to the self-assembly to the elasticity mechanism. *Pathol. Biol. (Paris)* 53(7): 383–389.
24. Debelle, L., Tamburro, A.M. 1999. Elastin: molecular description and function. *Int. J. Biochem. Cell Biol.* 31(2): 261–272.
25. Ma, K., Wang, K. 2003. Malleable conformation of the elastic PEVK segment of titin: Non-co-operative interconversion of polyproline II helix, beta-turn and unordered structures. *Biochem. J.* 374(Pt. 3): 687–695.
26. Rosenbloom, J., Abrams, W.R., Mecham, R. 1993. Extracellular matrix 4: The elastic fiber. *FASEB J.* 7(13): 1208–1218.
27. Floquet, N., Pepe, A., Dauchez, M., Bochicchio, B., Tamburro, A.M., Alix, A.J. 2005. Structure and modeling studies of the carboxy-terminus region of human tropoelastin. *Matrix Biol.* 24(4): 271–282.
28. Tamburro, A.M., De Stradis, A., D'Alessio, L. 1995. Fractal aspects of elastin supramolecular organization. *J. Biomol. Struct. Dyn.* 12(6): 1161–1172.
29. Pepe, A., Bochicchio, B., Tamburro, A.M. 2007. Supramolecular organization of elastin and elastin-related nanostructured biopolymers. *Nanomedicine* 2(2): 203–218.
30. Tamburro, A.M. 1995. The supramolecular structures of elastin and related synthetic polypeptides: Scale invariant weaving. *Macrocyclic and Supramolecular Chemistry in Italy.* G. Savelli, Centro Stampa Università di Perugia: 265–268.
31. D'Alessio, L., Tamburro, A.M., De Stradis, A. 1999. Observation of fractal structures in the supramolecular organization of protein molecules. *Fractals in Engineering, Proceedings*: 130–137.
32. Bressan, G.M., Argos, P., Stanley, K.K. 1987. Repeating structure of chick tropoelastin revealed by complementary DNA cloning. *Biochemistry* 26(6): 1497–1503.
33. Long, M.M., King, V.J., Prasad, K.U., Freeman, B.A., Urry, D.W. 1989. Elastin repeat peptides as chemoattractants for bovine aortic endothelial cells. *J. Cell. Physiol.* 140(3): 512–518.
34. Castiglione Morelli, M.A., Bisaccia, F., Spisani, S., De Biasi, M., Traniello, S., Tamburro, A.M. 1997. Structure-activity relationships for some elastin-derived peptide chemoattractants. *J. Pept. Res.* 49(6): 492–499.
35. Lograno, M.D., Bisaccia, F., Ostuni, A., Daniele, E., Tamburro, A.M. 1998. Identification of elastin peptides with vasorelaxant activity on rat thoracic aorta. *Int. J. Biochem. Cell. Biol.* 30(4): 497–503.
36. Brassart, B., Fuchs, P., Huet, E., Alix, A.J., Wallach, J., Tamburro, A.M., Delacoux, F., Haye, B., Emonard, H., Hornebeck, W., Debelle, L. 2001. Conformational dependence of collagenase (matrix metalloproteinase-1) up-regulation by elastin peptides in cultured fibroblasts. *J. Biol. Chem.* 276(7): 5222–5227.

37. Pepe, A., Armenente, M.R., Bochicchio, B., Tamburro, A.M. 2009. Formatio of nanosturctures by self-assembly of an elastin peptide. *Soft Matter* 5(1): 104–113.

38. Tamburro, A.M., Pepe, A., Bochicchio, B. 2006. Localizing alpha-helices in human tropoelastin: Assembly of the elastin "puzzle." *Biochemistry* 45(31): 9518–9530.

39. Woody, R.W. 1995. Circular dichroism. *Methods Enzymol.* 246: 34–71.

40. Woody, R.W. 1992. Circular dichroism and conformation of unordered polypeptides. *Adv. Biophys. Chem.* 2: 37–79.

41. Bochicchio, B., Tamburro, A.M. 2002. Polyproline II structure in proteins: Identification by chiroptical spectroscopies, stability, and functions. *Chirality* 14(10): 782–792.

42. Perczel, A., Hollosi, M. 1996. Turns. *Circular Dichroism: Conformational Analysis of Biomolecules.* G.D. Fasman. New York, Plenum: 285–380.

43. Perczel, A., Hollosi, M., Sandor, P., Fasman, G.D. 1993. The evaluation of type I and type II beta-turn mixtures. Circular dichroism, NMR and molecular dynamics studies. *Int. J. Pept. Protein Res.* 41(3): 223–236.

44. Bürgi, R., Pitera, J., van Gunsteren, W.F. 2001. Assessing the effect of conformational averaging on the measured values of observables. *J. Biomol. NMR.* 19(4): 305–320.

45. Bochicchio, B., Pepe, A., Tamburro, A.M. 2001. On (GGLGY) synthetic repeating sequences of lamprin and analogous sequences. *Matrix Biol.* 20(4): 243–250.

46. Bochicchio, B., Pepe, A., Tamburro, A.M. 2008. Investigating by CD the molecular mechanism of elasticity of elastomeric proteins. *Chirality* 20(9): 985–994.

47. Bochicchio, B., Floquet, N., Pepe, A., Alix, A.J., Tamburro, A.M. 2004. Dissection of human tropoelastin: Solution structure, dynamics and self-assembly of the exon 5 peptide. *Chem. Eur. J.* 10(13): 3166–3176.

48. Villani, V., Tamburro, A.M. 1995. Conformational modeling of elastin tetrapeptide Boc-Gly-Leu-Gly-Gly-NMe by molecular dynamics simulations with improvements to the thermalization procedure. *J. Biomol. Struct. Dyn.* 12(6): 1173–1202.

49. Bochicchio, B., Ait-Ali, A., Tamburro, A.M., Alix, A.J. 2004. Spectroscopic evidence revealing polyproline II structure in hydrophobic, putatively elastomeric sequences encoded by specific exons of human tropoelastin. *Biopolymers* 73(4): 484–493.

50. Pappu, R.V., Rose, G.D. 2002. A simple model for polyproline II structure in unfolded states of alanine-based peptides. *Protein Sci* 11(10): 2437–2455.

51. Conway, K.A., Harper, J.D., Lansbury, P.T., Jr. 2000. Fibrils formed *in vitro* from alpha-synuclein and two mutant forms linked to Parkinson's disease are typical amyloid. *Biochemistry* 39(10): 2552–2563.

52. Tamburro, A.M. 1981. *Elastin: Molecular and Supramolecular Structure. Connective Tissue Research: Chemistry, Biology, and Physiology.* New York, A. R. Liss.

53. Megret, C., Lamure, A., Pieraggi, M.T., Lacabanne, C., Guantieri, V., Tamburro, A.M. 1993. Solid-state studies on synthetic fragments and analogues of elastin. *Int. J. Biol. Macromol.* 15(5): 305–312.

54. Bochicchio, B., Pepe, A., Tamburro, A.M. 2007. Elastic fibers and amyloid deposition in vascular tissues. *Future Neurology* 2: 523–537.

55. Partridge, S.M., Davis, H.F., Adair, G.S. 1955. The chemistry of connective tissues. 2. Soluble proteins derived from partial hydrolysis of elastin. *Biochem. J.* 61(1): 11–21.

56. Bellingham, C.M., Woodhouse, K.A., Robson, P., Rothstein, S.J., Keeley, F.W. 2001. Self-aggregation characteristics of recombinantly expressed human elastin polypeptides. *Biochim. Biophys. Acta* 1550(1): 6–19.
57. Vrhovski, B., Jensen, S., Weiss, A.S. 1997. Coacervation characteristics of recombinant human tropoelastin. *Eur. J. Biochem.* 250(1): 92–98.
58. Mithieux, S.M., Rasko, J.E., Weiss, A.S. 2004. Synthetic elastin hydrogels derived from massive elastic assemblies of self-organized human protein monomers. *Biomaterials* 25(20): 4921–4927.
59. Bochicchio, B., Jimenez-Oronoz, F., Pepe, A., Blanco, M., Sandberg, L.B., Tamburro, A.M. 2005. Synthesis of and structural studies on repeating sequences of abductin. *Macromol. Biosci.* 5(6): 502–511.
60. Pepe, A., Guerra, D., Bochicchio, B., Quaglino, D., Gheduzzi, D., Pasquali Ronchetti, I., Tamburro, A.M. 2005. Dissection of human tropoelastin: Supramolecular organization of polypeptide sequences coded by particular exons. *Matrix Biol.* 24(2): 96–109.
61. Gotte, L., Giro, M.G., Volpin, D., Horne, R.W. 1974. The ultrastructural organization of elastin. *J. Ultrastruct. Res.* 46(1): 23–33.
62. Volpin, D., Pasquali-Ronchetti, I., Urry, D.W., Gotte, L. 1976. Banded fibers in high temperature coacervates of elastin peptides. *J. Biol. Chem.* 251(21): 6871–6873.
63. Bressan, G.M., Castellani, I., Giro, M.G., Volpin, D., Fornieri, C., Pasquali Ronchetti, I. 1983. Banded fibers in tropoelastin coacervates at physiological temperatures. *J. Ultrastruct. Res.* 82(3): 335–340.
64. Pepe, A., Flamia, R., Guerra, D., Quaglino, D., Bochicchio, B., Pasquali Ronchetti, I., Tamburro, A.M. 2008. Exon 26-coded polypeptide: An isolated hydrophobic domain of human tropoelastin able to self-assemble *in vitro*. *Matrix Biol.* 27(5): 441–450.
65. Bisaccia, F., Morelli, M.A., De Biasi, M., Traniello, S., Spisani, S., Tamburro, A.M. 1994. Migration of monocytes in the presence of elastolytic fragments of elastin and in synthetic derivates. Structure-activity relationships. *Int. J. Pept. Protein Res.* 44(4): 332–341.
66. Floquet, N., Hery-Huynh, S., Dauchez, M., Derreumaux, P., Tamburro, A.M., Alix, A.J. 2004. Structural characterization of VGVAPG, an elastin-derived peptide. *Biopolymers* 76(3): 266–280.
67. Kozel, B.A., Wachi, H., Davis, E.C., Mecham, R.P. 2003. Domains in tropoelastin that mediate elastin deposition *in vitro* and *in vivo*. *J. Biol. Chem.* 278(20): 18491–18498.
68. Tamburro, A.M., Pepe, A., Bochicchio, B., Quaglino, D., Ronchetti, I.P. 2005b. Supramolecular amyloid-like assembly of the polypeptide sequence coded by exon 30 of human tropoelastin. *J. Biol. Chem.* 280(4): 2682–2690.
69. Bochicchio, B., Pepe, A., Flamia, R., Lorusso, M., Tamburro, A.M. 2007. Investigating the amyloidogenic nanostructured sequences of elastin: Sequence encoded by exon 28 of human tropoelastin gene. *Biomacromolecules* 8(11): 3478–3486.
70. Ostuni, A., Bochicchio, B., Armentano, F., Bisaccia, F., Tamburro, A.M. 2007. Molecular and supramolecular structural studies on human tropoelastin sequence. *Biophys. J.* 93(10): 3640–3651.

71. Fan, K., Nagle, W.A. 2002. Amyloid associated with elastin-staining laminar aggregates in the lungs of patients diagnosed with acute respiratory distress syndrome. *BMC Pulm. Med.* 2: 5.

72. Guantieri, V., Tamburro, A.M., Gordini, D.D. 1983. Interactions of human and bovine elastins with lipids: Their proteolysis by elastase. *Connect. Tissue Res.* 12(1): 79–83.

73. Claire, M., Jacotot, B., Robert, L. 1976. Characterization of lipids associated with macromolecules of the intercellular matrix of human aorta. *Connect. Tissue Res.* 4(2): 61–71.

74. Getie, M., Schmelzer, C.E., Neubert, R.H. 2005. Characterization of peptides resulting from digestion of human skin elastin with elastase. *Proteins* 61(3): 649–657.

75. Rodriguez-Cabello, J.C., Prieto, S., Arias, F.J., Reguera, J., Ribeiro, A. 2006. Nanobiotechnological approach to engineered biomaterial design: The example of elastin-like polymers. *Nanomedicine* 1(3): 267–280.

76. Bochicchio, B., Pepe, A., Tamburro, A.M. 2005a. Circular dichroism studies on repeating polypeptide sequences of abductin. *Chirality* 17(7): 364–372.

77. Bochicchio, B., Jimenez-Oronoz, F., Pepe, A., Blanco, M., Sandberg, L.B., Tamburro, A.M. 2005b. Synthesis of and structural studies on repeating sequences of abductin. *Macromol. Biosci.* 5(6): 502–511.

10

Elastin and Elastin-Based Polymers

Daniela Quaglino, Deanna Guerra, and Ivonne Pasquali-Ronchetti
University of Modena and Reggio Emilia, Italy

CONTENTS

10.1 Introduction

Elastin is an insoluble, elastomeric, extracellular matrix protein that provides resilience and deformability to tissues. The elasticity is critical for the functioning of several tissues, such as arterial vessels, lungs, skin, esophagus, and stomach [1,2]. Tropoelastin (TE) is the soluble precursor of elastin and a vital building block in elastic fiber formation [3]. It has been shown that tropoelastin is formed intracellularly and then cross-linked extracellularly giving rise to the elastic fibers [4]. Elastic fibers consist of two distinct components: elastin (an insoluble polymer of 70 kDa tropoelastin monomers) and microfibrils (10 nm unbranching fibrillin-containing fibrils) [5]. During development, the microfibrils form linear bundles that appear to act as a scaffold for the deposition and orientation of tropoelastin monomers.

10.1.1 Elastin and Elastic Fibers

Elastin is the main component of the elastic fibers and is responsible for the elasticity of biological tissues. The number and size of elastic fibers are strictly related to the elastic modulus of tissues, intended as the capacity of tissues to regain their original shape when relaxed after stretching. Elastic fibers are immersed in a composite three-dimensional matrix made of a variety of proteins, glycoproteins, and glycosaminoglycans that take part of and contribute to tissue plasticity and deformability of soft connective tissues. Elastic fibers are mainly responsible for connective tissue elastic recoil.

From an evolutionary point of view, elastin is a rather recent protein. In vertebrates, it is mainly produced by fibroblasts and smooth muscle cells. Its synthesis is very active during fetal life and after birth up to puberty. Then, the elastin gene is almost completely silenced, and it has been calculated that the elastin renewal would take about 70 years. Therefore, the elastin and the elastic fibers deposited in the extracellular space have to persist throughout the life span. Elastin molecules are linked into a polymer by intermolecular cross-links, called desmosines, that resist several proteolytic enzymes.

Elastin is a major component of vessels, is well represented in lung and skin, and plays a peculiar role in certain organs, such as in the Bruch's membrane of the retina, where it forms a three-dimensional elastic network sustaining the retinal vessels.

Although affected by the aging process, elastin is one of the most resistant proteins, and it is still present and functioning in very old persons, even in the absence of a real renewal [6]. Therefore, for its resistance and elastic properties as well for the possibility to form three-dimensional networks, elastin and elastin-based polymers have been studied by a number of researchers as ideal materials for tissue engineering.

10.1.1.1 Structural Organization of Elastic Fibers

Elastic fibers are composed of elastin and of a series of proteins, glycoproteins, and proteoglycans that probably vary depending on the animal species, on the tissue, and on the age and health conditions [6,7].

Elastic fibers can be seen by light microscopy as long, branched, and interconnected strips, 0.5 to 2 μm large and several micrometers long, forming a three-dimensional network. In skin and lung, such strips are rather thin, and the network is loose. In the vessels and in the Bruch membrane, elastin is organized into lamellae that form a rather compact network. Elastic

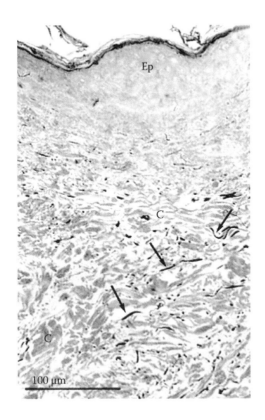

FIGURE 10.1
Light microscopy of a semithin section of human skin stained with toluidine blue. Under the epidermis (Ep), the dermis is mainly composed of collagen bundles (C, pale gray) and dark thin elastic fibers (arrows).

fibers can be stained red by resorcin-fucsin, intense blue by toluidine blue, or black by hematoxylin and potassium iodide Verhoeff's stain. In all cases, they appear as homogeneous strips interconnected with collagen bundles and cells. The other matrix constituents, such as proteoglycans and glyco-proteins, can be hardly seen by light microscopy (Figure 10.1).

In transmission electron microscopy (TEM), elastic fibers appear to be made of a homogeneous, slightly electron opaque material organized into branched, interconnected fibers. Figure 10.2 illustrates elastic fibers in the human dermis, as they appear by TEM, and how they change with age [6]. In Figure 10.2a, small amounts of homogeneous elastin are surrounded by linear microfibrils that are present during elastogenesis and form an efficient scaffold for elastin deposition. Figure 10.2b shows an adult elastic fiber made of electron transparent material containing microfibrillar remnants, and Figure 10.2c illustrates an old elastic fiber containing electron dense deposits made of degenerated elastin and ion precipitates. Therefore, TEM allows the visualization of at least some of the proteins and glycoproteins associated

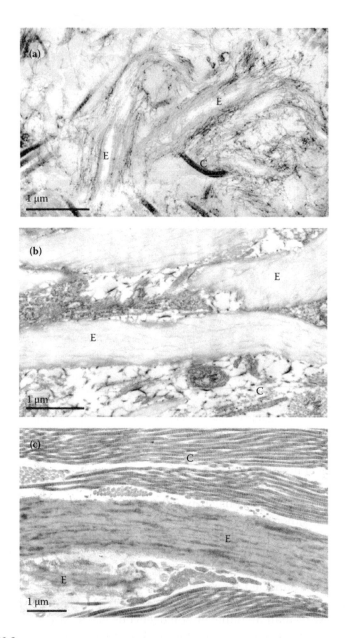

FIGURE 10.2
Transmission electron microscopy showing age-dependent changes of elastic fibers (E) in the human dermis. In newborns (a) fibers are characterized by small amounts of amorphous elastin surrounded by a thick coat of microfibrils. In adults (b), amorphous elastin is predominant and few microfibril remnants are visible at the periphery of the fibers. In old individuals (c), elastic fibers contain numerous electron-dense strips. Collagen fibrils (C) are also visible.

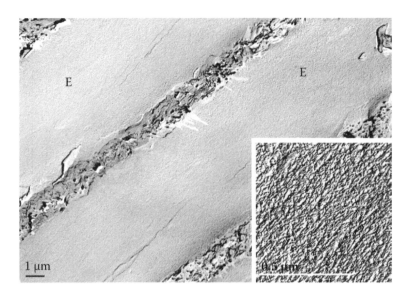

FIGURE 10.3
Bovine nuchal ligament elastic fibers (E) observed by transmission electron microscopy after freeze-fracturing and rotary shadowing. Using this technique, the amorphous component of the elastic fiber appears to form beaded filaments arranged into a three-dimensional network, as clearly visualized at higher magnification (insert).

with elastic fibers in the form of electron dense spots interspersed within the amorphous elastin.

However, the amorphous organization of elastin revealed by conventional TEM derives from the hydrophobic nature of the protein and from treatments necessary to perform the observation. Actually, when elastic fibers are frozen in liquid nitrogen and fractured [8,9], they appear composed of beaded filaments arranged into a three-dimensional network (Figure 10.3). When the fibers are stretched before freezing, these filaments appear oriented in the direction of the force applied, and this suggests that at least part of the increment in length of fibers upon stretching is due to aligment of filaments (Figure 10.3, insert). The organization into beaded filaments may indicate that elastin molecules do not have a random organization within the polymer, and beads might represent hydrophobic domains. This organization represents the supramolecular arrangement of the elastin molecules that, by contrast, when analyzed at a molecular level, reveal amorphous behavior.

By differential scanning calorimetry (DSC) (Q2000, TA Instruments), a *heat-cool-heat* thermal profile has been used to identify whether elastin has amorphous or crystalline structure. A very low temperature has been used to determine if there were any transitions in the lower temperature region. After determining that there were no events in the lower temperature region, all the other samples were exposed to an initial start temperature

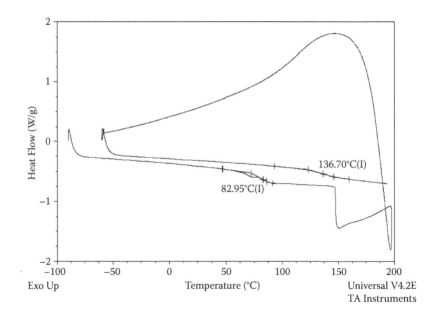

FIGURE 10.4
Elastin fiber transition from differential scanning calorimeter (DSC). Shows amorphous nature and no melting peak.

of –25°C, and glass transition temperature (*Tg*) was determined only after from the second heat cycle. The first heat cycle contained a very broad endothermic peak that might have been caused by evaporation of moisture from the elastin sample or from ineffective contact of the sample with the bottom of the pan. The exothermal peak (Figure 10.4) in the cool cycle was not due to any transitions in the material structure (such as crystallization) but was due to the heat flow associated with the rapid cooling of the DSC cell. From Figure 10.4, *Tg* of elastin is around 136.70°C, and it has been shown that there is no transition of melting point in the heat cycle which depicts amorphous behavior of elastin at the molecular level (Figure 10.4).

The beaded supramolecular organization of elastin molecules within the fibers has also been revealed by scanning force microscopy on fresh-frozen and fractured elastic fibers [8], as well as on sections of resin-embedded elastic fibers [10]. Figure 10.5 shows an elastic fiber visualized by atomic force microscopy on sections of the apparently amorphous material described in Figure 10.2. This confirms once more that elastin is organized into elastic fibers in the form of beaded filaments, and that the amorphous structure of elastin revealed by conventional transmission electron microscopy is an artifact due to the technical procedure. Moreover, by atomic force microscopy, it is possible to visualize bridges among beaded filaments (Figure 10.5b), as expected for a three-dimensional polymer.

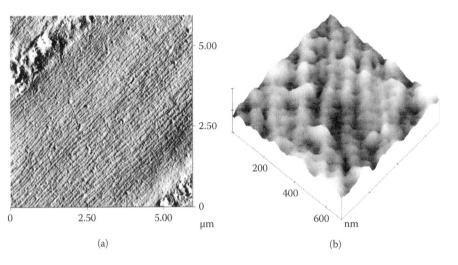

FIGURE 10.5
Epoxy embedded and sectioned elastic fibers of bovine nuchal ligament observed by atomic force microscopy. Elastic fibers are formed by beaded filaments (a), as also observed after freeze-fracture (Figure 10.3). At higher magnification (b), bridges among filaments can be appreciated.

10.2 Tropoelastin and Tropoelastin Aggregates

Tropoelastin is the protein precursor of elastin. It is a 70 kDa protein whose modular structure is described in detail by Tamburro et al. (see Chapter 9). It is made of alternate hydrophobic and hydrophilic domains and exhibits great tendency to aggregate in saline solution by increasing the temperature in the physiological range [11]. The process is called coacervation, and it is temperature reversible. Isolated and purified tropoelastin molecules in saline tend to aggregate by rising the temperature above 30°C, and the solution becomes cloudy. By increasing the temperature very slowly and analyzing the phenomenon by negative staining electron microscopy, it was observed that tropoelastin molecules aggregate into short sticks that become longer and longer and undergo lateral aggregation up to the formation of long bundles of filaments [12]. Several studies have been done in order to understand the driving forces sustaining the process and the involvement of hydrophobic and hydrophilic domains in the assembly of molecules into discrete filaments and of these latter into ordered bundles [11–13]. Tropoelastin is mainly produced by fibroblasts and smooth muscle cells in vertebrates and is secreted in the extracellular space, where it aggregates into small clumps that later interact with preexisting microfibrils, giving rise to the elastic fibers [14]. The final event is polymerization of tropoelastin molecules into an insoluble cross-linked network by a chemical reaction catalyzed by the extracellular enzyme lysyl oxidase. This enzyme induces the oxidative deamination of lysine residues on tropoelastin, creating the aldehydic groups necessary for

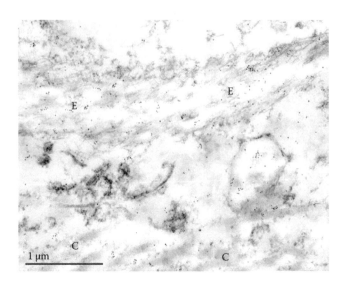

FIGURE 10.6
Immunoelectron microscopy for decorin, a small proteoglycan that is present in the extracellular matrix and interacts with collagen fibrils (C) as well as with elastic fibers (E) in the human dermis. The immunoreaction is visualized by the small black gold particles.

the formation of the ultimate desmosine cross-links either within the same molecule or among molecules. The process of tropoelastin aggregation and cross-linking is rather complex, and it is still not completely known whether other matrix molecules favor or inhibit the aggregation process and cross-linking [15].

Several matrix molecules have been observed to be present within normal elastic fibers; however, it is not known if these molecules are simply entrapped within the elastic polymer during its formation or if they really contribute to the formation or stability of fibers. It was described that glycosaminoglycans are present within normal elastic fibers (Figure 10.6) [16,17], and that these constituents may increase in pathological conditions, so perturbing the fiber organization.

Inhibition of the formation of intermolecular desmosine cross-links by treatment of animals with lathyrogens (Figure 10.7a) or in Menkes (Figure 10.7b), a genetic disorder due to mutations in the enzyme lysyl oxidase, induces the formation of abnormal elastic fibers containing a high quantity of glycosaminoglycans [18–20]. To explain the presence of these molecules within elastic fibers, it was suggested that negatively charged glycosaminoglycans would remain attached to the positively charged lysine groups on tropoelastin due to the inhibition or mutations of the cross-linking enzyme lysyl oxidase [19].

On the other hand, *in vitro* studies have shown that isolated and purified tropoelastin molecules interact with glycosaminoglycans, such as heparan sulfate, and this interaction has been demonstrated to favor tropoelastin aggregation [17]. Figure 10.8a shows an aggregate of purified tropoelastin

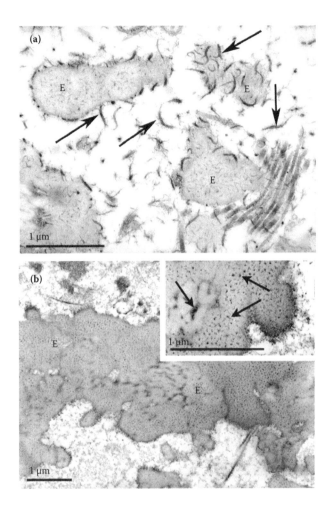

FIGURE 10.7
Transmission electron microscopy showing the interaction of glycosaminoglycans, visualized by cytochemical methods, with elastic fibers (E) after inhibition of intermolecular desmosine cross-links by lathyrogens in chicken's aorta (a) and in human aorta in Menkes disease (b). Glycosaminoglycan chains (arrows) are stained with alcian blue (a) or with toluidine blue (b and b insert) added to the fixative.

obtained by increasing the temperature above 30°C and observing after negative staining with uranyl acetate. The stain surrounds structures and provides evidence that tropoelastin molecules aggregate into filaments that form discrete bundles upon drying (Figure 10.8a) [12,21]. These filaments are similar to those seen on sections of natural elastic fibers in bovine nuchal ligament after prolonged chemical fixation with osmium tetroxide (Figure 10.8b and Figure 10.8c). Tropoelastin, in the presence of heparan sulfate at 30°C (Figure 10.8d), forms aggregates similar to those in the absence of heparan sulfate, when observed by negative staining (compare Figure 10.8a and Figure 10.8d). However, Figure 10.8e and Figure 10.8f demonstrate that when the same samples are resin embedded and sectioned, the three-dimensional network formed by tropoelastin molecules is looser in the presence of heparan sulfate (compare Figure 10.8e and Figure 10.8f). Therefore, heparan sulfate interacts with tropoelastin molecules and is incorporated into the tropoelastin coacervate. Moreover, as revealed by scanning calorimetry, it induces a significant reduction of the coacervation temperature of tropoelastin molecules [17].

Similar data were obtained by Weiss and coworkers [22]. They observed that negatively charged glycosaminoglycans interact with the positively

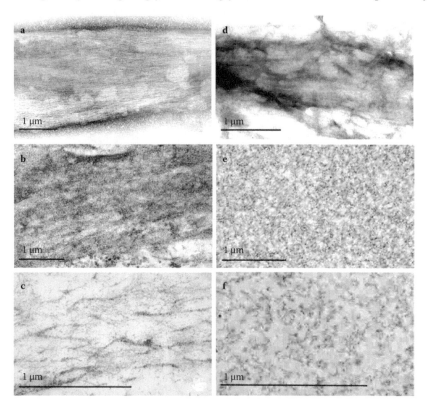

FIGURE 10.8

Transmission electron microscopy of tropoelastin and elastin in different experimental conditions. Purified tropoelastin, when incubated in physiological solution at 35°C (a), forms bundles of organized filaments, revealed by negative staining. The visualization of similar elongated and longitudinally oriented elastin filaments can be seen within elastic fibers of bovine nuchal ligament after prolonged fixation with osmium tetroxide (27 h, panel b; 4 days, panel c) and embedding into epoxy resin. The addition of heparan sulfate to purified tropoelastin molecules determines the formation of aggregates similar to those in panel a but at lower temperatures [16]. Purified tropoelastin coacervates, as those in panels a and d, when chemically fixed and embedded in epoxy resin, reveal that heparan sulfate (f) induces the formation of a looser tropoelastin network compared to samples incubated in the absence of the glycosaminoglycan (e).

charged lysine residues of purified tropoelastin molecules, so promoting the coacervation of tropoelastin in a temperature- and concentration-dependent manner. Interactions between tropoelastin and sulfated glycosaminogly-cans have also been observed by Mecham and coworkers who suggested that elastin assembly in the extracellular space may be influenced by sulfated proteoglycans in the matrix [14]. More specifically, there are observations suggesting that cell surface sulfated glycosaminoglycans, and in particular, plasma membrane–associated heparan sulfates, bind to tropoelastin through the carboxyl terminus of the protein [23], and that these elastin–heparan–sulfate interactions may be mediated by elastin-associated microfibrils [24]. Therefore, whatever their role, sulfated glycosaminoglycans, and in particular heparan sulfate, make contacts with tropoelastin either *in vivo* and *in vitro*, and this should be taken into account in any attempt to use elastin-derived polymers for biological purposes and nanodevices.

We recently conducted a series of experiments aimed at investigating the effect of added heparan sulfates on the production and deposition of elastic fibers by *in vitro* cultured human fibroblasts. Results show that exogenous heparan sulfate does not modify the quantity of elastin production, whereas it perturbs the organization of neosynthesized elastic fibers by favoring the formation of clumps of elastin which only later aggregate into real fibers (Figure 10.9) (unpublished data). Similarly, Buczek-Thomas and coworkers [23] showed that depletion of sulfated glycosaminoglycans by pharmacological treatments diminished the incorporation of tropoelastin into the extracellular matrix by pulmonary fibroblasts *in vitro*, suggesting that sulfated proteoglycans play a central role in modulating the extracellular matrix deposition and, in particular, the elastic fiber network [25].

Recombinant human tropoelastin can be converted to elastin *in vitro* by deamination of lysyl residues with the bifunctional cross-linker bis (sulfo-succinimidyl) suberate (BS^3). The tropoelastin molecule has about 35 lysine residues that can potentially be used for intra- and intermolecular cross-links (desmosines). BS^3 is a suberate cross-linker. It is an amine-reactive molecule, and it cross-links with amine groups of lysine residues within the same molecule or adjacent molecules (Figure 10.10). In the case of BS^3, the reactive moieties are NHS-esters [3].

10.3 Polymers Containing Tropoelastin and Tropoelastin-Derived Motives

10.3.1 Molecular Conformation and Properties

The organization of the elastin molecule into alternate hydrophobic and hydrophilic domains stimulated studies tending to elucidate the specific role

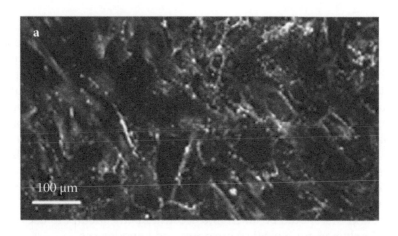

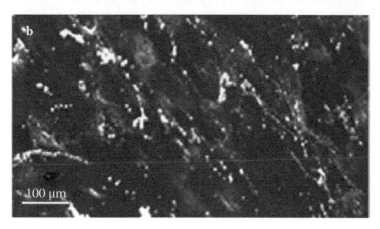

FIGURE 10.9
Tropoelastin produced *in vitro* by human dermal fibroblasts is visualized by specific fluorescent polyclonal antibodies (white spots). The addition of heparan sulfate to the culture medium (b) favors the formation of bigger and more numerous elastin aggregates compared to controls (a).

of domains implicated in the assembly of molecules into fibers, or in the interactions with cells or with other extracellular constituents. Attention has also been paid to the conformation characteristics of domains in order to explain the elastic behavior in water of the entire molecule. It is worth mentioning studies of Tamburro's group, who synthesized and analyzed by circular dichroism (CD) and nuclear magnetic resonance (NMR) different elastin peptides, demonstrating that peptides with amino acid sequences corresponding to the various tropoelastin domains exhibit peculiar chemical–physical characteristics and, depending on environmental conditions, form aggregates with distinctive supramolecular organization and properties [26–29].

As an example, Figure 10.11 illustrates the aggregation properties of peptides derived from exons 20 (Figure 10.11a) and 30 (Figure 10.11b) of the elastin

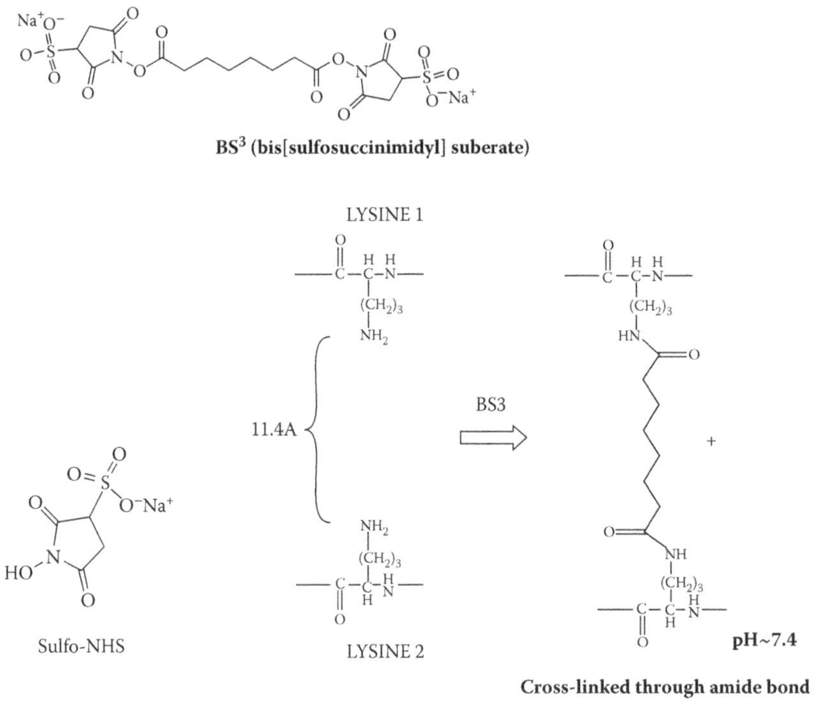

BS³ (bis[sulfosuccinimidyl] suberate)

Cross-linked through amide bond

FIGURE 10.10
BS³ cross-linking and isolation of cross-link pairs. Tropoelastin monomers were cross-linked with BS³ during coacervation. (From Wise, S.G., Mithieux, S.M., Raftery, M.J., Weiss, A.S. 2005. *J Struct Biol* 149:273–281. With permission.)

gene. The amino acid sequence and the physico-chemical characteristics of these two peptides are described by Tamburro et al. in this book. The two peptides aggregate in completely different ways, because peptide 20 forms bundles of filaments similar to those of the whole tropoelastin molecule [30], whereas peptide 30 forms single long and discrete sticks and amyloid-like structures [31]. Because during aging and in pathological conditions elastin is degraded by proteases, elastin-derived peptides may manifest *in vivo* these same properties and may induce subsequent physiological events depending on their supramolecular organization.

10.3.2 Interactions with Cells and with Cellular Components

At first, Urry's group reported the biological effects of elastin-derived peptides and showed that the nonapeptide Ala-Gly-Val-Pro-Gly-Phe-Gly-Val-Gly and the hexapeptide Val-Gly-Val-Ala-Pro-Gly, sequences that are present several times in the elastin molecule, were chemotactic for fibroblasts, aortic endothelial cells, and monocytes [32–34]. This means that cells *in vitro* move toward the highest concentration of peptides.

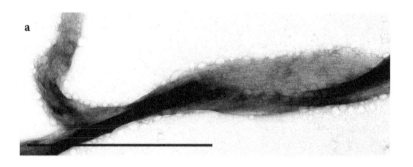

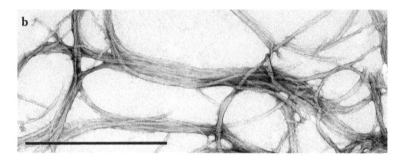

FIGURE 10.11
Negative staining of tropoelastin peptides incubated in physiological buffer at 50°C. Peptides derived from sequences corresponding to exon 20 of the elastin gene (a) form bundles of filaments similar to those obtained with the whole tropoelastin molecule (Figure 10.8a). By contrast, peptides derived from sequences corresponding to exon 30 of the elastin gene (b) form long and rigid sticks.

The chemotactic properties of elastin-derived peptides were defined by analyzing various hexapeptides and pentapeptides typical of elastin in an attempt to identify the smallest and most active peptides exhibiting biological properties [35,36]. Several biological molecules have chemo-attractant properties for living cells *in vitro* and, very likely, *in vivo*. These same substances are able to bind receptors on the surface of attracted and migrated cells. A large number of cell types were shown to have receptors for elastin, tropoelastin, and tropoelastin-derived peptides, which may be used to capture and degrade elastin [37].

10.3.3 Biological Activities of Tropoelastin and of Elastin-Derived Peptides

10.3.3.1 Proliferation

Soluble peptides and elastin-derived peptides stimulate the proliferation of mesenchymal cells, such as smooth muscle cells, by interacting with the 67 kDa cell membrane elastin binding protein (EBP), and generating intracellular signals with sequential activation of tyrosine kinases and of cyclin-dependent kinases leading to cell proliferation [38]. A similar mechanism would seem to operate in astrocytoma cells where malignant cell proliferation might be sustained by the cellular aberrant expression of elastin, and by binding of elastin fragments to the cell EBP receptors [39].

This physiological effect of elastin-derived peptides was also observed in lower animal species, namely in ciliates, which do not produce elastin, but on which elastin peptides were shown to exert chemotactic effect and to stimulate cell proliferation [36].

10.3.3.2 Inflammation

In the formulation of nano-biocomposites to be used as implants into living beings, one of the main concerns should be to check for activation of the inflammatory process and of the immune response. These processes may alter the implant and would favor its retention by encapsulation or its rejection.

The inflammatory process is characterized by recruitment and activation of macrophages, production of molecules with proteolytic activities and of cytokines with stimulatory effects on macrophages, lymphocytes, and fibroblasts. In turn, proteolysis may favor elastin breakdown leading to the formation of elastin fragments [40]. The inflammatory response that accompanies vessel degeneration, for instance, has been attributed to the elastin-derived peptides produced during the elastic fiber destruction occurring in this pathological process [41]. In turn, elastin fragments have been shown to elicit the immune response with production of antibodies and further destruction of the elastin polymer. In this context, it has been observed that there are differences in the antibodies produced against tropoelastin or against the alpha-elastin fragments derived from degradation of the elastin polymer [42] and that cross-linked domains behave differently than noncross-linked regions in antibody responses [43].

Simple elastin-derived peptides, like VGVAPG, VGVPG, and VGAPG, were shown to exhibit chemotactic activity (i.e., stimulate macrophages to migrate toward the higher concentration of the peptide), and it was also observed that migration was rather selective, implying the existence of different chemotactic receptors on the cell membrane of monocytes/macrophages for different peptides [35]. The same group investigated the chemotactic activity of exon 26A, an exon that is not expressed in humans in normal conditions, whereas it is expressed in some pathological processes, such as in hypertension. Interestingly, sequences derived from exon 26A exhibited a conformation-dependent chemotactic activity for monocytes/macrophages, indicating that the biological activity of elastin-derived peptides mostly depends on their structural organization [26,44]. Domains derived from the same exon 26A were also shown to have an effect on the vascular tone, suggesting for them a vasorelaxant role [45], similar to that described for peptides Val-Gly-Val-Ala-Pro-Gly and Val-Gly-Val-Hyp-Gly [46]. Peptides typical of the elastin molecule have been chemically synthesized and modified in their sequence structure in order to abolish or to augment their biological activities with interesting results [47].

As a whole, elastin-derived peptides have been shown to elicit a series of cellular responses, from elastase release, to free radical production, low-density lipoprotein (LDL) oxidation and cholesterol production [48–50]. More specifically, elastin-derived peptides have been shown to stimulate the production of elastase-like enzymes in macrophages [51] and in fibroblasts [40]. However, not all elastin peptides would seem to stimulate inflammatory cells in the same way, with some peptides being more active than others in the production of free radicals or in elastase release or in intracellular free calcium metabolism [51,52]. This datum could be important in the evaluation of the best domain of the elastin molecule to be selected and used for specific purposes.

It was recently observed that elastin-derived peptides, in the concentration range that can be reached in the inflammatory process, may regulate the inflammatory response. Actually, elastin-derived peptides were shown to interact with monocytes leading to downregulation of the expression of proinflammatory cytokines, such as TNF-alpha, IL-1-beta and IL-6. By contrast, the same elastin-derived peptides did not modify the production of the anti-inflammatory cytokines IL-10 and TGF-beta [53]. These data would suggest that certain elastin-derived peptides elicit an anti-inflammatory response. The effect of elastin-derived peptides on the inflammatory response should be better investigated in any attempt to use elastin-derived fragments for biocomposites or artificial tissues.

10.3.3.3 Pleiotropic Role of Elastin-Derived Peptides

From the analysis of the multiple biological effects of elastin-derived peptides, it emerges that elastin and its derivatives may have a role in cancer progression (cell multiplication) and metastasis (cell migration) as well as in inflammation and fibrosis (chemotaxis and activation of macrophages and fibroblasts). Due to the pleiotropic effects of elastic peptides, the hypothesis has been put forward that elastin-derived peptides may be considered as matrikines—that is, molecular fragments/domains exerting biological activities and modulating signaling pathways on several cell types [54], similarly to cytokines. In a recent report, the elastin-derived hexapeptide Val-Gly-Val-Ala-Pro-Gly was shown to protect the heart from ischemia/reperfusion injury by inhibiting cell reactions leading to cell death [55].

Elastin-derived peptides were shown to upregulate the production of proMT1-MMP, a membrane-associated metalloprotease that degrades the extracellular matrix favoring tissue neovascularization. The angiogenic effect of elastin peptides could also be important in promoting cancer growth and metastasis [56]. This effect has to be considered when elastin or elastin-derived peptides are chosen as constituents of biomaterials.

10.4 Elastin-Derived Polymers

10.4.1 Nanodispositives

An interesting utilization of elastin-derived peptides is linked to their molecular conformation in water. Their high hydrophobicity induces peculiar conformations that are sensitive to various environmental conditions, such as temperature and ionic composition. In this context, an elastin-like polypeptide has been conjugated with nitroxide, and the product has been used to measure changes in the temperature as a function of changes in the electron paramagnetic resonance spectrum of the polypeptide–nitroxide conjugate [57].

10.4.2 Delivery Systems

As already mentioned, elastin-derived and elastin-like peptides exhibit pleiotropic properties. They can form rigid or gel-like aggregates, they can be recognized by cell surface receptors and degraded by macrophages and by extracellular enzymes, and they can modulate the inflammatory reaction. For all these reasons and for their physico-chemical properties, several elastin-like peptides have been synthesized for various purposes [58]. Among these, it is worth mentioning the design of copolymers made of silk and elastin-like peptides for drug and DNA delivery [59]. Recently, by using recombinant techniques, Haider and coworkers were able to synthesize silk and elastin-like protein polymers forming hydrogels with defined physical and behavioral characteristics depending on the polymer composition [60]. Similarly, the influence of environmental conditions on the swelling ratio and stability of an engineered silk-elastin-like copolymer was investigated in order to have better control of the hydrogel eventually used as the delivery agent [61,62].

Elastin-like and elastin-derived polypeptides have been used to form complexes with drugs in order to retard degradation and cell efflux of the chemicals. Interestingly, conjugation of elastin-derived peptides with doxorubicin, a chemical used in the treatment of cancer cells, may lead to a dual advantage. First, elastin peptides would inhibit the drug efflux from cancer cells, so prolonging the toxic effect of the chemical. Second, elastin peptides would increase the drug concentration due to the gelation properties of the elastin peptide moiety of the complex [63]. Moreover, elastin-derived peptides conjugated with doxorubicin were shown to have different distribution within the cellular compartments, suggesting new approaches of intervention toward tumor cells [64].

By using recombinant techniques, mosaic proteins can be obtained to specifically target cells. Elastin-derived peptides able to coacervate by increasing the temperature have been linked to HIV-1 tat protein to facilitate the cellular uptake of a toxic doxorubicin derivative. Interestingly, this complex was found to be much more active in killing uterine sarcoma cells than the simple antitumor drug [65]. Basically, the higher efficiency of the complex was due to the more efficient cellular uptake due to the virus tat protein, to the slower efflux of the drug chemically linked to the elastin-derived peptide, and to the major concentration of the toxic drug due to coacervation of the elastin peptide moiety at physiological temperature [58,61].

10.4.3 Scaffolds for Tissue Equivalents

A series of tissue equivalents was prepared by using natural matrices seeded with mesenchymal cells of different origins, in the presence of growth factors, in the attempt to induce cell differentiation and production of specialized tissue substitutes [66]. For these purposes, elastin-rich natural matrices, such as heart valve or vessel wall matrices, have largely been used. However, results were mostly unsuccessful due to calcification of the elastic component of implanted tissues [67]. It was shown that tropoelastin per se or some of its degradation derivatives may inhibit calcium precipitation because the addition of soluble recombinant tropoelastin or of the peptide VGVAPG inhibits calcification in an *in vitro* system created to investigate the mechanisms of calcium precipitation on elastic fibers [68]. However, the process of calcification is rather complex and, *in vivo*, is controlled by specific inhibitors, such as matrix Gla protein (MGP) and fetuin [69], whose expression and balance are under different and complex regulation mechanisms [70].

Elastin-derived sequences have been shown to be very promising, as they allow the production of proteinaceous polymers largely biocompatible and with self-assembling properties [71]. Moreover, gene recombinant techniques allow the production of protein with the desired physical, chemical, and biological functions. In this context, elastin sequences have been used, alone or in combination with other protein sequences, to obtain a chimera protein with the expected structure and function [72,73]. Interestingly, proteins can be produced by assembling the properties of several molecules. As an example, a single protein molecule may contain hydrophobic domains of tropoelastin with elastic and self-assembling properties, alanine and lysine-rich domains for cross-linking purposes, and fibronectin-derived cell adhesion domains for cell attachment to the polymer [74].

The field of polyfunctional and biocompatible materials is very active today, and several theoretical studies and practical applications of polymers and tissues containing elastin-derived sequences have been published [75,76].

One of the most-used materials for building up scaffolds for tissue engineering is poly(D,L-lactic acid), and the interactions between cells and the PDLLA surfaces can be improved by covering PDLLA with a film of protein with adhesive properties for specific cell types. As an example, silk fibroin has been shown to ameliorate the adhesive, proliferation, and differentiation properties of osteoblasts seeded on silk-covered PDLLA compared with PDLLA alone [77]. Moreover, elastin-derived peptides, used to cover the surface of polyethylene glycol terephthalate, were shown to improve the duplication and the phenotypic expression of endothelial cells seeded on the composite compared with the polymer alone [76].

Several studies have been conducted on the characterization of cross-linked polymers basically made of collagen, but in which elastin and other extracellular matrix components were incorporated. In the great majority of these studies, elastin and elastin-derived peptides were added to increase elasticity and plasticity, whereas glycosaminoglycans were included to increase the water-binding capacity of the scaffold [78]. The main goal of these studies is to produce biomaterials with chemical, physical, and biological characteristics that may be controlled and that mimic as much as possible natural scaffolds and tissues. The porosity of the scaffold is of paramount importance, as seeded cells have to spread, migrate, and orient themselves depending on the cell type and on the structure of the scaffold [79].

Furthermore, apart from problems derived from the chemical nature and structural organization of the polymer, another big problem is how to standardize the industrial production of such materials and how to operate during cell culturing. Electrospinning of a mixture of collagen and elastin, stabilized by cross-linking with EDC (N-(3-dimethyl aminopropyl)-N-ethylcarbodiimide hydrochloride) and NHS (N-hydroxysuccinimide) and connected to a rotating tubular mandrel collector, made possible the production of biomaterials with high thermal stability and allowed smooth muscle cells to grow to a confluent layer [79,80]. This kind of scaffold was specifically studied for production of biomaterials suitable for vessel substitution.

The presence of elastin and of elastin-derived peptides is also important in scaffolds designed to repair wounds and burns. Actually, given the low expression of the elastin gene in physiological wound repair [81], elastin is theoretically one of the components that should already be present in a scaffold used to prepare biomaterials for wound healing. Solubilized elastin, incorporated into an insoluble collagen three-dimensional scaffold, was shown to stimulate angiogenesis and elastic fiber formation when the biomatrix was inserted subcutaneously in young rats. This was revealed by immunostaining of sections of the implanted material with antibodies specific for collagen type IV, indicating the synthesis of vessel basal membranes, and of the two main components of the elastic fibers, namely, elastin and fibrillin 1 [82].

Furthermore, elastin synthesis can be promoted by growth factors and cytokines that stimulate cell proliferation and differentiation into matrix synthesizing cells. Sales and coworkers induced endothelial progenitor cells, seeded on a poly (glycolic acid)/poly (4-hydroxybutyrate) scaffold, to produce collagen and elastin by treatment with TGF-beta1 [83].

10.4.4 The Skin Equivalent Model

Apart from the use of scaffolds containing elastin or elastin-derived domains, another goal is to produce a scaffold where elastin may be efficiently produced and elastic fibers deposited. As already mentioned, elastin synthesis and renewal are practically null in adults. Moreover, fibroblasts and smooth muscle cells produce very little elastin during wound repair and in the *in vitro* systems [81]. It should be worthwhile to find out how to induce an efficient elastic fiber production by cells grown into an artificial scaffold. This would lead to the production of artificial tissues useful for implants in vessels, skin, and other elastic tissues.

The skin equivalent tissue is an example. Fibroblasts can be grown in a three-dimensional matrix made of collagen, chitosan, and glycosaminoglycans [84]. Fibroblasts invade the polymer and start to produce extracellular matrix molecules, among them collagen. If, after 3 weeks of culture, keratinocytes are seeded on the top of the artificial dermis and grown at the air–liquid interface for another 5 to 6 weeks, the extracellular matrix produced by fibroblasts underneath is characterized by the occurrence of elastic fibers (Figure 10.12). This indicates that interactions between keratinocytes and fibroblasts are necessary to induce the expression of elastin and of elastin-associated proteins, among them LOX (lysyl oxidase) and LOXL (lysyl oxidase L), that are essential for the maturation and organization of the elastin network [85,86].

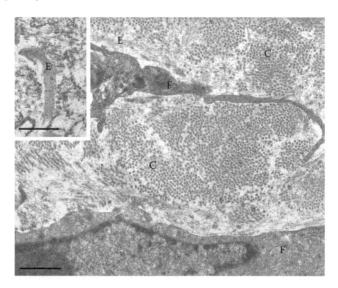

FIGURE 10.12

Transmission electron microscopy of the extracellular matrix produced by human dermal fibroblasts grown into a chitosan-glycosaminoglycan-based scaffold covered by keratinocytes. The cross talk between fibroblasts (F) and keratinocytes allows the deposition of amorphous elastic fibers (E) beside collagen bundles (C).

10.5 Summary

Tropoelastin, elastin, and elastin-derived peptides are very interesting molecules for the production of biomaterials. Because of its atypical structure and physical-chemical properties, based on alternate hydrophobic and hydrophilic domains, tropoelastin as such and each fragment of the tropoelastin molecule can be used for different and specific purposes. Actually, tropoelastin has domains that can be selected for their elasticity, self-assembly properties, interactions with cells, artificial polymers, and extracellular matrix constituents, for the activation of metabolic pathways. The use of elastin-derived peptides for the production of scaffolds for tissue engineering is a wide and still unrevealed field. Interestingly, hydrogels can also be produced for cell delivery of DNA and of drugs, so becoming a promising tool in tumor chemotherapy.

Acknowledgments

Studies in the author's laboratory were supported by grants from the Italian MIUR and from EU (Telastar, Geneskin, and Elastage).

References

1. Vrhovski, B., Weiss, A.S. 1998. Biochemistry of tropoelastin. *Eur J Biochem* 258: 1–18.
2. Mithieux, S.M., Rasko, J.E.J., Weiss, A.S. 2004. Synthetic elastin hydrogels derived from massive elastic assemblies of self-organized human protein monomers. *Biomaterials* 25: 4921–4927.
3. Wise, S.G., Mithieux, S.M., Raftery, M.J., Weiss, A.S. 2005. Specificity in the coacervation of tropoelastin: Solvent exposed lysines. *J Struct Biol* 149: 273–281.
4. Mecham, R.P., Heuser, J.E. 1991. The elastic fiber (Chapter 3). In *Cell Biology and Extracellular Matrix*, 2nd edition. Hay, E.D., Ed., Plenum Press, New York, pp. 79–110.
5. Muiznieks, L.D., Jensen, S.A., Weiss, A.S. 2003. Structural changes and facilitated association of tropoelastin. *Arch Biochem Biophys* 410: 317–323.
6. Pasquali-Ronchetti, I., Baccarani-Contri, M. 1997. Elastic fiber during development and aging. *Microsc Res Tech* 38: 428–435.
7. Pasquali-Ronchetti, I., Fornieri, C., Baccarani-Contri, M., Quaglino, D. 1995. Ultrastructure of elastin. *Ciba Found Symp* 192: 31-42; discussion 42–50.

8. Pasquali-Ronchetti, I., Fornieri, C., Baccarani-Contri, M., Volpin, D. 1979. The ultrastructure of elastin revealed by freeze-fracture electron microscopy. *Micron* 10: 89–99.

9. Fornieri, C., Pasquali-Ronchetti, I., Edman, A.C., Sjöström, M. 1982. Contribution of cryotechniques to the study of elastin ultrastructure. *J Microsc* 126: 87–93.

10. Pasquali-Ronchetti, I., Alessandrini, A., Baccarani-Contri, M., Fornieri, C., Mori, G., Quaglino, D., Valdrè, U. 1998. The elastic fiber organization by scanning force microscopy. *Matrix Biol* 17: 75–83.

11. Cox, B.A., Starcher, B.C., Urry, D.W. 1974. Coacervation of tropoelastin results in fiber formation. *J Biol Chem* 249: 997–998.

12. Bressan, G.M., Pasquali Ronchetti, I., Fornieri, C., Mattioli, F., Castellani, I., Volpin, D. 1986. Relevance of aggregation properties of tropoelastin to the assembly and structure of elastin fibers. *J Ultrastruct Mol Struct Res* 94: 209–216.

13. Urry, D.W., Long, M.M. 1977. On the conformation, coacervation and function of polymeric models of elastin. *Adv Exp Med Biol* 79: 685–714.

14. Kozel, B.A., Ciliberto, C.H., Mecham, R.P. 2004. Deposition of tropoelastin into the extracellular matrix requires a competent elastic fiber scaffold but not live cells. *Matrix Biol* 23: 23–34.

15. Hirai, M., Horiguchi, M., Ohbayashi, T., Kita, T., Chien, K.R., Nakamura, T. 2007. Latent TGF-beta-binding protein 2 binds to DANCE/fibulin-5 and regulates elastic fiber assembly. *EMBO J* 26: 3283–3295.

16. Baccarani Contri, M., Vincenzi, D., Cicchetti, F., Mori, G., Pasquali Ronchetti, I. 1990. Immunochemical localization of proteoglycans within normal elastin fibre. *Eur J Cell Biol* 53: 305–312.

17. Gheduzzi, D., Guerra, D., Bochicchio, B., et al. 2005. Heparan sulphate interacts with tropoelastin, with some tropoelastin peptides and is present in human dermis elastic fibers. *Matrix Biol* 24: 15–25.

18. Baccarani-Contri, M., Fornieri, C., Pasquali-Ronchetti, I. 1985. Elastin proteoglycans association revealed by cytochemical methods. *Connect Tissue Res* 13: 237–249.

19. Fornieri, C., Baccarani-Contri, M., Quaglino, D., Pasquali-Ronchetti, I. 1987. Lysyl oxidase activity and elastin-glycosaminoglycans interactions in growing chick and rat aortas. *J Cell Biol* 105: 1463–1469.

20. Pasquali-Ronchetti, I., Baccarani-Contri, M., Young, R.D., Vogel, A., Steinmann, B., Royce, P.M. 1994. Ultrastructural analysis of skin and aorta from a patient with Menkes disease. *Exp Mol Pathol* 61: 36–57.

21. Volpin, D., Pasquali-Ronchetti, I. 1977. The ultrastructure of high-temperature coacervates from elastin. *J Ultrastruct Res* 61: 295–302.

22. Wu, W.J., Vrhovski, B., Weiss, A.S. 1999. Glycosaminoglycans mediate the coacervation of human tropoelastin through dominant charge interactions involving lysine side chains. *J Biol Chem* 274: 21719–21724.

23. Broekelmann, T.J., Kozel, B.A., Ishibashi, H., et al. 2005. Tropoelastin interacts with cell-surface glycosaminoglycans via its COOh-terminal domain. *J Biol Chem* 280: 40939–40947.

24. Cain, S.A., Baldock, C., Gallagher, J., et al. 2005. Fibrillin-1 interactions with heparin. Implications for microfibril and elastic fiber assembly. *J Biol Chem* 280: 30526–30537.

25. Buczek-Thomas, J.A., Chu, C.L., Rich, C.B., Stone, P.J., Foster, J.A., Nugent, M.A. 2002. Heparan sulphate depletion within pulnomary fibroblasts: Implications for elastigenesis and repair. *J Cell Physiol* 192: 294–303.

26. Bisaccia, F., Castiglione-Morelli, M.A., Spisani, S., et al. 1998. The amino acid sequence coded by the rarely expressed exon 26° of human elastin contains a stable beta-turn with chemotactic activity for monocytes. *Biochemistry* 37: 11128–11135.

27. Tamburro, A.M., Bochicchio, B., Pepe, A. 2003. Dissection of human tropoelastin: Exon-by exon chemical synthesis and related conformational studies. *Biochemistry* 42: 13347–13362.

28. Tamburro, A.M., Bochicchio, B., Pepe, A. 2005. The dissection of human tropoelastin: From the molecular structure to the self-assembly to the elasticity mechanism. *Pathol Biol* 53: 383–389.

29. Tamburro, A.M., Pepe, A., Bochicchio, B. 2006. Localizing alpha-elices in human tropoelastin: Assembly of the elastin puzzle. *Biochemistry* 45: 9518–9530.

30. Pepe, A., Guerra, D., Bochicchio, B., Quaglino, D., Pasquali Ronchetti, I., Tamburo, A.M. 2005. Dissection of human tropoelastin self-assembly: Supramolecular organization of polypeptide sequences coded by particular exons. *Matrix Biol* 24: 96–109.

31. Tamburro, A.M., Pepe, A., Bochicchio, B., Quaglino, D., Pasquali Ronchetti, I. 2005. Supramolecular, amyloid-like organization of the polypeptide sequence coded by Exon 30 of human tropoelastin. *J Biol Chem* 280: 2682–2690.

32. Senior, R.M., Griffin, G.L., Mecham, R.P., Wrenn, D.S., Prasad, K.U., Urry, D.W. 1984. Val-Gly-Val-Ala-Pro-Gly, a repeating peptide in elastin, is chemotactic for fibroblasts and monocytes. *J Cell Biol* 99: 870–874.

33. Long, M.M., King, V.J., Prasad, K.U., Urry, D.W. 1988. Chemotaxis of fibroblasts toward nonapeptide of elastin. *Biochim Biophys Acta* 968: 300–311.

34. Long, M.M., King, V.J., Prasad, K.U., Freeman, B.A., Urry, D.W. 1989. Elastin repeat peptides as chemoattractants for bovine aortic endothelial cells. *J Cell Physiol* 140: 512–518.

35. Castiglione-Morelli, M.A., Bisaccia, F., Spisani, S., De Blasi, M., Traniello, S., Tamburro, A.M. 1997. Structure-activity relationships for some elastin-derived peptides chemoattractants. *J Pept Res* 49: 492–499.

36. Kohidai, L., Kun, L., Eva, P., et al. 2004. Cell-physiological effects of elastin derived (VGVAPG)n oligomers in a unicellular model system. *J Pept Sci* 10: 427–438.

37. Jung, S., Hinek, A., Tsugu, A., et al. 1999. Astrocytoma cell interaction with elastin substrates: Implications for astrocytoma invasive potential. *Glia* 25: 179–189.

38. Mochizuki, S., Brassart, B., Hinek, A. 2002. Signaling pathways transduced through the elastin receptor facilitate proliferation of arterial smooth muscle cells. *J Biol Chem* 277: 44854–44863.

39. Jung, S., Rutka, J.T., Hinek, A. 1998. Tropoelastin and elastin degradation products promote proliferation of human astrocytoma cell lines. *J Neuropathol Exp Neurol* 57: 439–448.

40. Gminski, J., Weglarz, L., Drozdz, M., Sulkowski, P., Goss, M. 1991a. Modulation of elastase-like activity in fibroblasts stimulated with elastin peptides. *Biochem Med Metab Biol* 45: 254–257.

41. Hance, K.A., Tataria, M., Ziporin, S.J., Lee, J.K., Thompson, R.W. 2002. Monocyte chemotactic activity in human abdominal aortic aneurisms: Role of elastin degradation peptides and the 67-kDa cell surface elastin receptor. *J Vasc Surg* 35: 254–261.

42. Daynes, R.A., Thomas, M., Alvarez, V.L., Sandberg, L.B. 1977. The antigenicity of soluble porcine elastins: I. Measurement of antibody by a radioimmunoessay. *Connect Tissue Res* 5: 75–82.

43. Mecham, R.P., Lange, G. 1982. Antigenicity of elastin: Characterization of major antigenic determinants on purified insoluble elastin. *Biochemistry* 21: 669–673.

44. Maeda, I., Mizoiri, N., Briones, M.P., Okamoto, K. 2007. Induction of macrophage migration through lactose-insensitive receptor by elastin-derived nonapeptides and their analog. *J Pept Sci* 13: 263–268.

45. Ostuni, A., Lograno, M.D., Gasparro, A.M., Bisaccia, F., Tamburro, A.M. 2002. Novel properties of peptides derived from the sequnce coded by exon 26A of human elastin. *Int J Biochem Cell Biol* 34: 130–135.

46. Lograno, M.D., Bisaccia, F., Ostuni, A., Daniele, E., Tamburro, A.M. 1998. Identification of elastin peptides with vasorelaxant activity on rat thoracic aorta. *Int J Biochem Cell Physiol* 30: 497–503.

47. Bisaccia, F., Morelli, M.A., DeBiasi, M., Traniello, S., Spisani, S., Tamburro, A.M. 1994. Migration of monocytes in the presence of elastolytic fragments of elastin and in synthetic derivates. Structure-activity relationships. *Int J Pept Protein Res* 44: 332–341.

48. Fulop, T. Jr., Jacob, M.P., Khalil, A., Wallach, J., Robert, L. 1998. Biological effects of elastin peptides. *Pathol Biol (Paris)* 46: 497–506.

49. Fulop, T. Jr., Larbi, A.A., Robert, L., Khalil, A. 2005. Elastin peptide induced oxidation of LDL by phagocytic cells. *Pathol Biol (Paris)*. 53: 416–423.

50. Antonicelli, F., Bellon, G., Debelle, L., Hornebeck, W. 2007. Elastin-elastases and inflammation-aging. *Curr Top Dev Biol* 79: 99–155.

51. Hauck, M., Seres, I., Kiss, I., et al. 1995. Effects of synthesized elastin peptides on human leukocytes. *Biochem Mol Biol Int* 37: 45–55.

52. Gminski, J., Weglarz, L., Drozdz, M., Goss, M. 1991b. Pharmacological modulation of the antioxidant enzyme activities and the concentration of peroxidation products in fibroblasts stimulated with elastin peptides. *Gen Pharmacol* 22: 495–497.

53. Baranek, T., Debret, R., Antonicelli, F., et al. 2007. Elastin receptor (spliced galactosidase) occupancy by elastin peptides counteracts proinflammatory cytokine expression in lipopolysaccharide-stimulated human monocytes through NF-kappaB down regulation. *J Immunol* 179: 6184–6192.

54. Duca, L., Floquet, N., Alix, A.J., Haye, B., Debelle, L. 2004. Elastin as a matrikine. *Crit Rev Oncol Hematol* 49: 235–244.

55. Robinet, A., Millart, H., Oszust, F., Hornebeck, W., Bellon, G. 2007. Binding of elastin peptides to S-Gal protects the heart against ischemia/reperfusion injury by triggering the RISK pathway. *FASEB J* 21: 1968–1978.

56. Robinet, A., Fahem, A., Cauchard, J.H., et al. 2005. Elastin-derived peptides enhance angiogenesis by promoting endothelial cell migration and tubulogenesis through upregulation of MT1-MMP. *J Cell Sci* 118: 343–356.

57. Dreher, M.R., Elas, M., Ichikawa, K., et al. 2004. Nitroxide conjugate of a thermally responsive elastin-like polypeptide for noninvasive thermometry. *Med Phys* 31: 2755–2762.

58. Chilkoti, A., Dreher, M.R., Meyer, D.E., Raucher, D. 2002. Targeted drug delivery by thermally responsive polymers. *Adv Drug Deliv Rev* 54: 613–630.
59. Megeed, Z., Cappello, J., Ghandehari, H. 2002. Genetically engineered silk-elastinlike protein polymers for controlled drug delivery. *Adv Drug Deliv Rev* 54: 1075–1091.
60. Haider, M., Leung, V., Ferrari, F., et al. 2005. Molecular engineering of silk-elastinlike polymers for matrix-mediated gene delivery: Biosynthesis and characterization. *Mol Pharm* 2: 139–150.
61. Meyer, D.E., Kong, G.A., Dewhirst, M.W., Zalutsky, M.R., Chilkoti, A. 2001. Targeting a genetically engineered elastin-like polypeptide to solid tumors by local hyperthermia. *Cancer Res* 61: 1548–1554.
62. Dinerman, A.A., Cappello, J., Ghandehari, H., Hoag, S.W. 2002. Swelling behavior of a genetically engineered silk-elastinlike protein polymer hydrogel. *Biomaterials* 23: 4203–4210.
63. Bidwell, G.L., 3rd, Fokt, I., Priebe, W., Raucher, D. 2007a. Development of elastin-like polypeptide for thermally targeted delivery of doxorubicin. *Biochem Pharmacol* 73: 620–631.
64. Dreher, M.R., Raucher, D., Balu, N., Michael-Colvin, O., Ludeman, S.M., Chilkoti, A. 2003. Evaluation of an elastin-like polypeptide-doxorubicin conjugate for cancer therapy. *J Control Release* 91: 31–43.
65. Bidwell, G.L., 3rd, Davis, A.N., Fokt, I., Priebe, W., Raucher, D. 2007b. A thermally targeted elastin-like polypeptide-doxorubicin conjugate overcomes drug resistance. *Invest New Drugs* 25: 313–326.
66. Narine, K., De Wever, O., Van Valckenborgh, D., et al. 2006. Growth factor modulation of fibroblast proliferation, differentiation, and invasion: Implication for tissue valve engineering. *Tissue Eng* 12: 2707–2716.
67. Paule, W.J., Bernick, S.B., Nimni, M.E. 1992. Calcification of implanted vascular tissues associated with elastin in an experimental animal model. *J Biomed Mater Res* 26: 1169–1177.
68. Wachi, H., Sugitani, H., Murata, H., Nakazawa, J., Mecham, R.P., Seyama, Y. 2004. Tropoelastin inhibits vascular calcification via 67-kDa elastin binding protein in cultured bovine aortic smooth muscle cells. *J Atheroscl Thromb* 11: 159–166.
69. Fiore, C.E., Celotta, G., Politi, G.G., et al. 2007. Association of high alpha2-Heremans-Schmid glycoprotein/fetuin concentration in serum and intima-media thickness in patients with atherosclerotic vascular disease and low bone mass. *Atherosclerosis* 195: 110–115.
70. Gheduzzi, D., Boraldi, F., Annovi, G., et al. 2007. Matrix Gla Protein (MGP) in Pseudoxanthoma Elasticum (PXE) and in genetic disorders with PXE-like clinical manifestations. *Lab Invest* 87: 998–1008.
71. Arias, F.J., Remoto, W., Martin, S., Lòpez, I., Rodriguez-Cabello, J.C. 2006. Tailored recombinant elastin-like polymers for advanced biomedical and nano(bio)technological applications. *Biotechnol Lett* 28: 687–695.
72. Rodriguez-Cabello, J.C. 2004. Smart elastin-like polymers. *Adv Exp Med Biol* 553: 45–57.
73. Rodriguez-Cabello, J.C., Prieto, S., Arias, F.J., Reguera, J., Ribeiro, A. 2006. Nanobiotechnological approach to engineered biomaterial design: The example of elastin-like polymers. *Nanobiomed* 1: 267–280.

74. Girotti, A., Reguera, J., Rodriguez-Cabello, J.C., Arias, F.P., Alonso, M., Matestera, A. 2004. Design and bioproduction of a recombinant multi(bio)functional elastin-like protein polymer containing cell adhesion sequences for tissue engineering purposes. *J Mater Sci Mater Med* 15: 479–484.

75. Dutoya, S., Levebvre, F., Deminières, C., et al. 1998. Unexpected original property of elastin-derived proteins: Spontaneous tight coupling with natural and synthetic polymers. *Biomaterials* 19: 147–155.

76. Dutoya, S., Verna, A., Lefebvre, F., Rabaud, M. 2000. Elastin-derived protein coating onto poly(ethylene terephthalate). Technical, microstructural and biological studies. *Biomaterials* 21: 1521–1529.

77. Cai, K., Yao, K., Lin, S., et al. 2002. Poly(D,L-lactic acid) surfaces modified by silk fibroin: Effects on the culture of osteoblast *in vitro*. *Biomaterials* 23: 1153–1160.

78. Daamen, W.F., Van Moerkerk, H.T., Hafmans, T., et al. 2003. Preparation and evaluation of molecularly-defined collagen-elastin-glycosaminoglycan scaffolds for tissue engineering. *Biomaterials* 24: 4001–4009.

79. Buttafoco, L., Kolkman, N.G., Engbers-Buijtenhuijs, P., et al. 2006. Electrospinning of collagen and elastin for tissue engineering applications. *Biomaterials* 27: 724–734.

80. Davidson, J.M., Giro, G.M., Quaglino, D. 1992. Elastin repair. In *Wound Healing*. Cohen, I.K., Diegelman, R., and Lindblad, W., Eds., Saunders, Baltimore, pp. 223–236.

81. Engbers-Buijtenhuijs, P., Buttafoco, L., Poot, A.A., et al. 2006. Biological characterization of vascular grafts cultured in a bioreactor. *Biomaterials* 27: 2390–2397.

82. Daamen, W.F., Nillesen, S.T., Wismans, R.G., Reinhardt, D.P., Hafmans, T., Veerkamp, J.H., Van Kuppevelt, T.H. 2008. A biomaterial composed of collagen and solubilized elastin enhances angiogenesis and elastic fiber formation without calcification. *Tissue Eng Part A* 14: 349–360.

83. Sales, V.L., Engelmayr, G.C. Jr., Mettler, B.A., Johnson, J.A. Jr., Sacks, M.S., Mayer, J.E. Jr. 2006. Transforming growth factor-beta1 modulates extracellular matrix production, proliferation and apoptosis of endothelial progenitor cells in tissue-engineering scaffolds. *Circulation* 114: 1193–1199.

84. Shahabeddin, L., Berthod, F., Damour, O., Collombel, C. 1990. Characterization of skin reconstructed on a chitosan-cross-linked collagen-glycosaminoglycan matrix. *Skin Pharmacol* 3: 107–114.

85. Duplan-Perrat, F., Damour, O., Montrocher, C., et al. 2000. Keratinocytes influence the maturation and organization of the elastin network in a skin equivalent. *J Invest Dermatol* 114: 365–370.

86. Noblesse, E., Cenizo, V., Bouez, C., et al. 2004. Lysyl oxidase-like and lysyl oxidase are present in the dermis and epidermis of a skin equivalent and in human skin and are associated to elastic fibers. *J Invest Dermatol* 122: 621–630.

11

PLA-Based Bio- and Nanocomposites

Johannes Ganster and Hans-Peter Fink

Fraunhofer Institute for Applied Polymer Research, Germany

CONTENTS

11.1 Introduction

Polylactic acid (PLA) (for example, see Reference [1]) is one of the most important bio-based and biodegradable thermoplastic polymers commercially available today. Production capacities are located in the United States (Nature Works, Minnetonka, Minnesota) and Asia (Unitika, Mitsui Chemicals, Japan), and a 60,000 metric ton per year plant is due to come on stream in Guben, Germany, in 2011. The main industrial method used to synthesize PLA is ring-opening polymerization of dilactide, which is usually obtained by fermentation of corn starch. There are three stereoisomers of dilactide—L-lactide, D-lactide, and *meso*-lactide, as shown in Figure 11.1.

Most polymers are produced from L-lactide with varying degrees of purity, leading to a certain percentage of *meso* sequences in the polymer chain. This has consequences for crystallization and mechanical properties, decreases glass transition temperature and melting point, and increases biodegradability. PLA in its amorphous state is transparent and resembles unmodified polystyrene in terms of high stiffness and low impact strength.

The traditional applications for PLA (and its copolymers) have been mainly in the medical sector (bioabsorbable implants, sutures, etc.) (for example, see Reference [2]), (food) packaging (films, trays, bags, etc. [3,4]), and agriculture (mulch films, bags [5]). To improve material properties for known applications and also open up new fields of use, for example in the durable goods sector, PLA polymers have been reinforced with various fibers and other

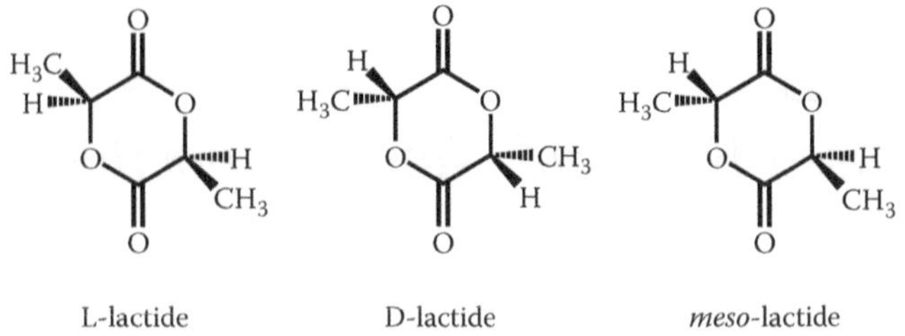

L-lactide   D-lactide   *meso*-lactide

FIGURE 11.1
Stereoisomers of dilactide.

fillers, including nanoparticles. Biobased, biodegradable reinforcing fibers are obviously of particular interest as they make it possible to create "green" composites based entirely on renewable resources.

In the first section, attempts to improve PLA properties by reinforcing with lignocellulosic fibers (both wood based and from annual plants) and man-made cellulosic fibers (rayon) are outlined. Our own results with thermomechanical pulp fibers and rayon are reported and compared with findings in the current literature on this topic. Throughout this section, the mechanical properties of composites obtained from melt compounding are mainly considered, because this is the most industrially relevant technology for fusible polymers owing to the avoidance of solvent-related problems.

As an alternative to reinforcement with cellulose-based fibers in the microscale range, nanoscale reinforcement is discussed in the second section. These fillers are not normally biodegradable (except for cellulose nanoparticles), but just a few percent often suffice to produce remarkable changes in properties. The current literature is reviewed, again mostly from the viewpoint of mechanical (and thermomechanical) properties. Finally, a concluding comparison is made between micro- and nanoscale reinforcement.

11.2 Bio-Based Composites

11.2.1 Wood-Based and Natural Fibers

Truly bio-based PLA composites can only be manufactured using a reinforcing phase based on renewable raw materials. Here the obvious choice would be cellulosic and lignocellulosic fibers. A typical example of PLA (NatureWorks 7000D, M_n = 80,000 g/mol, M_w = 150,000 g/mol from gel permeation chromatography (GPC), optical purity 92%) reinforced with spruce fibers from thermomechanical pulping (TMP) is given in Figure 11.2, where tensile strength and modulus are shown for injection-molded composites of varying fiber content [6].

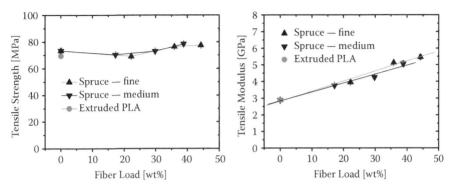

FIGURE 11.2
Tensile strength (left) and modulus (right) of spruce thermomechanical pulp fiber-reinforced polylactic acid composites as a function of fiber content.

Obviously, PLA strength is neither decreased nor improved by the addition of up to 45 wt% of fiber. By contrast, stiffness is almost doubled. At the same time, tensile elongation at break and (unnotched) Charpy impact strength are reduced considerably, as shown in Figure 11.3.

This reduction, particularly with respect to impact properties, is a general problem with wood and plant fiber-reinforced injection-molded composites and can be overcome by the use of spun cellulose fibers (see below).

The fiber fractions used in the composites were obtained from a commercial TMP product by sieving through appropriate meshes, giving fiber length number averages of 0.9 mm and 2.9 mm for the fine and medium fractions, respectively. Fiber width number averages were 32 μm and 38 μm, respectively, giving aspect ratios (length to width) of 28 and 76, respectively. However, from Figure 11.2 and Figure 11.3, no clear influence of the precompounding fiber length or aspect ratio on composite properties can be

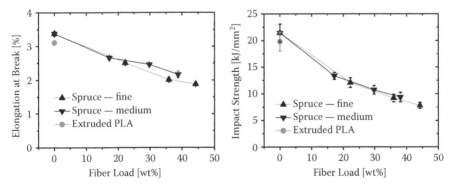

FIGURE 11.3
Elongation at break (left) and Charpy unnotched impact strength (right) of spruce thermomechanical pulp fiber-reinforced polylactic acid composites as a function of fiber content.

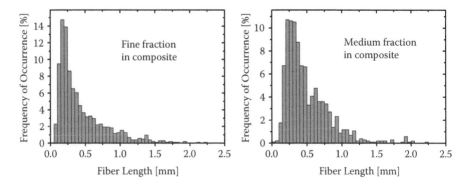

FIGURE 11.4
Fiber length distribution of spruce thermomechanical pulping fine (left) and medium (right) fractions in injection-molded polypropylene composites with 20% fiber content.

seen. Obviously, fiber lengths are reduced during the compounding process. This is shown in Figure 11.4 for fine and medium fractions Soxhlet extracted from, in this case, polypropylene (PP) composites and measured by optical microscopy with a sample of 1000 single fibers.

Despite the fact that the fibers in the medium fraction are, on average, three times the length of the fibers in the fine fraction before processing, after extrusion, the number averages are almost the same (i.e., 0.56 mm and 0.44 mm, respectively). The aspect ratio is reduced in both cases to about 20.

Scanning electron microscopy (SEM) cryofracture surfaces of composites with 20% fine fraction are shown in Figure 11.5, where the overview (left) demonstrates the general homogeneity of the composites, and the detail (right) can be used to judge fiber/matrix adhesion. There is no simple pullout of fibers. Fiber stumps are seen for fibers oriented in the loading direction (lower left corner), while perpendicularly oriented fibers are split lengthwise, revealing the internal fibrillar structure (center and upper right corner). This

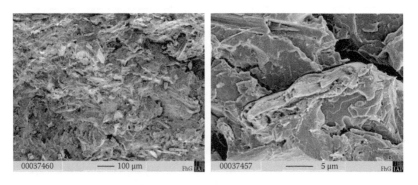

FIGURE 11.5
Scanning electron microscopy cryofracture surfaces of spruce thermomechanical pulp and polylactic acid composites.

internal structure with weak interaction between the fibrils is likely to cause a reduction in impact strength.

Other attempts to improve mechanical properties by (ligno)cellulosic fiber reinforcement through melt compounding are reported in the literature, with results showing the same overall tendencies. Huda et al. [7] improved the tensile modulus of a PLA (Biomer L 9000, Biomer, Germany) with a weight-average molecular weight of 20,000 g/mol (M_n = 10,000) from 2.7 GPa to 6.7 GPa by using 30% of a certain grade of recycled newspaper fiber. Strength was not changed significantly, and notched Izod impact strength, for another fiber grade, showed a linear decrease from 25 J/m (unfilled) to 15 J/m (30% fiber content). Only with chopped glass fibers were the authors able to improve tensile strength by 27% and notched Izod impact strength by 50% [8].

Three types of cellulosic reinforcement were investigated by Mathew et al. [9] with a PLA having a molecular weight of M_w = 97,000 g/mol (Pollait, Fortum Oil, Porvoo, Finland). Static tensile and dynamic mechanical analysis (DMA) was performed on composites containing 25 wt% of microcrystalline cellulose (MCC), low-lignin cellulose fibers (CF, Terracel, Rayonier, United States), and wood flour (WF). Tensile strength dropped from 50 MPa for the unreinforced PLA to 36 MPa, 45 MPa, and 45 MPa, for composites with MCC, CF, and WF, respectively, while the modulus was increased from 3.6 GPa to 5.0 GPa, 6.0 GPa, and 6.3 GPa. The relatively poor performance of MCC is attributed to the surface topography (smoothness) and low aspect ratio (length to width) of the filler.

Attempts have been made by the same Norwegian group to go one step further from MMC to nanoscaled cellulose reinforcement [10,11]. The difficulties encountered here mostly concerned the aggregation tendency of these kinds of particles. So, solution-based methods were used. In Reference [10], a cellulose nanowhisker suspension in LiCl/DMAc was fed into the extruder with subsequent removal of the solvent by a venting system in the extrusion process. In Reference [11], a complete solution method followed by film casting was pursued. Both studies were focused on structural features, and a straightforward improvement in mechanical properties was not reported. Iwatake et al. [12] improved the tensile strength of 300 μm thick compression-molded PLA films from 50 MPa to 70 MPa by reinforcing with 10% microfibrillated cellulose (mostly consisting of nanofibers) via a solution method. Higher fiber contents decreased this property again. Impact data are not reported.

Non-wood-based cellulosic fibers, namely flax [13], hemp [14], kenaf [15], and bamboo [16], have also been tested for their reinforcing capabilities in PLA. Again, tensile modulus increased while strength diminished. When impact data were reported, values were lower for the composites than for the unreinforced polymer, with one exception. For kenaf [15], improvements in notched Charpy impact strength of 65% were obtained with 30% fiber content (from a reported 7.4 J/m to 12.2 J/m, probably 1.8 kJ/m^2 to 3 kJ/m^2).

TABLE 11.1

Diameter and Mechanical Properties of Cellulose Man-Made Fibers from Single-Fiber Measurements

Material	Diameter [μm]	Strength [MPa] Average	s.d.[a]	Elongation at Break [%] Average	s.d.	Modulus [GPa] Average	s.d.
Cordenka 700	12	830	30	13	2	20	1.5
Enka viscose	15	310	15	24	1	11	0.5
Viscose sliver	10	340	40	12	4	11	1
NewCell	10	600	30	9	1	31	3
Tencel sliver	11	550	40	11	2	23	2
Carbamate	11	360	15	8	1	22	1

[a] Standard deviation.

11.2.2 Man-Made Cellulosic Fibers

A biobased, biodegradable alternative to plant- and wood-based fibrous PLA reinforcement is offered by the use of cellulosic man-made fibers. To produce these fibers, cellulose (in the specially pure form of dissolving pulp) is completely dissolved in an appropriate solvent, either as it is (nonderivative methods, like N-methylmorpholine-N-oxide [NMMO]) or as a cellulose derivative (derivative methods, like viscose or cellulose carbamate). Fibers are spun from solution by wet spinning or dry jet-wet spinning processes into filament yarn or staple fiber. A well-defined, homogeneous structure is obtained, which does not have the drawbacks of an internal composite structure as is the case for the reinforcements discussed above. Typical diameters and mechanical properties from single fiber tests [17] are shown in Table 11.1.

Cordenka 700 is a technical high-tenacity fiber used to reinforce high-speed tires (Cordenka GmbH Obernburg, Germany), and Enka viscose is a typical textile fiber (Enka GmbH & Co. KG, Oberbruch, Germany). Viscose and Tencel sliver originate from staple fiber production and NewCell from an NMMO filament process. Carbamate is a cellulose filament spun in this institute via the CARBACELL process [18], an environment-friendly alternative to the viscose process using urea as the derivatizing agent.

These kinds of fibers have a high elongation at break value and, especially in the case of rayon tire cord yarn, a high level of mechanical properties (Table 11.1). Their reinforcing capabilities have been intensively investigated [17,19–21] for polypropylene (PP), the most commonly used thermoplastic in the automotive sector, with results matching those for short glass fiber–reinforced PP and polycarbonate/acrylonitrile-butadiene styrol polymer blends (PC/ABS). In addition to improving strength and modulus, rayon tire cord yarn proved to be an excellent impact modifier, both at room and low temperature, owing to its high elongation at break and structural homogeneity.

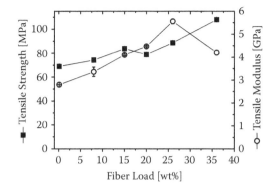

FIGURE 11.6
Tensile strength and modulus of cellulose rayon-reinforced polylactic acid as a function of fiber content.

Soon it became clear that these properties could also be used to improve the notoriously low impact strength of PLA [19].

In Figure 11.6, the tensile strength and modulus of rayon tire cord yarn (Cordenka 700)-reinforced PLA (the same as used above) are shown as a function of fiber content. The composites were manufactured in the way described in Reference [19] using a pultrusion technique with subsequent melt homogenization. A clear improvement in tensile strength is seen with increasing fiber content, starting from 70 MPa for the unreinforced PLA and reaching 108 MPa for 36% fiber content. This is clearly different from the behavior with wood-based or natural fibers, where in many cases a decrease in strength results from the introduction of these fillers, because they act as weak points owing to the internal composite structure with poor lateral adhesion. Modulus also increases from 3 GPa to almost 6 GPa at 25% fiber content, followed by a decrease for the high fiber content of 36%. This is probably a result of a somewhat reduced fiber orientation due to fiber entanglements caused by the high fiber concentration.

Charpy unnotched and notched impact strengths measured at room temperature are shown in Figure 11.7. Here the advantages of using a man-made cellulose fiber become evident.

Unnotched impact strength increases from 20 kJ/m^2 to almost 70 kJ/m^2, and the notched value from 2 kJ/m^2 to 8 kJ/m^2 for the unreinforced PLA and 36% composite, respectively.

This dependency is also maintained at lower temperatures, as demonstrated for the notched values at –18°C in Figure 11.8.

Fiber length distribution in a test bar injection molded from the 25% rayon-PLA composite is shown in Figure 11.9.

The initial length after pelletizing in the pultrusion step was roughly 4 mm, so that the homogenization step together with the second pelletizing

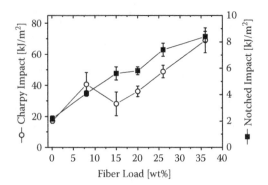

FIGURE 11.7
Unnotched and notched Charpy impact strength of cellulose rayon-reinforced polylactic acid as a function of fiber content at room temperature.

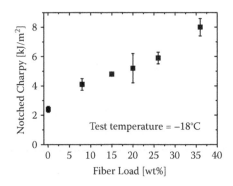

FIGURE 11.8
Notched Charpy impact strength of cellulose rayon-reinforced polylactic acid as a function of fiber content at –18°C.

step reduced fiber length considerably, an effect known from rayon–PP composites [19]. However, compared with the analogous distribution for TMP fibers (Figure 11.4), the maximum is shifted toward longer fibers and a longer tail is evident, indicating that the fibers are less damaged by the processing procedure. Moreover, with a number average fiber length of 1.14 mm and a fiber diameter of 12 μm, the average aspect ratio is almost 100 and thus far greater than for the wood-based compounds.

Cryofracture surface morphologies obtained by SEM are shown in Figure 11.10.

In the overview micrograph (left) demonstrating the homogeneity of the composite, fiber stumps, pullouts, and fractured matrix parts are visible. In the detail picture (right), the typical rayon fracture morphology is revealed, and small gaps are seen between the fiber and the matrix. Fiber coupling is not yet optimal, but a really uncoupled system has far more extensive fiber pullout and appears macroscopically like a "paintbrush."

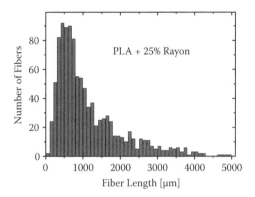

FIGURE 11.9
Fiber length distribution of cellulose rayon-reinforced polylactic acid in an injection-molded composite with 25% fiber content.

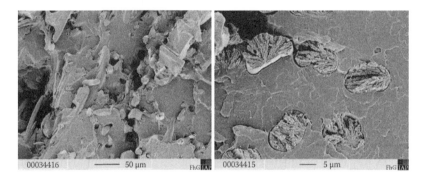

FIGURE 11.10
Scanning electron microscopy cryofracture surfaces of cellulose rayon-polylactic acid composites.

Similar results were obtained for a different PLA type by Bax and Müssig [22]. They slightly increased tensile strength from 50 MPa for the unreinforced polymer to 58 MPa for the polymer with 30% Cordenka fiber addition. However, Charpy unnotched impact strength was raised from 16 kJ/m^2 to about 70 kJ/m^2, fully corroborating the findings reported above.

Altogether, it can be concluded that with wood-based, natural, and manmade cellulosic fibers, PLA can be reinforced to create a fully biobased, biodegradable composite with improved mechanical properties. For wood-based and natural fibers, stiffness is increased considerably, while strength remains roughly the same (and is sometimes reduced). Impact properties, already very low for unreinforced PLA, are further reduced. On the other hand, man-made cellulose fibers lead to systematic, albeit moderate, gains in strength along with a remarkable improvement in impact properties, so overcoming an inherent weakness in this bioplastic.

11.3 Nanocomposites

In greening the twenty-first century materials world, biodegradable polymers and their layered silicate nanocomposites are hailed as a new class of materials with tremendous potential [23]. In addition to layered silicates, other fillers have been incorporated in nano dimensions by various processes, including carbon nanotubes [24–28], layered titanates [29], titanium dioxide [30], silica [31,32], synthetic mica [33], and calcium carbonate [34]. Compared to other types of reinforcement, where 20% to 40% filler is needed, nanofillers offer the possibility of achieving the desired effect with very low filler contents, often in the range of a few percent. Owing to the very small filler dimensions, the interface between the matrix and filler is large, even at low filler concentration, and filler properties can be largely transferred to the matrix.

Owing to the vast amount of scientific literature on PLA combined with nanofillers, only selected papers reporting tensile or bending properties, including strength data, and dealing with melt compounding techniques are discussed as follows.

Naturally occurring layered silicates have gained the most attention in recent years. Here, microsized particles consisting of nanoscale silica layers (with foreign ions) are intercalated by or exfoliated in the matrix polymer. To facilitate this process, the inorganic ions keeping together the silicate layers and compensating the net charge of the layers are exchanged for larger organic ions, giving rise to organomodified layered silicates, often modified bentonites or montmorillonites (MMTs).

In 2003, Ray et al. published a series of papers [35–38] comparing the structure and properties of injection-molded PLA composites filled with five types of organomodified clay. A PLA with 1.1% to 1.7% D content (optical purity 96.6% to 97.8%), $M_w = 177{,}000$ g/mol, and $M_w/M_n = 1.58$ supplied by Unitika Co., Japan, was used in the investigations.

In terms of the flexural strength and strain at break of annealed (crystallized) samples, the authors found an optimum clay concentration of 4% for montmorillonite modified with octadecyltrimethylammonium cations [35] and 7% for synthetic fluorine mica modified with N-(cocoalkyl)-N,N-[bis(2-hydroxyethyl)]-N-methylammonium cations [36]. At the 4% level chosen for comparison between the clay types [37,38], flexural strength was improved from 86 MPa for PLA to more than 130 MPa for MMTs with either octadecyltrimethylammonium or octadecylammonium ions as modifier. Increases in moduli were moderate. Saponite clay, MMT modified with dioctadecyldimethylammonium, and synthetic mica performed less well. The latter, however, was shown to decrease the oxygen permeability of amorphous films to 35% of the value for pure PLA at mica contents from 10% up to 20%, while the other clays increased the oxygen barrier only slightly. This is certainly due in part to the larger lateral layer size for mica of 200 to 300 nm compared with 50 to 150 nm for the other clays [37]. The heat distortion (HDT) values of annealed samples were also increased by the use of clays, especially at load levels of around 1 MPa (HDT-B), where unfilled PLA has an HDT value of 75°C, while the clay-filled systems show values of 100°C to 110°C at 7% clay content. Impact values are not reported, but an increase in strain at break from 1.9% to 3.9% is observed for the MMT nanocomposites, indicating slightly increased ductility [37].

However, brittleness is still a problem for PLA-based composites, and it remains to be seen whether layered silicates can help in this respect. One hint in this direction was given recently by Pogodina et al. [39]. They used Nature Works PLA with an optical purity of 92%, Mw = 108,000 g/mol, and M_w/M_n = 1.7 melt compounded with commercial MMT modified with methyl bis-2-hydroxyethyl hydrogenated tallow ammonium (Cloisite 30B, Southern Clay Products, Gonzales, Texas). A clay content of 3% did not change the 60 MPa tensile strength of the unreinforced polymer but increased tensile elongation from 5% to 15%, while the modulus was increased from 1.9 GPa to roughly 2.3 GPa. Unfortunately, no impact data are given for this high-elongation composite.

Similarly, Jiang et al. [34] found improved elongation at break for Nature Works PLA 4032 D (Mw = 207,000 g/mol, M_w/M_n = 1.74) and MMT modified with the same cation. They reached approximately 12% elongation for 2.5% clay content, starting from 5% elongation. With nano-sized precipitated calcium carbonate, similar elongation values were obtained, although with higher concentrations of 5% and 7.5% filler. Again, failure behavior in short-term tests (i.e., impact conditions) was not investigated.

Another way of improving impact properties is, of course, to introduce impact modifiers such as polyethylene glycol (PEG) into the matrix. However, this leads to considerable reductions in strength and stiffness. This effect might be offset by the use of nanoclay reinforcement.

Tanoue et al. [40] used 5% PEG with a molecular weight of 2000 and 300,000 to 500,000 g/mol to plasticize a commercial PLA (LACEA H-100, Mitsui Chemicals, Japan, M_w = 140,000 g/mol) and added 5% of two types of organomodified clay (dimethylstearylammonium and dimethyldistearylammonium). Only with the latter and 5% of the PEG-2000 were they able to increase tensile elongation from 1.4% (PLA) to 1.9% (composite). The modulus remained the same, but tensile strength increased from 29 MPa to 35 MPa. No values for the plasticized PLA are given, so the action of the clays in the plasticized matrix remains unclear, and only a modest increase in elongation (ductility) was observed compared with pure PLA. No impact values are reported.

Much stronger effects are observed by Shibata et al. [41] for the same PLA type plasticized with diglycerine tetraacetate and reinforced with n-octadecylammonium (ODA)-modified MMT. Following the addition of 10% plasticizer, tensile strength was reduced from 65 MPa to 45 MPa and modulus from 2.6 GPa to 2.5 GPa, while elongation was not appreciably increased. However, after adding 3% ODA-modified clay, tensile elongation jumped to more than 200%, while strength and modulus did not change much. The published stress–strain curve for this composite shows a large drop in stress to half the peak value after the yield point at approximately 3% elongation. The unplasticized ODA-MMT composite also displays an increased elongation at break value of around 20% (numbers are not given in the paper and values are estimated from the graphs) versus 3% to 4% for the pure PLA. Again, impact behavior is not considered.

Carbon nanotubes (CNTs) have also been used for improving the mechanical properties of PLA [25,26]. Again, these fillers cannot be used without modification, and pretreatment is necessary to functionalize the pure carbon structure and facilitate filler–matrix interactions. This effect can be enhanced by additional modification of the matrix. Kuan et al. [26] grafted PLA with γ-chloropropyltriethoxysilane and modified the CNTs with vinyltrimethoxysilane. Wu and Liao [25] grafted PLA with acrylic acid and functionalized the CNTs with 6-hydroxyhexylester. In the latter case, tensile strength was improved from 49 MPa to 62 MPa with 1% filler content. In the former case, with 4% filler, strength was increased from 57 MPa to 62 MPa, while, following a water crosslinking reaction after 7 h, these values changed to 63 MPa and 69 MPa, respectively. In neither case are other mechanical characteristics reported.

11.4 Summary

Reinforcement of PLA with wood-based and plant fibers, man-made cellulose fibers, and nanostructures such as nanoclays or carbon nanotubes can lead to considerable improvements in mechanical properties. Generally speaking, natural fibers improve the stiffness and, in rare cases, strength of the composite, while impact properties, already very low for pure PLA, are further reduced. The problem of maintaining the bio-based, biodegradable nature of the reinforcement can be overcome by using man-made cellulose fibers (e.g., rayon tire cord yarn). Impact properties are dramatically improved along with strength and stiffness, an effect not observed with conventional impact modifiers.

Nanoscale reinforcements, albeit not biodegradable, can be successfully used at very low concentrations (1% to, say, 4%) to improve stiffness, strength, and tensile elongation. To the authors' knowledge, impact behavior is not reported in the literature. It remains to be seen whether composites with increased tensile elongation (i.e., increased ductility at low testing speeds) have reduced brittleness under impact conditions.

References

1. Södergård, A., Stolt, M., 2002. Properties of lactic acid based polymers and their correlation with composition. *Prog. Polym. Sci.* 27: 1123–1163.
2. Middleton, J.C., Tipton, A.J., 2000. Synthetic biodegradable polymers as orthopedic devices. *Biomaterials* 21: 2235–2246.
3. Auras, R., Harte, B., Selke, S., 2004. An overview of polylactides as packaging materials. *Macromol. Biosci.* 4: 835–864.
4. Sinclair, R.G., 1996. The case for polylactic acid as a commodity packaging plastic. *J. Macromol. Sci., Part A* 33: 585–597.
5. Drumright, R.E., Gruber, P.R., Henton, D.E., 2000. Polylactic acid technology. *Adv. Mat.* 12: 1841–1846.
6. Horbens, M., 2008. Orientierende Untersuchungen zum Einsatz von Holzfasern in Holz-Polymer-Verbunden. Diploma thesis, Technical University Dresden and Fraunhofer Institute for Applied Polymer Research, Potsdam, Germany.
7. Huda, M.S., Mohanty, A.K., Drzal, L.T., Schut, E., Misra, M., 2005. "Green" composites from recycled cellulose and poly(lactic acid): Physico-mechanical and morphological properties evaluation. *J. Mat. Sci.* 40: 4221–4229.
8. Huda, M.S., Drzal, L.T., Mohanty, A.K., Misra, M., 2006. Chopped glass and recycled newspaper as reinforcement fibers in injection molded poly(lactic acid) (PLA) composites: A comparative study. *Comp. Sci. Technol.* 66: 1813–1824.
9. Mathew, A.P., Oksman, K., Sain, M., 2006. Mechanical properties of biodegradable composites from poly lactic acid (PLA) and microcrystalline cellulose (MCC). *J. Appl. Polym. Sci.* 97: 2014–2025.
10. Oksman, K., Mathew, A.P., Bodeson, D., Kvien, I., 2006. Manufacturing process of cellulose whiskers/polylactic acid nanocomposites. *Comp. Sci. Technol.* 66: 2776–2784.
11. Petersson, L., Kvien, I., Oksman, K., 2007. Structure and thermal properties of poly(lactic acid)/cellulose whiskers nanocomposite materials. *Comp. Sci. Technol.* 67: 2535–2544.
12. Iwatake, A., Nogi, M., Yano, H., 2008. Cellulose nanofiber-reinforced polylactic acid. *Comp. Sci. Technol.* 68: 2103–2106.
13. Oksman, K., Skrifvars, M., Selin, J.-F., 2003. Natural fibers as reinforcement in polylactic acid (PLA) composites. *Comp. Sci. Technol.* 63: 1317–1324.
14. Masirek, R., Kulinski, Z., Chionna, D., Pirkowska, E., Pracella, M., 2007. Composites of poly(L-lactide) with hemp fibers: Morphology and thermal and mechanical properties. *J. Appl. Polym. Sci.* 105: 255–268.
15. Garcia, M., Garmendia, I., Gareia, J., 2008. Influence of natural fiber type in eco-composites. *J. Appl. Polym. Sci.* 107: 2994–3004.
16. Lee, S.-H., Wang, S., 2006. Biodegradable polymers/bamboo fiber biocomposite with biobased coupling agent. *Composites: Part A* 37: 80–91.
17. Ganster, J., Fink, H.-P., 2006. Novel cellulose fiber reinforced thermoplastic materials. *Cellulose* 13: 271–280.
18. Voges, M., Brück, M., Gensrich, J., Fink, H.-P., 2002. The CARBACELL process—An environmentally friendly alternative for cellulose man-made fiber production. *ipw/Das Papier*, issue 4 T74.

19. Ganster, J., Fink, H.-P., Pinnow, M., 2006. High tenacity man-made cellulose fiber reinforced thermoplastics—Injection molding compounds with polypropylene and alternative matrices. *Composites: Part A* 37: 1796–1804.
20. Fink, H.-P., Ganster, J., 2006. Novel thermoplastic composites from commodity polymers and man-made cellulose fibers. *Macromol. Symp.* 244: 107–118.
21. Ganster, J., Fink, H.-P., Uihlein, K., Zimmerer, B., 2008. Cellulose man-made fiber reinforced polypropylene—Correlations between fiber and composite properties. *Cellulose* 15: 561–569.
22. Bax, B., Müssig, J., 2008. Impact and tensile properties of PLA/Cordenka and PLA/flax composites. *Comp. Sci. Technol.* 68: 1601–1607.
23. Ray, S.S., Bousmina, M., 2005. Biodegradable polymers and their layered silicate nanocomposites: In greening the 21st century materials world. *Prog. Mat. Sci.* 50: 962–1079.
24. Kim, H.-S., Park, B.H., Yoon, J.-S., Jin, H.-J., 2007. Thermal and electrical properties of poly(L-lactide)-graft-multiwalled carbon nanotube composites. *Europ. Polym. J.* 43: 1729–1735.
25. Wu, C.-S., Liao, H.-T., 2007. Study on the preparation and characterization of biodegradable polylactide/multi-walled carbon nanotube nanocomposites. *Polymer* 48: 4449–4458.
26. Kuan, C.-F., Chen, C.-H., Kuan, H.-C., Lin, K.-C., Chiang, C.-L., Peng, H.-C., 2008. Multi-walled carbon nanotube reinforced poly(L-lactide acid) nanocomposites enhanced by water-crosslinking reaction. *J. Phys. Chem. Sol.* 69: 1399–1402.
27. Wu, D., Wu, L., Zhang, M., Zhao, Y., 2008. Viscoelasticity and thermal stability of polylactide composites with various functionalized carbon nanotubes. *Polym. Degr. Stabil.* 93: 1577–1584.
28. Chiu, W.-M., Chang, Y.-A., Kuo, H.-Y., Lin, M.-H., Wen, H.-C., 2008. A study of carbon nanotubes/biodegradable plastic polylactic acid composites. *J. Appl. Polym. Sci.* 108: 3024–3030.
29. Hiroi, R., Ray, S.S., Okamoto, M., Shiroi, T., 2004. Organically modified layered titanate: A new nanofiller to improve the performance of biodegradable polylactide. *Macromol. Rapid Commun.* 25: 1359–1364.
30. Liao, H.-T., Wu, C.-S., 2008. New biodegradable blends prepared from polylactide, titanium tetraisopropylate, and starch. *J. Appl. Polym. Sci.* 108: 2280–2289.
31. Yan, S., Yin, J., Yang, J., Chen, X., 2007. Structural characteristics and thermal properties of platicized poly(L-lactide)-silica nanocomposites synthesized by the sol-gel-method. *Mat. Lett.* 61: 2683–2686.
32. Wu, C.S., Liao, H.-T., 2008. Modification of biodegradable polylactide by silica and wood flour through a sol-gel process. *J. Appl. Polym. Sci.* 109: 2128–2138.
33. Chang, J.-H., An, Y.U., Cho, D., Giannelis, E.P., 2003. Poly(lactic acid) nanocomposites: Comparison of their properties with montmorillonite and synthetic mica (II). *Polymer* 44: 3715–3720.
34. Jiang, L., Zhang, J., Wolcott, M.P., 2007. Comparison of polylactide/nano-sized calcium carbonate and polypactide/montmorillonite composites: Reinforcing effects and toughening mechanisms. *Polymer* 48: 7632–7644.
35. Ray, SS, Yamada, K., Okamoto, M., Ueda K., 2003. New polylactide-layered silicate nanocomposites. 2. Concurrent improvements of material properties, biodegradability and melt rheology. *Polymer* 44: 857–866.

36. Ray, S.S., Yamada, K., Okamoto, M., Ogami, A., Ueda, K., 2003. New polylactide/layered silicate nanocomposites. 3. High performance biodegradable materials. *Chem. Mater.* 15: 1456–1465.
37. Ray, S.S., Okamoto, M., 2003. Biodegradable polylactide and its nanocomposites: Opening a new dimension for plastics and composites. *Macromol. Rapid Commun.* 24: 815–840.
38. Ray, S.S., Yamada, K., Okamoto, M., Jujimoto, Y., Ogami, A., Ueda, K., 2003. New polylactide/layered silicate nanocomposites. 5. Designing of materials with desired properties. *Polymer* 44: 6633–6646.
39. Pogodina, N.V., Cercle, C., Averous, L., Thomann, R., Bouquey, M., Muller, R., 2008. Processing and characterization of biodegradable polymer nanocomposites: Detection of dispersion state. *Rheol. Acta* 47: 543–553.
40. Tanoue, S., Hasook, A., Iemoto, Y., Unryu, T., 2006. Preparation of poly(lactic acid)/poly(ethylene glycol)/organoclay nanocomposites by melt compounding. *Polym. Comp.* 27: 256–263.
41. Shibata, M., Someya, Y., Orihara, M., Miyoshi, M., 2006. Thermal and mechanical properties of plasticized poly(L-lactide) nanocomposites with organo-modified montmorillonites. *J. Appl. Polym. Sci.* 99: 2594–2602.

12

Nanomaterials Formulation and Toxicity Impact

Khalid Lafdi

University of Dayton Research Institute, Ohio

CONTENTS

12.1 Introduction

Novel materials are continually being developed to improve performance within the fields of engineering and communications. Due to advances in manufacture and characterization methods, nanosized materials have emerged as powerful tools to support progress in these technologies. Nanomaterials are roughly defined as a material with a base constituent that is between 1 $\times$ 10^{-9} and 100 $\times$ 10^{-9} meters in length and at least one property that deviates from the value for the equivalent bulk material. Unique properties exhibited by nanomaterials that differ from bulk materials of the same composition include electrical and thermal conductivity, mechanical performance, and optical sensitivity. Changes in these properties at the nanoscale are the basis of the unique nature and value of nanomaterials in novel technologies [1].

Nanotechnology is growing at an exponential rate and is predicted to change the face of the world in which we live. According to Lux Research, the value of products incorporating nanomaterials worldwide was $13 billion in 2004, $32 billion in 2005, and nearly $50 billion in 2006. It was estimated that by the summer of 2007, there were as many as 500 consumer products that contained nanomaterials [2]. Some examples of these products include self-cleaning windows, sunscreens, bandages, antiwrinkle face creams, stain-resistant and static-free clothing, and lightweight sports equipment with enhanced strength.

In addition to consumer products, research for the development of more advanced nanostructures, coatings, adhesives, and devices is rapidly progressing. Some research directions include using semiconductor nanoparticles as fluorescent biolabels, bioconjugated gold nanoparticles for cancer treatment, nanoshell-polymer composites for drug delivery systems, and nanoparticle additives to decrease diesel emissions [3,4].

Of all the materials that can be synthesized and manipulated at the nanoscale, carbon-based nanomaterials arguably offer the most promise in future applications. No other materials exhibit such a wide range of variations in structures defined by extended networks of high-strength carbon–carbon bonds. Proposed applications of carbon-based nanomaterials include nanoporous filters, catalyst supports, coatings, field emitters, molecular electronics, biomedical devices, energy storage devices, adhesives, and advanced composites for the aerospace and automotive industries [5–7].

The beneficial aspects of nanomaterials are vast; however, there is a severe lack of information concerning potential human health risks upon exposure. Reasons for concern about the potential toxicity of nanomaterials are derived from the same size-related properties that make nanomaterials promising candidates in many technologies. For example, reduction in size and corresponding increase in specific surface area and surface energy may cause nanomaterials to be more biologically active [8,9]. Nanomaterials are smaller than eukaryotic cells and most organelles, which means they may be taken up within these structures, potentially interfering with cellular processes. Even though the ability of nanomaterials to pass through cell membranes and tissues or be distributed throughout the lymphatic and circulatory systems may be beneficial in some applications, such as targeted drug delivery, it could also result in unintended hazardous consequences.

In the past, failure to regulate the toxicity of materials before their extensive implementation in commercial products and industrial applications led to hazardous consequences. For example, throughout history, asbestos fibers were widely used in cements, insulations, coatings, and various other applications due to their high tensile strength, flexibility, and thermal resistance. After several decades of use in the United States, epidemiological and clinical studies exposed their hazardous health effects. Regulations on the production of asbestos fibers were set by the Occupational Safety and Health Administration (OSHA) in 1972, which included safety protocols for workers, employee involvement in safety and health management programs, and worksite analysis. However, by the time these regulations were put into effect, thousands of premature deaths had already occurred, and thousands of cases of asbestos-related lung diseases had been discovered [10,11].

Despite warning signs from past experiences and attempts from toxicologists to raise awareness of their potential toxicity, nanomaterials continue to be implemented in new technologies without strict regulation. If a threat to human health is discovered after several decades of use, it may be too late to set standards for workers who have already been exposed. In order to prevent repeating history, the effects of nanomaterials on health and the environment must be proactively explored, and regulatory agencies must act quickly to implement new policies before epidemiological and clinical data become available. Therefore, initial policies must be based on toxicological studies.

Although nanotechnology has promised invaluable progress in science and technology, it is up to the scientific community to predict unknown outcomes of exposure to nanomaterials. The establishment of test procedures to ensure safe manufacture and use of nanomaterials in the marketplace is urgently required and achievable. Because nanomaterials can vary with respect to composition, size, shape, surface chemistry, and crystal structure, it is not appropriate to establish general safety regulations for all nanomaterials. Detailed characterization and risk assessment of each property must be performed through toxicity studies, and methods for prediction must be established in order to implement proper safety practices.

The focus of this chapter is to investigate the toxicity of nanomaterials at the cellular level using an *in vitro* model in two phases. First, the effect of particle dimension on the toxicity of carbon-based nanomaterials was investigated. Second, the effect of impurities was investigated by comparing the toxicity of clean MWCNT (CULOT) to MWCNT containing metal impurities, and the effect of surface area and surface charges on the toxicity of carbon-based nanomaterials was evaluated using materials with low to high surface area and low impurities. To address the effect of impurity composition, the relative toxicity of metal catalyst impurities commonly used in the synthesis of carbon-based nanomaterials was studied. Finally, the data were used to develop a toxicity factor for predicting the effect of exposure to carbon-based nanomaterials. The basis of this study stems from the lack of information regarding occupational risks and health implications of exposure to carbon-based nanomaterials, and the results aim to contribute to a method for predicting the toxicity of nanomaterial toxicity to biological systems.

12.2 Nanomaterials and Nanotechnology

A nanomaterial is roughly defined as a material with a base constituent between 1×10^{-9} and 100×10^{-9} meters in length that exhibits at least one property that deviates from the value for the equivalent bulk material [1]. Properties that can be different between bulk materials and their nanosized counterparts include electrical and thermal conductivity, mechanical properties, optical behavior, and surface reactivity. Such size-dependent properties are the basis of the unique nature of nanomaterials and their value in novel technologies.

Materials that are manipulated at the nanoscale include metals, semiconductors, ceramics, polymers, glass, carbon, biomolecules, and composites. The method of synthesis used for a particular nanomaterial depends on its chemical nature and application. Some of the common methods used for nanomaterials synthesis include gas phase, sol-gel, sonochemical, microemulsion, and high-energy ball milling [12].

Gas-phase synthesis can be used to manufacture almost all types of materials and is one of the best techniques to produce size monodispersity in samples. In this method, a starting material is vaporized and condensed in an inert-gas environment. Combustion flame, laser ablation, chemical vapor deposition (CVD), and electrospray are all examples of gas-phase synthesis.

Sol-gel synthesis methods are often used to produce metal, semiconductor, and ceramic nanomaterials. This type of method involves a wet chemical synthesis approach by gelation, precipitation, and hydrothermal treatment. Various techniques can be used for size and stability control of resulting nanoparticles, such as introduction of a doping agent and heat treatment [12].

Sonochemical synthesis is used to generate metal, alloy, semiconductor, and carbide nanoparticles and is useful for producing a large volume of material. In this method, an acoustic cavitation process results in growth and implosive collapse of bubbles within a liquid where sudden changes in temperature and pressure lead to destruction of the precursor and formation of nanoparticles [13].

Microemulsion synthesis can be used to generate metallic, semiconductor, magnetic, and superconductor nanoparticles and is an ideal method for large-scale production. In this method, two immiscible phases are present with a surfactant. The surfactant molecules form nanosized micelles at the interface between the two immiscible phases. Precursor molecules are added, and after pressure treatments, nanoparticles are formed within the surfactant [12].

High-energy ball milling has been used for the generation of magnetic and catalytic nanoparticles. It uses a grinder to grind materials into extremely fine powder. Drawbacks to this method include low surface area and highly polydisperse size distribution, as well as the partially amorphous state of the resulting powders [12].

For some applications, nanomaterials are synthesized to contain surface modifications or use a patterned-growth method. Purification steps are often carried out after synthesis for removal of catalyst and growth support contaminants. Nanotechnology is driven by the idea that the ability to manipulate nanomaterials by advanced synthesis methods and surface modifications will lead to tailorable and enhanced properties. Continuous innovation in synthesis and characterization of nanomaterials has led to a more advanced understanding of the relationship between nanostructure and properties, and the exploitation of properties inherent to materials at the nanoscale has resulted in rapid growth in the field of nanotechnology.

Some materials are more promising than others due to diversity of structure and properties; however, semiconductor, metal, ceramic, polymer, glass, lipid, composite, and carbon nanomaterials have all found a place in nanotechnology applications.

Silicon-based nanomaterials have been shown to be useful because they are stable at a wide range of pH, ion strength, and temperature, allowing them to be resistant to varying chemical environments. Additionally, the surface of silica nanoparticles can be modified to contain functional groups, making them versatile for various biological applications [43,44]. Silica can be used as a shell for magnetic materials, and silica nanoparticles can be doped with dye molecules to generate luminescent nanoparticles that can be used in biodetection applications. Dye-doped silica nanoparticles (DSNPs) are often produced via a water-in-oil microemulsion technique and consist of silica spheres encapsulating thousands of dye molecules.

DSNPs exhibit increased intensity compared to traditional dye molecules and reduced photobleaching from molecule decomposition. DSNPs have been proposed for replacement of traditional fluorophores in various areas of biology research including labeling of proteins, DNA, genes, pathogens, subcellular structures, cells, and tissues. Due to their luminescent properties in conjunction with nanoscale properties, they can be used to study transport of nanomaterials within biological systems, as well as mechanisms involved in the uptake of nanomaterials by cells and tissues. DSNPs have also been shown to be successful candidates for replacement of biomolecules used as tags for quantifying proteins and sensors for noninvasive monitoring of intracellular pH changes [45,46]. Quantum dots, which are commonly composed of cadmium selenide, are another form of luminescent nanoparticles that have been shown to be useful in biodetection and sensing applications. In some cases, quantum dots are encapsulated with silica to increase their stability and allow for surface modification [43].

Metal-based nanoparticles have also been shown to be useful in many applications. For example, silver nanoparticles been proposed for use in bone cement and are implemented in many commercial products for their antimicrobial properties, and gold nanoparticles have been shown to be successful for differentiating healthy cells from cancer cells based on differences in metabolism between the two cell types [47,48]. Additionally, aluminum nanoparticles are being used by the U.S. Army in solid rocket fuel, and by the U.S. Naval Air Warfare Center to replace lead primers in artillery and wear-resistant coatings on propeller shafts [27,49]. Polymer-based core shell–type nanoparticles have been proposed for drug delivery due to their ability to encapsulate drug molecules and degrade at a controlled rate. Also, polymer shells can be functionalized with biomolecules for specific applications [50].

Some of the most promising nanomaterials are those composed of carbon. No other materials exhibit such a wide range of variation in extended networks of high-strength carbon–carbon bonds. This is due to the ability of carbon to form three different hybrid orbitals, including sp, sp2, and sp3, which leads to the ability to construct a wide variety of allotropes and structures. Allotropic forms of carbon in carbon-based nanomaterials include diamond, graphite, and fullerenes, which are composed of sp3, sp2, and a combination of sp and sp2 hybridized bonds, respectively.

Diamond is a naturally occurring allotrope of carbon that consists of sp3 hybridized bonds. In this structure, each carbon atom is covalently bonded to three other carbon atoms. Due to this structure, diamond is the hardest known naturally occurring material. Unlike graphite, the properties of diamond are isotropic. Pure diamond has a high electrical resistivity, low thermal resistance, and is colorless and transparent. As with most other materials, defects in the structure result in variations in these properties.

Nanosized diamond structures include nanodiamonds and diamond nanowhiskers. Nanodiamond is often synthesized by high-energy exothermic detonation methods, which result in solid spherical nanoparticles with average size of about 4 nm [14]. Diamond nanowhiskers have been formed using air plasma etching of polycrystalline diamond films and range from 60 to 300 nm in diameter [15].

Nanodiamond particles consist of a single-crystal diamond core surrounded by a shell consisting of functional groups. The core is composed of sp3 hybridized carbon, and the state of carbon atom hybridization in the shell is either sp3 or sp2. The composition of the shell can change via chemical modification. The properties of nanodiamonds can be controlled by modifying the shell with functional groups [16]. Unique properties of nanodiamonds include optical transparency, high specific surface area, chemical inertness, and hardness. Due to their low chemical reactivity and unique physical properties, nanodiamonds may be useful in a variety of biological applications, such as carriers for drugs, genes, or proteins, coatings for implantable materials, biosensors, and biomedical nanorobots. Also, due to their hardness and chemical stability, in addition to the ability to produce samples with uniform size distribution and controlled shell composition, nanodiamonds and diamond nanowhiskers have been proposed for use as field emitters in vacuum microelectronics [15].

When nanodiamonds are heated up to 1200 to 1800 K, they transform into onion-like carbon. The transition temperature is dependent on the particle size and occurs from the surface inward. Onion-like carbons can be transformed back into nanodiamond by electron irradiation, which begins in the center and grows to become single or polycrystalline nanodiamond. Additional sources of carbon nano-onions include synthesis by graphite arcing under deionized water, in which an electric current creates a discharge between two high-purity graphite electrodes [17].

Carbon nano-onions are spherical carbon structures with concentric rings resembling an onion and range from double- and triple- to multilayered structures. The amount of concentric shells tends to increase with pressure and deformation growth. The diameters of carbon nano-onions have been reported to range in size from 1 nm to 2 μm [17]. Carbon nano-onions have been found to exhibit very large specific surface area, indicating that they may be promising for gas storage applications [18].

One of the most well-known naturally occurring allotropes of carbon is graphite. Ideal graphite is composed of hexagonal planes of covalently bonded carbon atoms with sp2 hybridization, which are stacked in layers held together by van der Waals bonds. Between the layers, the van der Waals bonds are held by a delocalized electron that migrates from one side of the plane to the other. The distance between carbon atoms in adjacent layers is more than twice the distance between carbon atoms within the same hexagonal plane (Figure 12.1). Due to this structure, the properties of graphite are anisotropic; the electrical and thermal conductivity and mechanical strength are higher in the direction of the planes than in the direction perpendicular to the planes. Graphite layers are able to easily slide over each other due to weak van der Waals bonds, making it useful as a lubricant. Ideal graphite is chemically inert, but its chemical reactivity increases with impurities and defects.

Graphite crystallites can be combined to form many natural and synthetic carbon materials, such as exfoliated graphite, carbon black, and carbon fiber.

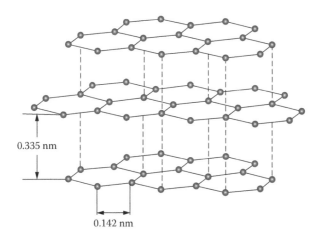

0.335 nm

0.142 nm

FIGURE 12.1
Ideal graphite.

Exfoliated graphite consists of graphite crystallites with interlayer spacing from 0.335 nm up to 50 nm. Exfoliated graphite can be synthesized by metal intercalation followed by exfoliation of graphene layers using a solvent or gas. Thermal shock can be used for drying and further separation of graphite sheets. The resulting material consists of high-aspect-ratio graphite nanoplatelets. Exfoliated graphite is used as reinforcement for high-strength carbon–carbon composites, in addition to replacement of asbestos in gaskets for automobiles, seals in vacuum furnaces, and thermal insulation materials [19,20]. Exfoliated graphite has also attracted attention because of its very high sorption capacity for spilled heavy oils or hydrogen storage [21,22].

Another synthetic carbon material composed of graphite crystallites is carbon black. Carbon black is hard and mainly crystalline with some amorphous particles and is distinguishable from carbonaceous soot, which is randomly formed in natural and industrial processes [23]. The size of carbon black particles varies from coarse to ultrafine ($\leq$100 nm), depending on the process used for synthesis. As the diameter of carbon black is reduced from 300 to 10 nm, the pH and degree of crystallinity increase [24]. Carbon black is manufactured using four main processes, including impingement, thermal decomposition, furnace, and lamp. Impingement produces the finest particle size with the highest surface area but also includes the highest volatile content. Thermal decomposition produces the coarsest particles with the lowest surface area, and the remaining two processes produce carbon black particles that are between fine and coarse in size. Carbon black is often used as filler in rubbers, polymers, and plastics as a reinforcement agent [24–27].

One of the most commonly used synthetic carbon materials in commercial applications is continuous carbon fiber (CF). Continuous CF is produced by the thermal decomposition of organic precursors, such as polyacrylonitrile (PAN), rayon, or pitch, and has been implemented since the 1960s in composite materials to improve mechanical and thermal performance. CF is on the order of 10 μm in diameter [28,29]. The size and morphology of typical PAN-based CF are shown in Figure 12.2.

Another class of CF includes those that are synthesized via CVD, or vapor grown through pyrolysis of carbon-containing gases on a metal catalyst. Although vapor growth can be used to produce fibers with the diameter of conventional CF, it can also be used to grow carbon nanofibers (CNFs), which exhibit nanomaterial characteristics. Depending on the processing conditions, CNFs can consist of stacks of cups or cone-shaped graphite sheets (Figure 12.3). In general, vapor-grown CNFs range in diameter from a few nanometers to hundreds of nanometers and in length from less than a micrometer to millimeters [21].

Individual graphite sheets, called graphene, can also form crystalline structures called fullerenes. Fullerenes are an allotrope of carbon that can consist of sp and sp2 hybridized bonds. One of the most common types of fullerene, which forms in natural and industrial processes, is the Buckminsterfullerene, or C60. C60 is a spherical structure composed of a carbon sheet containing 12

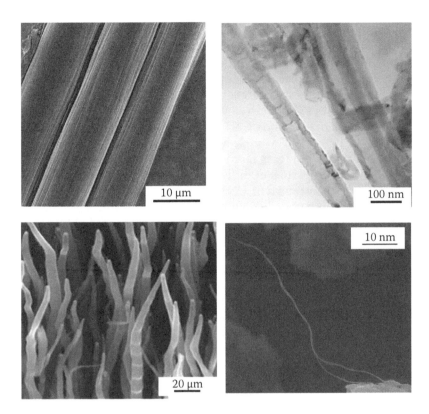

FIGURE 12.2
Scanning electon micrographs of carbon fibers, nanofiber, and multiwalled carbon nanotubes, and transmission electron micrographs of single-walled carbon nanotubes.

FIGURE 12.3
The carbon nanofiber.

pentagons and 20 hexagons and named based on the number of atoms in the molecule. Variations of fullerenes can be created by changing the number of hexagons. The diameter of a C60 fullerene molecule is close to 1 nm [21]. C60 and related fullerene molecules have been proposed for applications in medicine, electronics, energy storage, chemical sensors, and optical devices [30].

Carbon nanotubes (CNTs) are a cylindrical form of fullerene, which contain only sp2 hybridized bonds. CNTs were first recognized by Iijima of NEC Corporation in 1991 [31]. The angle at which the graphene layer of a CNT is wrapped to form a cylinder is represented by a pair of integers (n,m) called the chiral vector. Properties of individual CNTs depend on their diameter and chiral vector, and current CNT synthesis methods produce samples containing CNTs with various diameters and chirality. CNTs possess highly desirable electrical, mechanical, and thermal properties, are lightweight, and have the strongest tensile strength of all synthetic fibers [32].

CNTs are fabricated by arc discharge, laser evaporation, or CVD. In the arc discharge method, carbon located in a negative graphite electrode sublimates at high temperatures. CNTs produced by this method can be up to 50 µm in length. In laser ablation, a pulsed laser vaporizes a graphite target in a high-temperature reactor while an inert gas is introduced into the chamber. The CNTs develop on the cooler surfaces of the reactor, as the vaporized carbon condenses. This method has a higher yield than arc discharge or CVD methods (around 70%) but is the most expensive of the alternatives. In CVD, a hydrocarbon gas is decomposed at a high temperature, and carbon that is adsorbed onto metal catalyst particles (usually nickel, cobalt, or iron) grows in the form of a nanotube. CVD is a common method for commercial production due to the cost versus production rate advantage over the other processes. After synthesis, acid treatment can be used to remove amorphous carbon and free residual catalyst. Because purification steps also destroy CNTs, the removal of impurities must be balanced against defects. Current research is aimed at improving synthesis methods to increase yield and length of CNTs [21].

Two ideal forms of CNTs include multiwalled carbon nanotubes (MWCNTs) and single-walled carbon nanotubes (SWCNTs). MWCNTs consist of multiple graphene layers and contain between 5 and 50 walls and are about 10 to 50 nm in diameter. The spacing between walls is similar to that of graphite, and individual layers can slide over each other [33]. MWCNTs can be cylindrical, conical, or polyhedral, and have a Russian doll, spiral, or herringbone structure. MWCNTs grown via CVD are pictured in Figure 12.2 and Figure 12.3.

SWCNTs consist of a single graphene layer and are approximately 1 nm in diameter with a length that can be thousands of times the diameter. As a result of strong van der Waals forces, SWCNTs often self-organize into bundles, which can range in length up to several microns (Figure 12.2). SWCNTs are able to slide over each other within the bundles; however, electron beam irradiation can be used to create covalent bonds between the nanotubes to increase the Young's modulus and shear modulus [34]. The diameter of

SWCNTs is controlled by the size of metal nanoparticles from which they are grown and varies from 0.7 to 3 nm [35].

CNFs and CNTs exhibit many unique properties, which are being exploited in nanotechnology applications across a broad range of fields. Due to their high mechanical strength and low density, they have been implemented in a variety of composites, such as those used in sports equipment, automotive materials, and aerospace materials. Due to their light weight, large surface area, flexibility, and high thermal conductivity, CNTs have been used as reinforcement fillers in polymer composites. CNFs do not compare to CNTs in mechanical strength or density but are much less expensive and easier to synthesize with controlled structure and orientation [37]. As a result, CNFs are widely implemented in various low-weight composite materials to increase mechanical strength, as well as in energy storing devices, such as lithium batteries used for cell phones [1].

Due to their high surface area and hydrophobic nature, CNTs have been proposed for use in environmental technologies as adsorbents. Also, depending on their chirality, CNTs can be electrical conducting or semiconducting and are being implemented in various electronic technologies, such as superconductors, electrical wires, and quantum computers [1]. Incorporation of CNTs in construction materials such as concrete or structural plastics may allow for real-time monitoring of material integrity and quality [36].

For medicine and biology, nanomaterials are attractive because they are smaller than cells, which range in size from 10 to 100 μm, and are comparable in size to viruses (20 to 450 nm) and proteins (5 to 50 nm). Because of their size, SWCNTs have been proposed as infrared photosensitizers for cancer cells [6]. Additionally, due to the ability of acid-oxidized SWCNTs to spontaneously adsorb proteins on their sidewalls, protein–nanotube conjugates have been successfully developed for use as molecular transporters for protein delivery [7]. Because of their low density and ability to mimic biological compounds, CNFs, MWCNTs, and SWCNTs have been proposed for use in structural support for bone implants and neural tissue regeneration [38–42].

12.3 Nanomaterials as a Potential Hazard

The unique physicochemical properties of engineered nanomaterials are unquestionably valuable in many technologies; however, these properties are also cause for concern about their potential toxicological effects on biological systems. Due to their small size, individual nanomaterials and small agglomerates may be deposited deep within the lungs when inhaled, reaching areas that are not as easily accessed by larger materials. Also, their size may permit them to pass directly through tissues and cell membranes, allowing them to translocate from their initial site of exposure to other organs in the body.

In addition to their small size, the increased surface area per unit mass of nanomaterials increases their potential for interaction with biological components. A high specific surface area may lead to increased adsorption of physiological surfactants and macromolecules, which may alter the surface chemistry of the nanomaterial. One study showed that carbon and silica-based nanoparticles from combustion processes can adsorb components of lung surfactant, altering both the surface chemistry of the particles and their toxicological effects [51]. In addition, adsorption of macromolecules, such as proteins, may lead to denaturation and loss of function, potentially leading to a devastating cascade effect within the body. Additionally, nanomaterials may be able to mimic biological components and attach to cellular membrane receptors, disrupting the cell's normal metabolism and functions.

As materials are reduced in size, their radius of curvature increases due to the increased amount of uncoordinated surface atoms versus bulk atoms, which results in increased surface energy. This may lead to additional interaction with macromolecules, which may cause denaturization of macromolecules and inhibit the ability for cells and tissues to perform their normal functions. The combination of nanomaterial size, surface area, and surface energy characteristics are likely to initiate effects in biological systems that are not possible with the bulk form of the same materials. Therefore, it is of paramount importance to investigate the potential hazards of nanomaterial exposure [52,53].

In order to understand potential toxicity of nanomaterials, it is important to understand the potential routes of exposure and the body's natural defense mechanisms within target organ systems. Nanomaterials implemented in commercial and industrial products may enter the body through several ports but are most likely to come into contact with the skin, respiratory tract, and gastrointestinal (GI) tract, which are in direct contact with the environment [54].

12.3.1 Skin

Skin provides the most significant surface area for potential contamination and is the first line of defense against the outside environment. It is composed of two layers: the epidermis and the dermis. The epidermis is the top layer and is mostly composed of keratinocytes, which form sublayers in different stages of differentiation. The keratinocytes develop at the bottom, and as they move to the surface, they gradually express new protein markers and accumulate keratin proteins. When they reach the surface, they are shed from the surface as dead cells. In healthy skin, the epidermis provides protection for the underlying dermis, which makes up 90% of the skin and consists mostly of connective tissue and contains a rich supply of blood and tissue macrophages, lymph vessels, and different types of sensory nerve endings [55].

Materials absorbed by the skin have to pass through several cell layers before entering blood and lymph capillaries in the dermis. The rate-determining barrier in the dermal absorption of materials is the horny layer at the top of the epidermis, which is biologically inactive due to keratinization. Therefore, once materials pass through the horny layer, they are expected to pass through the other layers more rapidly [56]. Little evidence of nanomaterial absorption through healthy skin has been demonstrated. However, it was found that when normal skin is flexed through natural motions, such as wrist movements, micrometer-size fluorescent beads can penetrate through the epidermis to the dermis [57]. Additionally, quantum dots (37 nm) injected into the top layer of the skin (dermis) of female rats were found to move through the lymphatic system to regional lymph nodes, accumulate in the liver, kidney, spleen, and lymph nodes at the site of injection and in other parts of the body [58].

12.3.2 Respiratory Tract

The respiratory tract is a probable site of exposure when nanoparticles are suspended in air and subject to inhalation. The respiratory tract can be broken down into three main regions: nasopharyngeal, tracheobronchial, and alveolar. The nasopharyngeal region is further broken down into the naris (or nostrils), vestibule, which is unciliated, nasal cavities, which are ciliated and contain turbinates functioning to increase the surface-area-to-volume ratio, and pharynx. The nasal passages are separated by the nasal septum but are rejoined at the pharynx. The tracheobronchial region can be broken down into the trachea, bronchi, and bronchioles, which branch out in the upper lung. The alveolar region includes the deep lung where gas exchange occurs. Nanoparticles are generally considered to fall within the respirable size range (<2.5 μm), meaning they can deposit deep within the alveolar portion of the respiratory tract.

Much of the respiratory tract is protected by a layer of mucus with the primary functions of trapping inhaled particles and bacteria and acting as a diffusion barrier for deposited particles and gaseous air pollutants [59]. Just beneath the mucous layer lie ciliated epithelia cells, each containing 25 to 100 cilia [60]. In a healthy respiratory tract, the cilia beat continuously, transporting mucus and trapped particulates via the mucociliary clearance mechanism toward the pharynx, where they are swallowed. Materials deposited in unciliated areas of the lung are cleared by uptake into macrophages (phagocytotic cells) and migration to the mucociliary escalator. The mucociliary mechanism is one of the most important defense methods of the upper respiratory system [61]. The viscoelastic properties of mucus are the essential features enabling mucus to perform its protective functions, especially in relation to mucociliary clearance and retardation of diffusion. Significant changes in mucus composition due to bronchial diseases (e.g., asthma, chronic bronchitis, cystic fibrosis) or nanoparticle accumulation can lead to changes in viscosity and disrupt the mucociliary clearance mechanism [62]. Also, surface tension lowering of mucus due to altered composition can increase epithelial encounters, which will result in increased potential for uptake and translocation of deposited particles [63].

The site of deposition in the respiratory tract plays a significant role in the clearance and potential toxicity of nanomaterials [64]. Materials deposited on the unciliated anterior portion of the nose tend to remain at the site of deposition until removed by wiping or sneezing, while particles deposited in the ciliated portion of the nasal passage are generally cleared and swallowed within an hour of deposition. Materials deposited on the tracheobronchial airways are generally carried to the pharynx and swallowed within the first day. Clearance from the alveoli occurs mainly by macrophage phagocytosis and subsequent mucociliary clearance. However, this clearance mechanism occurs much more slowly, allowing more time for materials to diffuse into the epithelium or lymphatic circulation before they can be mechanically removed. Nanomaterials that persist in lung epithelium for many years can increase the risk of lung cancer [54]. In one study, titanium dioxide (TiO_2) nanoparticles (20 nm) were found to have a greater ability to diffuse into lung tissue after intratracheal instillation than fine particles. (Note that fine particles are generally defined as ≤ 2.5 μm and >0.10 μm based on the assumptions that particles in this size range are able to deposit in the deep lung upon inhalation and exhibit distinct properties from nanoparticles.) In some cases, the small size of nanomaterials may cause them to be overlooked by the body's immune system as a foreign substance, which can inhibit the body's ability to eliminate them from the body. The length of high-aspect-ratio nanomaterials, such as CNTs, may also inhibit the body's ability to efficiently perform its usual clearance functions, which may lead to accumulation. In the case of asbestos fibers, for example, incomplete phagocytosis by respiratory macrophages leads to accumulation and persistence within the lungs, which eventually leads to lung diseases and cancer [65]. Additionally, when macrophages containing phagocytosed nanoparticles were intratracheally instilled, interstitial access was inhibited [66]. This result indicates that nanoparticles either diffuse into lung tissue too rapidly to be effectively phagocytosed by macrophages, or that they are too small to be detected by the cells as foreign matter. A schematic of the clearance and translocation of nanomaterials deposited in the respiratory system is shown in Figure 12.4.

12.3.3 Gastrointestinal Tract

Nanomaterials cleared from the respiratory tract via the mucociliary escalator can be ingested into the GI tract. Alternatively, nanomaterials can be ingested directly, for example, if contained in food or water or if used as drugs or drug delivery devices. The GI tract is coated with a layer of mucus similar in composition to that which covers the respiratory tract. This mucosa layer of the GI tract functions in selective absorption of ingested substances [67].

In some cases, nanomaterials may be eliminated rapidly after exposure to the GI tract, while in others, they may be taken up into circulation and

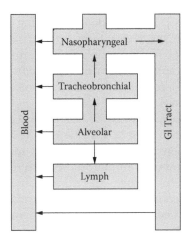

FIGURE 12.4
Translocation and clearance of inhaled nanomaterials.

accumulate in target organs (Figure 12.6). In one study, rats were dosed orally with radiolabeled functionalized C60 fullerenes. Nearly 98% were cleared in the feces within 48 h, and the rest were taken up into blood circulation and eliminated via urine [68]. In another study, TiO_2 particles administered to rats were found in the liver after being taken up into blood from the GI tract [69].

12.3.4 Circulatory System and Other Target Organ Systems

In addition to causing local injury to the skin, respiratory tract, or GI tract, nanomaterials that are able to diffuse through the epithelial layers may enter the bloodstream and circulate throughout the body, potentially affecting systemic organs [70]. Evidence of epithelial absorption and consequential systemic translocation of nanomaterials after inhalation has been shown in the hamster [71] and in humans [72]. Once in the bloodstream, systemic circulation and persistence are potential hazards. Target circulatory organs include the heart, liver, and kidneys. Critical organs, such as the brain, placenta, and testes, are protected by a barrier that prevents access of some chemicals and materials in the bloodstream [73]. However, some nanomaterials have been found to cross the blood–brain barrier [74]. This suggests they may also be able to cross into the testes or placenta and disrupt fetal development or be passed on to offspring.

12.4 Approaches for Assessing Toxicity

Toxicity is defined as the degree to which a substance or material poses a hazard to a biological system, including whole organisms, individual organ systems, tissues, or cells [75]. Methods for studying toxicity include *in vivo*, or within a living organism, and *in vitro*, or in artificial environments outside living organisms [76]. *In vivo* models include animals, such as mice, rats, hamsters, guinea pigs, rabbits, and primates. *In vitro* models include whole tissues (e.g., tissue slices), cellular systems (e.g., isolated cells), and subcellular systems (e.g., macromolecules).

When choosing a biological system to test, there are advantages and drawbacks to each method that must be considered. Whole animal studies have obvious advantages, such as realistic responses incorporating the body's defense mechanisms. However, animal studies also require high cost and are more complex than *in vitro* studies. Additionally, *in vivo* data from different species are not always consistent. For example, carbon black–induced lung tumors and lesions have been found to occur only in rats, but not in other laboratory species, such as mice, hamsters, guinea pigs, rabbits, and primates [77].

Although current *in vitro* methods cannot substitute for whole animal studies, they are useful for screening material toxicity and understanding mechanistic pathways that occur in the skin, lungs, and other target organs [78]. Additionally, when only one cell type is present in a culture, it is possible to specify with certainty the type of cell responsible for the observed results [79]. Acellular *in vitro* models can also be used to gain insight into the toxicity of materials. For example, the solubility of materials can be tested in physiological solutions to predict whether they have the potential to be biopersistent within the body [11].

For *in vitro* studies, cell lines are often cultured in a plate or dish, exposed to the chemical or material of interest, then analyzed using various toxicological endpoints. A culture of cells is considered primary if it is the first one after cells are isolated from an organism. Once a primary culture is subcultured, it becomes a cell line, and if a cell line is transformed and escapes from senescence control (a series of genes that regulate cell cycle progression), it becomes a continuous cell line. Continuous cell lines are often used for *in vitro* studies because they can be subcultured infinitely, whereas finite cell lines will eventually be extinct [80].

In order to interpret *in vitro* toxicity results, the fundamental structure of a typical eukaryotic cell must be understood. Cells are the basic functional units of life and control a host of intracellular and extracellular events. Animal cells are usually about 10 to 20 μm in diameter and are typically composed of one or more nuclei, which hold the cellular DNA, in addition to several membrane-bound organelles, which enable the cell to perform specific functions. Some of the most important organelles to be aware of when studying cytotoxicity include mitochondria, which produce energy for the cell, lysosomes, where cell digestion occurs, and endosomes, where materials are packaged and stored after being taken up into the cell. The Golgi apparatus modifies secretory products and packages them into vesicles to be exported, and ribosomes are very small organelles that initiate protein synthesis. Rough endoplasmic reticulum (ER) is covered by ribosomes and

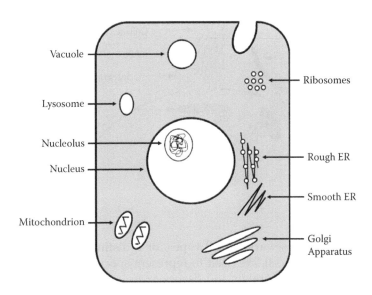

Vacuole

Ribosomes

Lysosome

Nucleolus

Rough ER

Nucleus

Smooth ER

Mitochondrion

Golgi
Apparatus

FIGURE 12.5
A typical animal cell.

functions in protein synthesis. Smooth ER performs synthesis of lipids, metabolism of carbohydrates, drug detoxification, and attachment of receptors on cell membrane proteins [81]. The plasma membrane is flexible and allows cells to move and be moved by changes in the internal cytoskeletal structure [82]. A schematic of a typical animal cell is shown in Figure 12.5.

The cytoskeleton of the cell is responsible for cellular structure and is composed of three main polymers: actin filaments, intermediate filaments, and microtubules [80]. Actin is one of the most abundant proteins in eukaryotic cells [83], and the actin cytoskeleton is composed of actin polymers and a variety of associated proteins. Actin is constantly assembling and disassembling and plays a role in cell shape and motility, cell division, endocytosis, secretion, organelle movement, and cell adhesion [83–85]. Of the cytoskeletal polymers, it is in the closest proximity to the plasma membrane, connecting to transmembrane proteins and signaling complexes located at the intracellular attachment sites and extracellular matrix adhesion sites. Integrins are the integral membrane adhesion receptors, which bind to membrane-associated proteins and extracellular matrix proteins [82]. Fibronectin and other extracellular proteins can influence cell activities, such as cytoskeleton remodeling and gene activity in the nucleus (Figure 12.6). Cell adhesion is a specific interest due to the fact that cell membrane interaction with nanomaterials can play a role in the function of actin, and cell–cell adhesion is required for tissue morphogenesis and homeostasis [86]. Actin genes have been highly conserved during evolution, and actin molecules from various organisms are functionally interchangeable *in vitro,* which means actin studies using mouse or rat cells can be directly related to human cells [85].

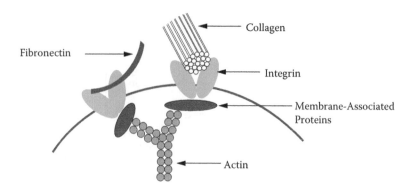

FIGURE 12.6
Extracellular matrix.

For *in vitro* studies, various cell types, or a distinct morphological and functional form of a cell, are used to represent specific target organs. Cells isolated from different organs and tissues express different genes based on their intended function. Basal cell toxicity describes toxicity that is common to all cells, whereas organ-specific toxicity refers to effects that are unique to cells derived from that organ [87].

In dermal and inhalation studies, target cells include fibroblasts, epithelial cells, and alveolar macrophages [76,88]. Fibroblasts are connective tissue cells that function in maintaining tissue structure and wound healing, and epithelial cells make up the skin and line the inner cavities of the body. The most commonly studied epithelial cells are lung epithelial cells, which line the respiratory tract, and keratinocytes, which represent the skin. Keratinocytes are involved in the earliest phase of epidermal repair and are an appropriate model with which to predict dermal toxicity. Characteristic features of keratinocytes include a large polygonal-shaped structure possessing a big vesicular nucleus, surrounded by a network of fibers and filaments [55].

Unlike fibroblasts and keratinocytes, alveolar macrophages are phagocytotic, meaning they function in the uptake of foreign materials [89]. Due to this function, phagocytotic cells may be more sensitive to nanomaterial exposure than nonphagocytotic cells. When macrophage-mediated clearance is impaired, even nontoxic materials may accumulate in the lungs leading to significant health effects. Studies have found that the clearance function of alveolar macrophages is fully impaired when particulate burdens are of volume size equivalent to 60% of the normal cell volume [90].

Nanomaterials that enter the bloodstream or lymphatic system and translocate to circulatory organs may be exposed to many other cell types, including cells in the kidney, liver, and brain, and representative cell lines from these organs are commonly used *in vitro* models. For studying the use of nanomaterials for orthopedic applications, osteoblasts, which are bone-forming cells, are often used. A partial list of commonly used cell lines for *in vitro* studies is shown in Table 12.1.

TABLE 12.1

Partial List of Commonly Used Cell Lines for *In Vitro* Studies

Cell Type	Organ	Primary Function
Keratinocytes	Skin	Provide barrier properties (most common cell type of the dermis)
Macrophages	Lung	Engulf particles and debris via phagocytosis
Astrocytes	Brain	Participate in wound healing and scarring
Fibroblasts	Lung, kidney	Provide structural framework and participate in wound healing
Osteoblasts	Bone	Excrete bone-forming material

Toxicity endpoints include assessment of cell viability after a specified exposure period, as well as various more detailed studies to determine the mechanisms involved in toxicity. The mechanism of cell death is one of the most important because it plays a role in the immune response. There are two main mechanisms of cell death: necrosis and apoptosis. Necrosis occurs due to massive cell injury, and results with distribution of cellular debris into the extracellular environment and inflammation [73]. Evidence of cell necrosis includes changes in cell morphology, plasma membrane rupture, cytoplasmic swelling, and dissolution of organelles [91]. Apoptosis is programmed cell death that occurs naturally to keep a balance between cell division and cell death and is triggered by physiological or abnormal signals. In apoptosis, the plasma membrane stays intact and is eventually engulfed by other cells. Evidence of apoptosis includes detachment from neighboring cells, rounded morphology, and cell size reduction. Abnormal signals, which can cause apoptosis, include oxidative stress caused by loss of balance between reactive oxygen species (ROS) and antioxidants, starvation due to lack of nutrients, or loss of access to the cell and lack of growth factors. Detachment of epithelial cells from their extracellular matrix has been found to induce apoptosis [92].

In general, toxicological effects are classified by the extent of injury, location of damaging affects, timeliness of the response, and the length of exposure. The extent of toxicity can be described as mild, moderate, serious, or lethal, which ranges from minor irritation to morbidity. The location of toxicological effects can include local or systemic depending on whether the substance or material affects only the tissue with which it first comes into contact or it is absorbed into the bloodstream, affecting other organs. The timeliness of a toxicological response can either be immediate or delayed, depending on whether effects are seen immediately or after a lag period. Carcinogenic substances and materials often do not produce noticeable effects until after a period of 20 to 30 years after initial exposure. The length of exposure to a substance or material can be defined as acute or chronic. Acute toxicity is an effect seen after short-term exposure (days or weeks), and chronic toxicity refers to effects seen after repeated exposures over a long period of time (months or years) [75].

Factors that affect toxicity include duration of exposure, dose, persistence at sites of interaction, biological role of affected sites, ability for defense mechanisms to repair or replace injured biological structures, nature and extent to which the substance or material is broken down or excreted from the body, and nature and extent of products released from injured cells that may cause cytotoxic or stimulatory effects on other cells [91].

Overall, the final goal of toxicity analysis is to establish a regulatory framework that protects workers, the general public, and the environment from impacts of new substances or materials. In general, results from toxicity studies are used to evaluate potential risks posed to humans as a result of exposure to a substance or material, as well as identify and understand cellular and molecular mechanisms by which they exert toxic effects. In addition to establishing exposure limits and protective protocols, identification of key properties associated with specific toxic effects will allow for toxicity prediction and the potential to engineer safer forms of substances and materials [93].

12.5 Nanomaterial Toxicity

Due to rapid advancement in the synthesis of nanomaterials, it is difficult to study the effects of each variation of material that is engineered. Therefore, it is necessary to establish a paradigm for predicting the toxicity of new materials. In order to achieve this, researchers must compile information about specific physicochemical properties of nanomaterials that drive toxicity [9]. Physicochemical properties that have been found to be important in nanomaterial toxicity studies include shape, size, agglomeration state, surface chemistry, surface charge, surface area, crystal structure, and chemical composition.

Many of the most promising nanomaterials in the field of nanotechnology are those that are carbon based, and several basic studies have been conducted to investigate their toxicity. Carbon-based nanomaterials can be synthesized with different crystal structure, shape, size, and surface chemistry, and each variation has the potential to exhibit very different toxicity. Carbon-based nanomaterials that have been studied for their potential toxicity include carbon black, nanodiamonds, nano-onions, C60 molecules, CNF, and CNT.

Carbon black, nanodiamonds, nano-onions, and C60 molecules represent sp2 and sp3 hybridized carbon-based nanomaterials with spherical shape. In general, conventional carbon black is known to produce minor irritation to the skin, eyes, and respiratory tract but is not considered a carcinogen. Short-term inhalation studies have found that nanosized carbon black causes inflammation in the lungs [94]. Evidence of size-related effects was observed in a study, where nanosized carbon black generated more oxidative stress *in vitro* than did microsized carbon black [95]. *In vivo*, smaller nanoparticles induced greater pulmonary effects after exposure to rats [94,96]. Few studies have been published to date on nanodiamond toxicity, but thus far no hazardous effects have been reported. In one study, nanodiamonds ranging in size from 2 to 10 nm did not induce toxicity or oxidative stress in various cell lines [97]. Nano-onions also have not been shown to induce toxicological effects. Specifically, in an *in vitro* study examining their effects on human skin fibroblasts, they were not found to elicit any toxicological responses [98].

In vitro, C60 induced cytotoxicity through ROS-mediated cell membrane lipid peroxidation [99,100]. *In vivo*, C60 molecules were found to produce toxicity in a fish model by generating oxidative stress and lipid peroxidation [101]. High concentrations (200 μg/mL) of C60 and C70 induced necrotic and apoptotic cell death and mortality in embryonic zebrafish, while C60 functionalized with hydroxyl groups produced less pronounced affects and did not produce apoptotic cell death [102]. C60 molecules functionalized with hydroxyl and carboxyl groups produced less toxicity and oxidative stress than those that were not functionalized in human dermal fibroblasts and human liver carcinoma cells [103]. However, *in vivo* studies did not reproduce those results, where neither underivatized nor functionalized C60 molecules exposed to rats via intratracheal instillation induced adverse effects after subchronic exposure [104]. Although the results are sometimes contradictory, they suggest that spherical carbon-based nanoparticles can elicit toxic responses *in vitro* and *in vivo*.

Other important carbon-based materials are those with lamellar or cylindrical shape and high aspect ratio, such as CF, CNF, and CNT. Carbon materials with high aspect ratio have been the focus of many toxicity studies due to their extensive use in nanotechnology applications. Additionally, previous toxic effects seen by fibrous materials, such as asbestos fibers and man-made vitreous fibers, raise concern about exposure to fibrous carbon-based materials. Unlike CNFs and CNTs, CFs have been used in applications for many decades without evidence of hazardous effects and usually are not respirable [105]. Furthermore, respirable CFs were reported to cause no adverse reactions in the rat lung treated in a chronic intratracheal instillation study [106]. Evidence thus far suggests that CNFs do not induce toxicity. For example, in a study on osteoblasts, CNFs with low metal content did not affect cell viability [107].

Of the high-aspect-ratio carbon materials, CNTs have been studied most extensively. Overall, the toxicity of CNTs is complex due to that fact that CNTs can be synthesized by different methods, which yield samples with varying chirality and purity. After synthesis, impurities are removed through many different techniques, which can drastically alter the properties of the final CNT sample. Therefore, different CNT samples are likely to elicit different toxicological responses. For example, in one study, HiPco® SWCNT samples (0.8 to 1.2 nm diameter; 800 nm long; 10% iron catalysts) decreased cell viability more drastically than SWCNT samples produced via arc discharge (1.2 to 1.5 nm diameter; 2 to 5 µm long; <1 wt% nickel and yttrium catalysts) [108].

Evidence that CNTs may be more toxic than carbon-based nanomaterials that do not have high aspect ratio has been shown *in vitro*. In one study, MWCNTs were found to be more toxic than carbon nano-onions to human skin fibroblasts [98]. In another, MWCNTs and SWCNTs were found to be more toxic than C60 after exposure to alveolar macrophages due to impairment of phagocytosis [109]. Also, MWCNTs and SWCNTs exhibited higher cytotoxicity to murine macrophages than carbon black at a dose of 0.01 mg/mL [110]. However, in another study, carbon black was found to exhibit higher toxicity to human tumor cell lines than MWCNTs [111]. These results emphasize the importance of detailed material characterization. For example, knowledge of the impurity content and size may provide insight into controversial results.

In vitro studies have found that CNTs can produce toxicity. Unmodified MWCNTs were found to cause toxicity to human epidermal keratinocytes [113], and unmodified SWCNTs exposed to human embryonic kidney cells caused cell apoptosis [114]. CNTs synthesized by catalytic CVD were found to be nontoxic to human umbilical vein endothelial cells when exposed at very low concentrations [115]. The concentrations used in this study were much lower than those used in traditional toxicity studies due to the assumption that CNTs would not accumulate to a high extent in the placenta region. Only one known human study has been performed to study the effect of carbon nanomaterials on dermal exposure. In this study, CNTs were administered to human volunteers with allergy susceptibility via skin patch and found to be nonirritant [116].

Several *in vivo* studies have been carried out to assess the toxicity of CNTs upon inhalation. Changes in the lung after CNT exposure include lung inflammation and fibrosis [117–120]. Fibrosis describes scarring that occurs due to the lung's attempt to repair inflammatory damage. The inflammation–fibrosis response in the lungs can be progressive and eventually impair the function of the lungs or may arrest at various stages [11]. In each of these studies, CNTs were more pathogenic than nanosized carbon black.

Lam et al. [117] demonstrated that a single intratracheal instillation in mice with three different types of SWNCT resulted in dose-dependent granulomas and inflammation. Each SWCNT sample contained different metal impurities, but this did not affect the induction of granulomas. Muller et al. [118] exposed rats intratracheally to whole (5.9 µm length) or ground (0.7 um length) MWCNTs and found longer unground CNTs to be more biopersistent after 60 days. This corresponds to results seen by asbestos fibers, where longer stable fibers are more likely to cause lung cancer than shorter fibers or those that dissolve in physiological solution [64]. Warheit et al. [119] showed SWCNTs containing impurities and amorphous carbon exposed to rats via intratracheal instillation caused inflammation and non-dose-dependent accumulation of granulomas. The lack of dose dependence was thought to be due to the route of exposure, which requires CNTs to be dispersed in a solution where they form agglomerates, and does not mimic realistic exposure scenarios. In an attempt to generate a more realistic exposure condition, Shvedova et al. [120] exposed SWCNTs to mice by pharyngeal aspiration, where the animal is forced to inhale a sample of materials that are placed on their tongue. SWCNTs produced lung inflammation, where well-dispersed CNTs produced milder affects than SWCNT aggregates. More realistic still, Mitchell et al. [121] exposed rats to MWCNTs using whole-body inhalation. The results of this study did not result in significant lung inflammation or tissue damage but did induce systemic immune function alterations.

Similar to the study with C60 molecules, there is evidence that functionalization may play a role in CNT toxicity. In one study, SWCNTs were found to inhibit human embryonic kidney cell growth by oxidative stress and induction of cell apoptosis depending on the degree of functionalization [122]. Additionally, adsorption of serum proteins on the surface of CNTs was found to modulate their toxicity and uptake *in vitro* [123].

Agglomeration is an additional parameter that has been shown to induce toxicity. The impact of SWCNTs on rat aortic smooth muscle cells was examined for concentrations of 0.0 to 0.1 mg/mL. Cell culture medium was filtered to remove the aggregate material and both unfiltered and filtered SWCNT samples were exposed to cells. In general, SWCNT aggregates were more toxic than filtered SWCNTs. However, samples induced equal cytotoxicity at the high dose [124].

Also, studies have indicated that metal catalysts may play a role in the toxic effect of CNTs on cells. In a study examining the effect of impurities in CNT, acute toxicity was not observed after exposure to CNTs, but intracellular ROS increased when exposed to unrefined CNTs versus refined. However, the refined CNTs still contained as much as 2.5% Ni/Co catalyst impurities and also had a high amount of oxygen on the surface, which could play a role in the development of oxidative stress [112]. In another study, refined SWCNTs had a greater impact on cell viability than unrefined, which was thought to be a result of increased dispersion and consequentially increased surface area. This increased surface area caused an increase in morphological changes and cell detachment. As expected, unrefined SWCNTs stimulated increased oxidative stress compared to refined SWCNTs [125]. In an additional study, SWCNTs were more toxic than MWCNTs to alveolar macrophages by impairing phagocytosis at low concentrations and inducing necrosis at high concentrations [109]. However, the SWCNTs used in this study contained a greater percentage of impurities than MWCNTs, which likely played a role in the observed effects.

Some of the most commonly used catalysts in CNT synthesis include cobalt (Co), nickel (Ni), iron (Fe), cobalt aluminum oxide (Co-Al_2O_3), nickel aluminum oxide (Ni-Al_2O_3), and iron aluminum oxide (Fe-Al_2O_3). No previous study has evaluated their relative toxicity; however, some studies have developed preliminary results regarding metal nanomaterial toxicity. For example, an *in vivo* study found that nanosized nickel and cobalt (20 nm) were able to induce significant inflammation in rat lungs after intratracheal instillation, where nickel was more toxic than cobalt and both particles had similar ability to induce oxidative stress [126]. In another study, nickel nanoparticles intratracheally instilled in rats were found to cause persistent inflammation after instillation of a small dose [127]. Additionally, iron nanoparticles with an average diameter of 72 nm were found to induce oxidative stress associated with a proinflammatory response in rats after intratracheal instillation at low doses (57 and 90 $\mu g/m^3$ for 3 days/6 h per day) [128]. This shows that when present at the nanoscale, as in the case of catalysts used in synthesis of CNTs, cobalt, nickel, and iron induce toxicological responses *in vivo*.

Aluminum oxide dust is generally considered nontoxic, but excessive inhalation has been associated with mild irritation. However, aluminum oxide nanoparticles have been found to produce greater pulmonary toxicity than fine aluminum oxide particles [126]. In another study, aluminum oxide nanoparticles were only found to have a marginal effect on the viability of alveolar macrophages and no effect on phagocytosis [129].

In addition to producing mild toxicity in their pure form, it is possible that in the presence of CNTs, the toxicity of metal catalysts can be significantly altered. Evidence of this phenomenon has been shown in the case of cobalt particles, where cobalt alone is unable to produce chronic lung disease in cobalt production workers, but when complexed with tungsten carbide to create hard metal, it is capable of inducing severe lung disease [130].

Further insight into the toxicity of carbon-based nanomaterials and their potential constituents can be obtained by reviewing data about other nanosized particles. As demonstrated with carbon-based nanomaterials, the toxic effects induced by materials containing similar elemental composition can vary drastically based on differences in size, crystal structure, and surface charge [131]. For example, inflammation in the rat lung after instillation of polystyrene particles of 64, 202, and 535 nm resulted in inflammatory indicators that increased as a function of increasing surface area. These results were confirmed *in vitro* where cytosolic calcium increased after exposure to nanoparticles only. Also, the polystyrene nanoparticles were found to induce an increased amount of ROS than the fine particles [132]. In another study, nanosized CB (14 nm), TiO_2 (20 nm), and latex (64 nm) particles were able to induce greater inflammation in instilled rats than their fine counterparts (CB 260 nm; TiO_2 250 nm; latex 202 nm) [133]. Titanium oxide particles of two sizes (20 nm and 250 nm diameter) of the same crystalline structure (anatase) exhibited similar mass deposition in the lower respiratory tract of rats exposed via inhalation. However, in the postexposure period of up to 1 year, sustained impairment of alveolar macrophages function and inflammation occurred, and a prolonged total pulmonary retention and increased translocation to the pulmonary interstitium of the TiO_2 nanoparticles was found [134]. This result is important because it indicates that certain nano-particles may persist in the lung for a long time after initial exposure, which could eventually lead to cancer.

The effects of crystal structure and surface charge have also been shown to play a role in toxicity. In a study examining the effect of nanoscale titania on cells in culture, toxicity was not dependent on size but correlated strongly to the crystal structure, where anatase TiO_2 was 100 times more toxic than an equivalent sample of rutile TiO_2 [135]. It is possible that the availability of a greater amount of surface atoms in anatase TiO_2 was the cause for the differ-ence. In another study, reduction in zeta potential corresponded to reduction in toxicity induced by asbestos fibers [136].

In addition to the size, crystal structure, and composition of the mate-rial, the atmosphere in which nanoparticles are inhaled can also affect their toxicity. Inhalation of acids does not usually initiate toxicological responses in the lungs; however, when exposed in combination with car-bon black nanoparticles, acids disrupted the functional integrity of lung defenses. Therefore, carbon black nanoparticles can carry toxic chemicals to the deep lung before being deposited, resulting in increased toxicity [137,138]. Additionally, carbon nanoparticles can induce slight inflamma-tory responses in rats, but coexposure with ozone agents increased the toxic response [139].

The mechanism of toxicity in many nanomaterial toxicity studies thus far has been found to include oxidative stress, plasma membrane leakage, and cell morphology changes. In one *in vitro* study, the mechanism of CNT toxic-ity was investigated more deeply, and it was found that SWCNTs decreased cell adhesive ability in a time- and dose-dependent manner and were found to downregulate genes associated with the first growth phase of the cell cycle and upregulate genes associated with apoptosis, as well as downregu-late expression of signal transduction–associated genes and adhesion-asso-ciated proteins, which are important to the extracellular matrix (e.g., collagen and fibronectin) [114]. Many studies have found that regardless of whether a nanomaterial induces direct toxic affects to cells, they often impair the func-tion of a particular cell type or tissue. Specifically, the phagocytotic function of macrophages has been found to be altered after exposure to various nano-materials. For example, Renwick et al. [140] found that titanium oxide in the fine and nanosize ranges were not toxic to the cells, but significantly reduced the ability of macrophages to phagocytize other particles. In another study, cell phagocytotic ability was significantly hindered by exposure to alumi-num nanoparticles at doses of 25 µg/mL but did not affect cellular viability at the same concentration.

Lundborg et al. [141] showed that exposure to fine and nano carbon particles at a concentration similar to that in urban areas impaired the phagocytotic function of alveolar macrophages, which was more prevalent for nanoparticles than their micron-sized counterparts. Furthermore, SWCNTs were shown to impair phagocytotic function more than MWCNTs and fullerenes at the same concentration [109]. In another study, fullerenes were taken up to a greater extent in alveolar macrophages than SWCNTs but were found to induce lower cytotoxicity than the high-aspect-ratio material [142]. This may be due to the length of SWCNT bundles versus the size of MWCNT agglomerates.

Nonphagocytotic cells have also been shown to uptake nanomaterials. For example, CNFs (60 to 100 nm) were found to be taken up into cells and enclosed in vacuoles visualized via transmission electron microscopy (TEM) [107]. Unmodified MWCNTs were found to localize within the free cytoplasm and cytoplasmic vacuoles of human epidermal keratinocytes [116]. SWCNT bundles have been shown to be encapsulated in endosomes inside human epithelial lung cells after 24 h exposure [143]. In another study; however, CNTs were only partially ingested by rat lung epithelial cells, which raises cause for concern [144]. In addition to cell uptake, adherence on the plasma membrane may also contribute to toxicity. In one study, it was suggested that SWCNT aggregates adhered to 3T3 fibroblast surfaces and likely inhibited mass transport across the cell membrane [145].

In some cases, the mechanisms involved in material toxicity may be predicted comparing properties with other materials that have known toxicity using structure–activity relationships. One of the materials with the most well-understood toxicity mechanisms is asbestos due to extensive epidemiological and clinical data. In the case of asbestos, the greatest risk of lung cancer is associated with fibers longer than 10 μm with diameters between 0.3 and 0.8 μm [146]. The induction of lung cancer due to long fibers is associated with particle ingestion by lung cells and consequential cell membrane permeability, leakage of proteolytic enzymes, and liberation of growth-promoting factors at the ingestion site, leading to inflammation, tissue necrosis, and collagen deposition and scar formation, which in turn can lead to cancer [65].

It is hypothesized that partial ingestion of other fibrous nanomaterials may cause similar effects. The biopersistence of fibers arises when they do not undergo chemical dissolution in the lung tissue. When fibers are treated *in vitro* with salt solutions typical of those found in biological systems, the fibers differ in their durability. Some fibers break into smaller fragments or dissolve, but other fibers may resist dissolution. CNTs are not likely soluble in a neutral pH and are being synthesized with increasing length. Furthermore, they tend to be contaminated with metals, which contribute to their toxicity. Based on toxicity information about other fiber types, accumulation of CNT agglomerates in the lung with a length of 10 μm or longer are likely to cause biopersistence and lead to hazardous effects [65].

Parametric studies carried out thus far have provided evidence that specific forms of nanomaterials are more toxic than others. Factors that have been shown to play a role include agglomeration state, crystal structure, size, aspect ratio, composition, and surface charge [119,131,135,136,141]. There is contradicting evidence whether functionalization of nanomaterials leads to increased biocompatibility [103,104]. What is lacking in many of these studies is detailed materials characterization and universal techniques to assess toxicity and toxicological mechanisms. Investigation into available characterization methods and their limitations will facilitate a better understanding for future directions in this area.

12.6 Nanomaterials Characterization and Cytotoxicity Endpoints

12.6.1 Nanomaterials Characterization

Many variations of nanomaterials are being synthesized for use in various applications. Additionally, nanomaterial samples may contain different size distributions, impurity content, and surface chemistry depending on the synthesis and purification methods employed. Incomplete characterization of nanomaterial properties has yielded many contradicting results in early nanotoxicity studies. Therefore, for accurate risk assessment, the first step in every toxicity study must be detailed materials characterization. Furthermore, before administration to a biological system, nanomaterials are often dispersed in water or biological media, which can lead to changes in some of these properties. Because of these changes, it is necessary to characterize nanomaterials in both the as-synthesized form and after dispersion to generate an accurate analysis of the toxicological effects of a specific nanomaterial sample and be able to compare these results across studies. Properties that may play a role in the toxic effects of nanomaterials include size, morphology, crystal structure, surface area, surface energy, composition, and surface chemistry.

For size and morphology in the as-synthesized form, transmission electron microscopy (TEM), scanning electron microscopy (SEM), and atomic force microscopy (AFM) are routinely used. The crystal structure of a material can be determined using X-ray diffraction (XRD). The Brunauer, Emmett, and Teller (BET) technique and inverse gas chromatography (IGC) can be used to estimate surface area and surface energy of as-synthesized nanomaterials, respectively. Some of the most common techniques to determine composition include X-ray fluorescence (XRF), atomic absorption spectroscopy (AA), and inductively coupled plasma (ICP) paired with optical emission spectroscopy (OES) or mass spectroscopy (MS). Surface chemistry can be evaluated using ultraviolet-visible spectroscopy (UV/VIS), Fourier transform infrared spectroscopy (FTIR), or Raman spectroscopy [147].

Once nanomaterials have been dispersed into solution, agglomeration may occur, leading to changes in size and morphology. The size and morphology of dispersed materials can be characterized using certain types of optical microscopy (OM) and environmental SEM. Size can also be evaluated

TABLE 12.2

Common Techniques for Nanomaterial Characterization

Physical or Chemical Property	Technique(s)	
	As Synthesized	After Dispersion
Size and morphology	TEM, SEM, AFM	DLS, OM, SEM
Elemental composition	XRF, AA, ICP-OES, ICP-MS	
Surface area	BET	
Surface energy	IGC	LDE
Surface chemistry	UV/VIS, FTIR, Raman	
Crystal structure	TEM, XRD	

Notes: AA, atomic absorption spectroscopy; AFM, atomic force microscopy; BET, Brunauer, Emmett, and Teller technique; DLS, dynamic light scattering; FTIR, Fourier transform infrared spectroscopy; ICP, inductively coupled plasma; IGC, inverse gas chromatography; LDE, laser doppler electrophoresis; MS, mass spectroscopy; OES, optical emission spectroscopy; OM, optical microscopy; SEM, scanning electron microscopy; TEM, transmission electron microscopy; UV/VIS, ultraviolet-visible spectroscopy; XRD, x-ray diffraction; and XRF, X-ray fluorescence.

using dynamic light scattering (DLS). The surface charge of nanomaterials in solution can be defined using laser doppler electrophoresis (LDE). Table 12.2 summarizes common characterization techniques used in nanotoxicity studies. Methods used for characterization in this thesis are further described in the following sections.

In TEM, electrons are accelerated by an electric field and transmitted through the sample. When the electrons that remain in the beam are detected, areas where electrons are scattered appear dark on a positive image, and a bright-field image is formed. When the diffracted electrons are detected, a dark-field image is formed. The maximum resolution of the TEM is limited by spherical and chromatic aberration, but high-resolution TEM can reach to below 1 nm when imaging atomic lattice planes [148]. Limitations of TEM are that samples must be thin (<100 nm) and dry (imaging is performed under vacuum).

In contrast to TEM where electrons are transmitted through the sample, SEM produces images by rastering a primary electron beam across the sample surface while detecting secondary electrons, which are emitted from the surface. In this technique, more electrons emitted from the side of a sample escape entry into the photomultiplier device than those from the top, which results in images with a characteristic three-dimensional appearance. In SEM, the electron beam is focused onto an area of only a few nanometers in diameter, which is significantly smaller than the wavelength of visible light. Therefore, the images obtained in an SEM provide greater resolution than light microscopy, but about an order of magnitude lower resolution than TEM [149]. Like TEM, SEM typically requires samples to be imaged under vacuum. However, thicker samples can be imaged [150].

XRF is a method for qualitative and semiquantitative element analysis and is based on the characteristic radiation emitted by ionized atoms of each element upon relaxation. When an atom is irradiated with X-rays, the atom is ionized (i.e., an electron from one of the inner shells is ejected). During relaxation (i.e., vacant place is filled by an electron from an outer shell), a photon is released. The radiation emitted, which is unique for each element, is dispersed into individual wavelengths and evaluated using a spectrometer. Semiquantitative analysis is made possible using software analysis programs, which calculate the XRF spectrum for a given radiation using standards. The accuracy of results depends on how closely the standards resemble the sample in composition and morphology with an average accuracy of ~10% [151]. In general, XRF can detect elements with atomic number greater than 11 [152].

ICP-OES (also called ICP-AES) is a method for quantitative analysis of known elements in a sample. ICP-OES is based on the principle that energy of emission is specific for each element. The sample is injected into a stream of argon gas where it is atomized by a nebulizer. The argon gas carries the atomized sample into the inductively generated argon plasma, where the elements are thermally excited. The spectrum emitted is transferred into a spectrometer where it is decomposed into the individual wavelengths and evaluated. The radiation intensities emitted by analysis samples are compared to elemental concentrations in calibration samples [153]. The sample may be liquid or solid, but solid samples must be digested in acid. ICP-OES can detect most elements, except refractory elements, which are more readily determined in a hotter plasma source, such as flame atomic absorption methods [154]. ICP-mass spectrometry (ICP-MS) is another method for quantitative elemental analysis, where the argon ICP decomposes the original sample into charged ions, which are transferred into the mass spectrometer. Rather than separating the light according to its wavelength as in ICP-OES, MS separates the ions according to their mass-to-charge ratio. Advantages of the ICP-MS are that standards are not required, it has better detection limits, and it has the ability to quantify elemental isotopic concentrations and ratios [155].

Raman spectroscopy is a convenient, nondestructive technique for identifying crystalline or molecular phases and obtaining structural information on noncrystalline solids. In this method, a laser beam illuminates the sample and interacts with phonons or other excitations, causing the energy of the laser photons to be shifted up or down. Wavelengths close to the laser line are filtered out while the rest of the light is dispersed onto a detector. When light impinges on a molecule and interacts with the electron cloud of the bonds of that molecule, the incident photon excites one of the electrons into a virtual state. For the spontaneous Raman effect, the molecule will be excited from the ground state to a virtual energy state and relax into a vibrational excited state, which generates Stokes Raman scattering. The inelastic light scattering measures the vibrational frequencies of the sample, which are given in wave numbers obtained from the difference between the wave number position of the laser line and the Raman line. Calculations are performed by a computer and only the difference wave numbers (cm^{-1}) are displayed [156].

In simple graphite, a single mode is observed at 1580 cm^{-1} (G band). The D band is usually located between 1330 and 1360 cm^{-1} and is a longitudinal optical phonon and is known as the disordered or defect mode because a defect is required to elastically scatter in order for the process to conserve momentum. The D band is present in all carbon allotropes, including sp2 and sp3 amorphous carbon. In CNTs, this band is activated from the first-order scattering process of sp2 carbons by the presence of in-plane substitutional hetero-atoms, vacancies, grain boundaries, or other defects, and by finite-size effects [157]. The graphitization degree is higher for smaller values for the ratio of I(D)/I(G).

XRD is a nondestructive analytical technique used to study material structure and is based on the elastic scattering of X-rays from structures that have long-range order. There have been extensive studies of carbon structure by XRD; however, the application of general diffraction theory to carbons, which are weak scatters of X-ray and have low-degree regularity in crystallinity, involves complex problems. The average internal distance of the layers is derived from the peak position 28 and the relationship between average crystalline size, La, and the width of the scattering peak, β, is well known as Scherrer's equation: La = $(\kappa\lambda)/(\beta\cos\theta)$, where κ is a constant, λ is the wavelength of X-rays. La is a mean value because it is anticipated that there are considerable distributions in the number of stacking layers and the dimension of each layer.

DLS is used to measure particle size in solution. The laser passes a beam, usually a red laser (633 nm), through the sample. The laser light is scattered, which is then picked up by a detector and sent through a photomultiplier tube and a signal amplifier. Information on particle position with respect to time is collected, and these data are used to calculate the hydrodynamic radius of the particle using the Stokes–Einstein equation and vector calculations. Using this technique, the particle size range that can be detected is 6 μm in size down to 1 nm. Samples in powder form must be dispersed into a solvent whose viscosity data are available. Sample concentration can range from 0.1 to 40 w/v%. About 1.5 mL of sample is needed for measurement, but with special cuvettes, sample size can be as small as 300 μL [158].

Although dispersed samples cannot be imaged directly in their dispersed environment, nanomaterials that have formed agglomerates in solution maintain their agglomerated form upon drying. Therefore, electron and optical microscopy methods can be employed to view these structures. Even though the limits for optical microscopy are much lower than for electron microscopy, general information about the size of large agglomerates can be achieved. In some cases, dark-field microscopy is useful for illuminating the materials against a dark background for better resolution.

Zeta potential (ZP) is an important and useful indicator of the electric charge that is acquired on the surface of materials in contact with liquid, which can be used to predict and control the stability of colloidal suspensions or emulsions. ZP is one of the main forces that mediate interparticle interactions: particles with a high zeta potential of the same charge sign, either positive or negative, will repel each other. For molecules and particles that are small enough, and of low enough density to remain in suspension, a high zeta potential (<–30 mV and >30 mV) will confer stability. Most particles dispersed in an aqueous system will acquire a surface charge, principally either by ionization of surface groups or adsorption of charged species. These surface charges modify the distribution of the surrounding ions, resulting in a layer around the particle that is different to the bulk solution. If the particle moves, this layer moves as part of the particle. The zeta potential is the potential at the point in this layer where it moves past the bulk solution. This is usually called the slipping plane. ZP is measured by applying an electric field across the dispersion. Particles within the dispersion with a ZP will migrate toward the electrode of opposite charge with a velocity proportional to the magnitude of the ZP. This velocity is measured using the technique of LDE. The frequency shift of an incident laser beam caused by these moving particles is measured as the particle mobility, and the ZP is calculated based on the Smoluchowzki or Huckel theories using the particle mobility and dispersant viscosity [158].

12.6.2 Cytotoxicity Endpoints

Once nanomaterial samples are well characterized and administered to cells for a predetermined exposure time, there are a wide range of endpoints available to assess toxicity. Each endpoint is designed to detect a specific marker for direct toxicity effects or mechanisms that may lead to toxicity. Techniques often used for assessment of toxicity endpoints include microscopy and imaging analysis, biochemical assays, and electrochemical assays.

Optical and electron microscopy are commonly used for visualizing cell morphology and nanomaterial interactions in cellular systems. Some forms of optical microscopy often used include bright field microscopy, dark field microscopy, fluorescence microscopy, and confocal microscopy. Each method carries trade-offs for analyses, which are described in the following sections.

Bright-field microscopy is useful for gaining insight into the overall morphology of cells exposed to nanomaterials at low magnification. Large agglomerates of nanomaterials can be seen using this method, but specific interactions between cells and nanomaterials cannot be resolved. An example of an image acquired via this method is shown in Figure 12.7.

Dark-field microscopy can be used to illuminate unstained nanomaterials and cell membranes, which appear as brightly illuminated objects on a dark background (Figure 12.8). Dark-field illumination removes unscattered light from the diffraction pattern formed at the rear focal plane of the objective. In situations where the refractive index is different from the surrounding medium or where refractive index gradients occur (as in the edge of a

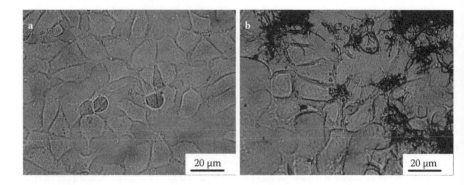

FIGURE 12.7
Mouse keratinocytes after a 24 h exposure acquired via bright-field microscopy: (a) control cells; and (b) cells exposed to 100 μg/mL carbon nanofiber.

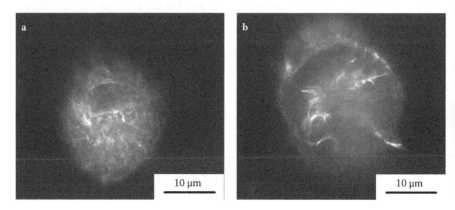

FIGURE 12.8
Rat alveolar macrophages imaged via URI method: (a) control; (b) 10 μg/mL carbon nanofiber.

membrane), light is refracted by the specimen. The resolving power of the objective is the same in dark-field illumination as found under bright-field conditions, but dark-field illumination allows for better contrast for revealing outlines, edges, and boundaries [159]. The CytoVivaTM150 Ultra-Resolution Imaging (URI) system (Aetos Technologies, Inc., Auburn, Alabama) is an attachment for a standard research transmission optical microscope, which allows for dark-field imaging. The URI system is currently growing as a widely used biological imaging tool due to its ability to image live biological specimens at high magnification using an oil lens [159,160].

Fluorescent microscopy can be used to image samples that have been tagged with fluorescent probes. Cell structures can be stained to examine location of cellular subunits and evaluate their integrity. This allows for qualitative analysis of cellular toxicity mechanisms, such as cell morphology changes and membrane potential disruption. Fluorescence microscopy

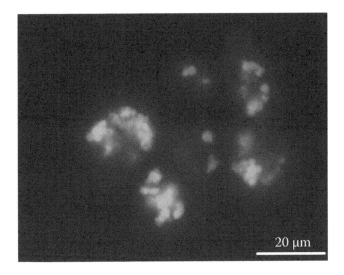

FIGURE 12.9
Rat alveolar macrophages exposed to 10 μg/mL dye-doped silica nanoparticles imaged via fluorescence microscopy.

is also a useful method for imaging fluorescent nanoparticles, such as the DSNP seen in Figure 12.9. In fluorescence microscopy, light is passed from the emitting source through the exciter filter. The exciter filter passes only the desired excitation wavelengths, which are reflected down through the objective to illuminate the specimen. Longer wavelength fluorescence emitted by the specimen passes back through the objective and dichroic mirror before finally being filtered by the emission filter.

Dual-mode fluorescence (DMF) is an attachment used in conjunction with the URI system (Aetos Technologies, Inc.), which allows simultaneous imaging of fluorescent and nonfluorescent structures. This method is useful because the nanomaterial agglomerates or cells can be illuminated using standard dark-field optics, and specific cell structures or nanoparticles that have been tagged by a fluorescent probe can be imaged using triple-pass emission and excitation filters. The versatility of the DMF system is shown in Figure 12.10. Note that in Figure 12.10a, the cells fluoresce while the nanomaterials are illuminated, and in Figure 12.10b, the nanomaterials fluoresce while the cell is illuminated.

Confocal microscopy allows for controlling depth of field and optical sectioning. There are two types: laser scanning confocal microscopy (LSCM) and spinning disk confocal microscopy (SDCM). LSCM yields high image quality, but the imaging frame rate is very slow (less than 3 frames/second), while SDCM can achieve very fast imaging (30 frames/second) with lower quality, which is preferred for dynamic observations [161].

The LSCM is composed of a pinhole aperture sitting conjugated to the focal plane, for only in-focus light passes through to the detector, while out-of-

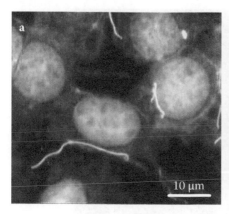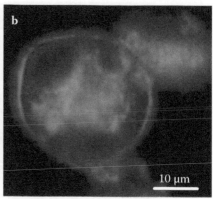

FIGURE 12.10
Two approaches for utilizing dual-mode fluoroscence to image nanomaterial interactions with cells: (a) Mouse keratinocytes stained with fluorophores to tag the nucleus and actin filaments after exposure to 10 µg/mL carbon nanofiber; and (b) rat alveolar macrophage after exposure to dye-doped silica nanoparticles.

focus light is blocked. One confocal pinhole per detector is used in most confocal systems [161]. The LSCM scans the sample sequentially point by point and line by line and assembles the pixel information to one image. SDCM uses the same principle as LSCM but uses an excitation source for excitation of many spots simultaneously. Furthermore, the use of an arc lamp as the excitation source in SDCM allows the excitation wavelength to be changed by simply changing the excitation filter.

In confocal microscopy, optical slices of the specimen are imaged with high contrast and high resolution in x, y, and z. By moving the focus plane, optical slices can be put together to build up a three-dimensional image (Figure 12.11). This method is optimal for determining three-dimensional structures of relatively thick live biological samples without destruction. The disadvantage of this method is that nonluminescent structures cannot be visualized, which makes conventional confocal microscopy unsuccessful for imaging carbon-based structures without fluorescence tagging. The overall resolution of confocal microscopy lies between 400 and 600 nm [162].

Electron microscopy can also be used to image biological samples after exposure to nanomaterials. There are many advantages to the high-resolution capabilities of electron microscopes, such as the ability to image individual nanomaterials within cellular structures. However, in many cases, extensive sample preparation is required. Fixing and drying procedures can affect the morphology of cells, which can lead to misanalysis of data. Additionally, samples for TEM must be <100 nm thick, so cells must be isolated by centrifugation and sliced into thin sections, which could cause forced uptake of nanomaterials. In SEM, less sample preparation is required; however, samples still must be imaged under a vacuum. The main disadvantage of using

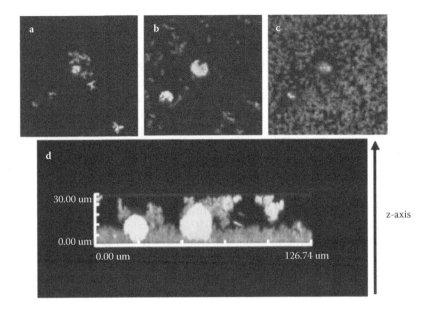

FIGURE 12.11
Rat alveolar macrophages after exposure to dye-doped silica nanoparticles imaged via confocal microscopy: (a) top slice in the xy plane; (b) middle slice in the xy plane; (c) bottom slice in the xy plane; and (d) stack of optical slices generating a projection in the xz direction.

SEM for biological samples, which are not electrically conductive, is that sample charging is a problem. Therefore, the sample must be dehydrated and coated with a thin (1 to 2 nm) carbon or metal film. The alternative is to use environmental scanning electron microscopy (ESEM). ESEM allows wet samples to be observed in low-pressure gaseous environments and in the presence of water vapor. However, resolution does not permit clear analysis of nanomaterial uptake, and live cell imaging is complicated.

Although there are many imaging techniques available, those that allow imaging of live samples are preferred. However, optical microscopy methods do not provide resolution high enough to image individual nanomaterials, so determining uptake of nonluminescent nanoparticles is complicated. It may be achievable, though, where in one study, the relative uptake of titanium oxide particles, quartz, and diesel particulates in alveolar macrophages was determined by measuring the right-angle light scatter of control and exposed cells [163]. This only provided a rough estimate of uptake, so a method is sought for analysis of nanomaterial uptake and interactions with biological samples with less rigorous preparation and more precise ability for measurement. In the meantime, the methods described above must be used in conjunction for adequate data.

Biochemical assays are commonly used to measure cytotoxicity *in vitro*, and the types of assays available are diverse. Assays are available to measure markers that indicate the number of dead cells (cytotoxicity assay), the number of live cells (viability assay), and the mechanism of cell death, such as apoptosis or necrosis or other biomarkers indicating mechanisms of toxicity.

Cell viability after exposure to potential toxins can be assessed using cell metabolism and membrane leakage. Common metabolic markers include ATP production and tetrazolium reduction. In tetrazolium-based assays, dehydrogenase enzymes in viable cells produces reducing compounds, such as NADH or NADPH, which pass their electrons to an intermediate electron transfer reagent that can reduce the tetrazolium product into a formazan product. At death, cells rapidly lose the ability to reduce tetrazolium products, so the production of the colored formazan product is proportional to the number of viable cells in culture.

One of the most commonly used tetrazolium-based methods is the MTT assay [164]. In this assay, the tetrazolium salt [3-(4,5-dimethylthiazol-2-yl)-2,5-diphenyltetrazolium bromide (MTT)] is incubated with live cells after exposure. During this incubation, the mitochondria of healthy cells cleave the ring of the MTT via succinate dehydrogenase, reducing it to a blue formazan salt. The resulting formazan salt must be extracted with alcohol. The absorbance of the final solution is read at 570 nm with a reference of 690 nm to eliminate error from scratches on the plastic or turbidity [164]. The final absorbance values are compared to negative control values, and cell viability is expressed as percent reduction or enhancement from control. It is important in biochemical assays to use both a positive and negative control to verify the probe is working properly. In the MTT assay, a chemical or nanomaterial with known toxic effects is often used as the positive control, and cells that have not been exposed the nanomaterial being studied are used as the negative control.

When using biochemical assays to evaluate nanomaterial toxicity, many challenges are faced. The most important is the potential adsorption of dyes on the nanomaterial surface [8]. Monteiro-Riviere et al. [165] found that carbon black interfered with biochemical assays through adsorption of biomarkers and hypothesized that other CNTs also have the capability to act as sinks for the nonspecific adsorption of organic molecules and macromolecules. In fact, cell viability experiments by Wörle-Knirsch et al. [143] found that SWCNT interfered with the MTT assay by adsorbing the formazan product and causing false positive for cytotoxicity. The results showed that when impurities were increased from 2.5 to 8%, adsorption of the formazan product decreased. This study showed that the reaction using mitochondrial dehydrogenases was unaffected by SWCNTs (verified using other cell viability assays relying on this reaction). Instead, the formazan salt produced by the MTT assay was adsorbed onto the surface of SWCNT, causing a false positive for cytotoxicity (determined using TEM images and comparison of results to other viability assays). However, in another study, Wick et al. [166] found that binding of MTT-formazan to MWCNTs was negligible. Some nanoparticles have been found to bind cytokines, which are a group of signaling compounds produced by cells to communicate with one another [167].

Due to these challenges, modifications to traditional protocols and extra control experiments are necessary when using colorimetric biochemical assays to assess cellular toxicity.

Apoptosis is a mechanism of cell death, in which an orderly series of biochemical events leads to death. One indicator of apoptosis is dissipation of mitochondrial membrane potential (MMP). A common method for determining the status of the mitochondrial polarization is to stain the mitochondria with dye, which is able to accumulate within the mitochondria of healthy cells. In this assay, dissipation of the MMP is marked by redistribution of the dye throughout the cell and culture media [168].

Apoptosis is sometimes initiated by oxidative stress, which has been found to be caused by exposure to various nanomaterials [52,94,101,103]. Oxidative stress occurs when there is an imbalance between damaging oxidants and protective antioxidants. Oxidants include hydrogen peroxide and hydroxyl radicals, which are collectively known as ROS. Release of ROS may cause cell membrane alterations and damage to macromolecules, leading to inflammation, cell injury, and death [145]. Additionally, many studies have indicated that oxygen radicals and other oxygen-derived species are important agents in aging and in many human diseases including cancer, multiple sclerosis, Parkinson's, autoimmune disease, and senile dementia [169].

A common assay to assess the state of oxidative stress of cells exposed to nanomaterials is the ROS assay. In this assay, a fluorescent probe 2′,7′-dichlorohydrofluorescein diacetate (DCHF-DA) is incubated with the cells prior to exposure, which reduces the probe via esterases to DCF. After specified time points of exposure to nanomaterials, the fluorescence intensity produced by DCF bound to ROS is measured with a 485 nm excitation filter and a 530 nm emission filter. The positive control to test the reactivity of the probe in this assay is most often hydrogen peroxide [170]. Therefore, it is likely that metal contaminants in carbon-based samples could lead to the production of ROS [116]. For example, iron impurities in asbestos were found to cause ROS generation after exposure to asbestos fibers [171].

Iron bioavailability is sensitive to partial oxidation, mechanical stress, and sample age [169]. Fe(III) alone appears to be unable to produce radicals; however, Fe(II) catalyzes the reduction of hydrogen peroxide or oxygen to ROS [88]. Additionally, Lund and Aust [172] found that there is a relationship between mobilized iron and DNA single-strand breaks. Therefore, the toxicity of a given material containing iron impurities could be predicted by careful analysis of its iron content that is reducible or in the reduced state and bioavailable at the surface.

Iron in CNT has been reported to be a mixture of α-Fe (zero-valent), γ-Fe, and carbide phases and much of the metal appears by TEM to be at least superficially encapsulated by carbon. The key question for the possible biological reactivity of these iron residues is whether some portion of the metal is accessible to reductants in biological media and thus able to participate in redox reactions. One assay to test the bioavailability of iron in a sample is to add a chemical such as ferrozine, which binds to mobilized Fe(II) forming a Fe(II)-ferrozine complex that can be measured spectrophotometrically [173].

Iron mobilization requires oxidation to a soluble form, which can occur through atmospheric exposure. The reaction between nanoscale zero-valent iron and dissolved dioxygen is reported to be kinetically limited with full oxidation reported after 60 days [31]. The early stages of oxidation in the atmosphere accelerate subsequent mobilization in solution and thus increase iron bioavailability and potential redox-mediated toxicity. Further oxidation eventually leads to formation of the more stable Fe_3O_4 or Fe_2O_3 phases and corresponding decreases in mobilization in the time range of 10 to 300 days.

Although assays for determining cell viability and cytotoxicity based on ATP production, tetrazolium reduction, and ROS generation are useful and cost effective, they have limits to the length of time to complete, resulting in loss of real-time response information. The time it takes for an assay to develop a measurable signal can be crucial to the accuracy and reduce the chance of false results caused by interaction of the test compound with the biological marker. Therefore, recent studies have examined real-time methods for assessing cell proliferation and morphology changes.

In one study, the real-time cell growth rate of human keratinocytes was monitored using image analysis software and was evaluated by measuring the average degree of confluence and cell concentration. This method eliminated the need to count cells directly using enzymatic digestion or a biochemical assay, both of which cannot be performed continuously [174]. In the future, it is reasonable to assume this method could be extrapolated to test the response of cellular growth rate after exposure to a potentially toxic chemical or material. In another, a human lung carcinoma epithelial cell line, normal human bronchial epithelial cells and normal human keratinocytes were fixed and stained after exposure to nanoparticles. Digital photographs were taken and an image processing program was used to measure colony surface areas [108].

A more thoroughly studied method for real-time analysis of cell proliferation in culture is through impedance measurements. The electrical properties of biological tissues and cells differ significantly depending on their structures, and the impedance increases with increasing presence of cells because they act as insulation and restrict current flow. Therefore, electrochemical impedance measurements can be made to monitor cell growth for several days without detectable electrical influence on the cells [175].

In an early study, real-time monitoring of cell proliferation was used by measuring the impedance of cells grown on interdigitated electrode structures and proved to be successful for detecting the influence of media serum components and toxicity of cadmium [175]. In another study, a linear relationship was found between capacitance and DNA content of single eukaryotic cells, which implies this technique is useful to detect changes in intracellular DNA associated with tumors. The results were found to be in agreement with those acquired using flow cytometry, which requires much more sample preparation [176].

In another study, human hepatocellular carcinoma cells were treated with cytotoxic agents, resulting in a decrease of impedance, indicating the detachment of dead cells. Using measurements of the impedance changes in conjunction with a relationship between the impedance and the cell growth curve, this method allows successful measurement of live or dead cells [177].

In a recent study, electrical impedance measurements were used to monitor cell adhesion of immortalized mouse fibroblasts that were cultured directly onto electrode structures. This method was also successfully used to detect cytotoxicity after treatment with different concentrations of a model toxicant [178]. A similar method was used to quantitatively monitor interaction of fibroblasts with extracellular matrix proteins [179].

Real-time methods for measurement of cell proliferation and adhesion have shown to be successful in monitoring *in vitro* toxicity. Although biochemical assays are currently the most commonly used method to study nanomaterial methods, they have been shown to cause issues with false signals due to adsorption of dyes onto nanomaterials and interference with absorbance and fluorescence measurements. Also, they are limited to endpoint analysis of toxic effects. Due to the disadvantages associated with biochemical assays, future studies examining the toxicity of nanomaterials should be aimed at incorporating impedance-based and other real-time methods to allow for more rapid assessment of cellular effects after exposure to nanomaterials. Another approach to this issue is to develop predictive models to limit the amount of studies necessary for toxicity analysis of new materials.

12.7 Carbon Nanomaterial Formulations

12.7.1 Effect of Particle Dimension on Toxicity

The diversity of nanomaterial properties has created the need to establish a paradigm for accurately predicting their toxicity. In order to create this paradigm, researchers must compile information about the specific physicochemical properties unique to nanomaterials which drive toxicity, such as shape, surface chemistry, and dimension [9]. In this study, the effect of particle dimension was tested using four carbon-based materials composed of graphitic carbon with similar symmetry and varying diameter. Because skin is the first line of defense and provides a significant area for exposure, material toxicity was assessed using a skin-derived cell line to represent dermal toxicity *in vitro* [180].

The four materials used in this study were CFs, CNFs, MWCNTs, and SWNCTs. The arrangement of the graphitic structure is different in each of these materials, resulting in diameters separated by about a factor of 10.

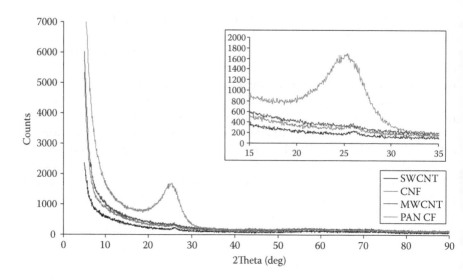

FIGURE 12.12
X-ray diffraction pattern for carbon fiber, carbon nanofiber, multiwalled carbon nanotube, and single-walled carbon nanotube.

The structure and diameter of each material was verified using TEM. The mean diameter of each individual CF, CNF, MWCNT, and SWCNT could be rounded to 10 µm, 100 nm, 10 nm, and 1 nm, respectively; however, SWCNT formed bundles that were 20 to 50 nm in diameter and up to several microns in length.

The XRD data show that there is a 002 peak at 2θ of 25 degrees seen in the spectra for CF which is indicative of typical turbostratic carbons. There is no signal at 112, which indicates that all the carbon samples consist of a two-dimensional carbon ordering structure (Figure 12.12).

Using a Raman technique for further characterization of the CNT samples (Table 12.3), both the I(G)/I(D) ratio and crystallite size are higher in the MWCNT sample than in SWCNT.

The state of dispersion and morphology of the nanomaterial agglomerates in solution were examined.

TABLE 12.3

Raman Spectroscopy Results for Carbon Nanotubes

Sample	I(G)/I(D)	La (nm)
MWCNT	2.170	9.2
SWCNT	1.159	5.0

Notes: MWCNT, multiwalled carbon nanotube; SWCNT, single-walled carbon nanotube.

TABLE 12.4

Impurity Content of Carbon Samples

Sample	Impurities (%)
CF	0
CNF	0.25
MWCNT	3.75
SWCNT	1.13

Notes: CF, carbon fiber; CNF, carbon nano-fiber; MWCNT, multiwalled carbon nanotube; SWCNT, single-walled carbon nanotube.

SEM micrographs show that the structure and agglomerates of CNF, MWCNT, and SWCNT differed drastically. In cell growth media, CNF appeared to interact with each other at the tips. This did not appear to affect the amount of surface that would be available to the cells but may have altered their overall size that could affect their ability to be taken up into cells. MWCNT agglomerates appeared to interact with cell media salts and proteins when dispersed in cell media. Dispersion in toxicity studies is often viewed as a challenge because material properties can change drastically in solution. In some cases, surfactants are used to facilitate dispersion; however, this does not accurately represent environmental exposure to nanomaterials. Materials that are exposed to the body will immediately be exposed to bodily fluids, especially in the case of inhalation. Therefore, CNF and CNT agglomerates formed in cell growth media were considered realistic compared to those that would be formed upon exposure to the body.

The impurity content was limited to iron, found using XRF, and the percent in each carbon sample was quantified using ICP-OES. The results are shown in Table 12.4.

Results from the iron mobilization assay confirmed the trend shown by these results. MWCNT contained about a fivefold increase in mobilized iron from control (culture media) and SWCNT contained about a threefold increase. Iron oxide nanoparticles contained about a ninefold increase (Figure 12.13). Studies using Fe II were carried out to quantify the amount of iron solubilized from the MWCNT sample, which was found to be 0.16 mg/L and did not change over a time period of 72 h (Figure 12.13 and Figure 12.14). The asterisk denotes significant difference from control based on the statistical analysis.

The zeta potential at the surface of each carbon material was determined at pH 7. CNF exhibited the most negative zeta potential out of all the materials, followed by MWCNT, then SWCNT (Table 12.5).

Confluency and overall morphology of HEL-30 cells were examined using bright-field microscopy at low magnification and dark-field microscopy using the URI system at high magnification (Figure 12.15). The tendency for

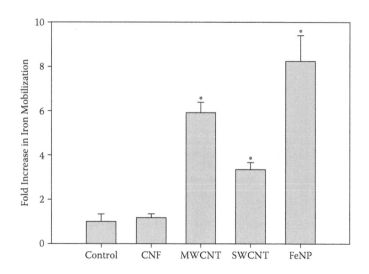

FIGURE 12.13
Fold increase of iron from growth media in carbon nanofiber, multiwalled carbon nanotube, single-walled carbon nanotube, and iron oxide nanoparticles suspended in growth media to a concentration of 0.1 mg/mL.

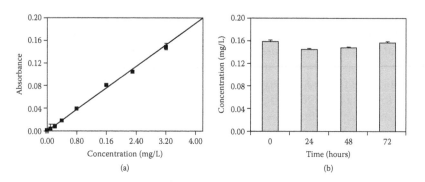

FIGURE 12.14
Iron mobilization quantification: (a) standard curve; (b) concentration of total iron ions in 0.1 mg/mL multiwalled carbon nanotube dispersion.

TABLE 12.5

Zeta Potential of Carbon Samples

Sample	Zeta Potential
CNF	−42.7
MWCNT	−39.0
SWCNT	−26.6

Notes: CNF, carbon nanofiber; MWCNT, multiwalled carbon nanotube; SWCNT, single-walled carbon nanotube.

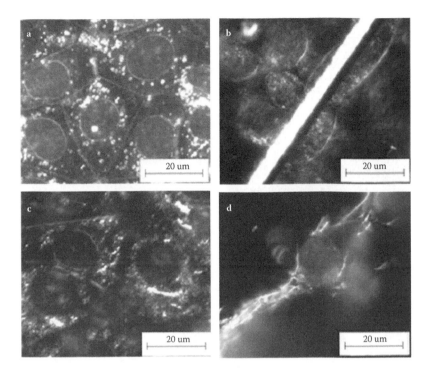

FIGURE 12.15
Cell morphology observed via the URI system: (a) control; (b) carbon fiber, 10 µg/mL; (c) carbon nanofiber, 10 µg/mL; and (d) single-walled carbon nanotube, 10 µg/mL.

growth of cells along CF was shown at this magnification. The cells exposed to CNF did not have any morphological changes when compared to the control. In an image of one isolated cell, SWCNT were bundled along the cell membrane. Cells exposed to both MWCNT and SWCNT changed their overall morphology and appeared to show membrane interaction. Due to the increased surface area of the SWCNT bundles over MWCNT agglomerates, there appeared to be increased membrane interaction after exposure to SWCNT (Figure 12.16).

MTT data were used to show cell viability after different exposures to each material at four concentrations (5, 10, 25, and 50 µg/mL) after four time points (12, 24, 48, and 72 h). After 24 h, CF and CNF did not cause significant toxicity, but cells exposed to MWCNT and SWCNT were significantly toxic compared to control cells (Figure 12.17a). After 72 h, all four materials presented no toxicity and no dose dependence (Figure 12.17b). MWCNT and SWCNT exhibited toxicity at all four concentrations up to 48 h, but cells completely recovered by the 72 h time point (Figure 12.17 and Figure 12.18).

In the negative control cells (Figure 12.19a), the MMP was unaffected, whereas in the positive control (Figure 12.19b), pockets of bright fluorescence signal can be seen, indicating the dye dissipated from the mitochondria.

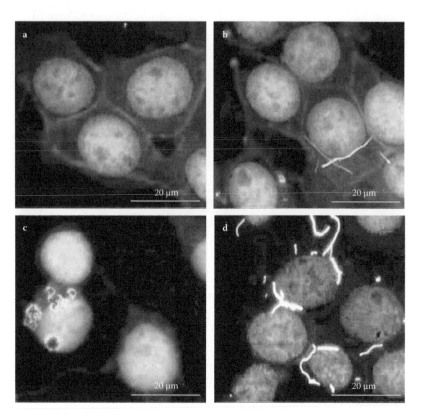

FIGURE 12.16
Cell morphology observed via the URI system coupled with dual-mode fluorescence module: (a) control; (b) carbon nanofiber, 10 µg/mL; (c) multiwalled carbon nanotube, 10 µg/mL; (d) single-walled carbon nanotube, 10 µg/mL.

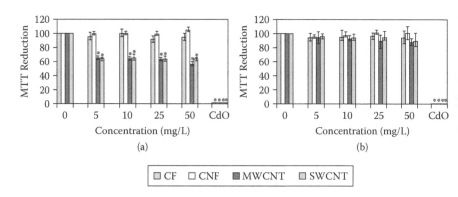

FIGURE 12.17
Cell viability determined via MTT assay after exposure to carbon materials at four concentrations (5, 10, 25, and 50 µg/mL): (a) 24 h exposure and (b) 72 h exposure.

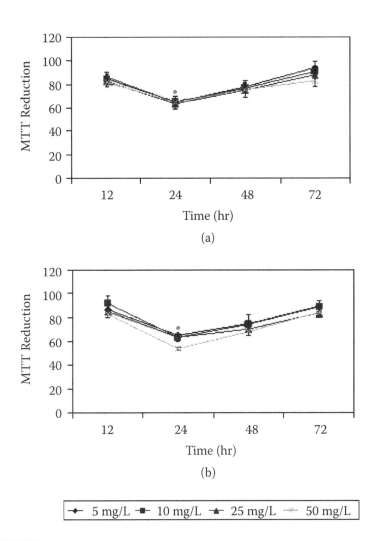

FIGURE 12.18
Cell viability over time determined via MTT assay after exposure to carbon materials at four concentrations (5, 10, 25, and 50 μg/mL) and four time points (12, 24, 48, 72 h): (a) multiwalled carbon nanotube and (b) single-walled carbon nanotube.

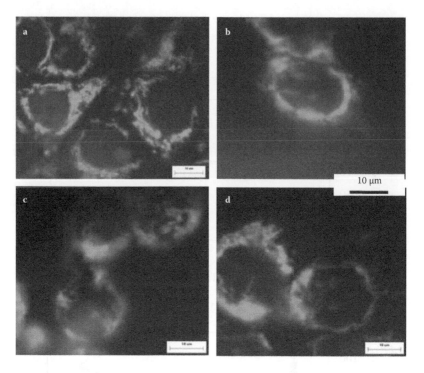

FIGURE 12.19
Mitochondrial membrane potential after 24 h exposure to carbon materials imaged via fluorescence microscopy: (a) control; (b) CdO, 2.5 µg/mL; (c) single-walled carbon nanotube, 50 µg/mL; (d) multiwalled carbon nanotube, 50 µg/mL.

However, the morphology changes of the cells exposed to MWCNT and SWCNT make it difficult to determine the extent of impact on MMP. It appears that cells exposed to both CNTs samples exhibited some dissipation of dye from the mitochondria, but whether this is indicative of cellular apoptosis is unclear (Figure 12.19c and Figure 12.19d).

ROS data were used to gauge the state of oxidative stress in cells after exposure to each material at four concentrations (5, 10, and 25 µg/mL) after 24 h. CF and CNF did not increase ROS significantly from control. SWCNT increased generation of ROS about three- to fourfold, while MWCNTs generated about three- to sevenfold ROS in a dose-dependent manner. Both MWCNT and SWCNT produced higher-fold ROS than the positive control, hydrogen peroxide (Figure 12.20a). The acellular ROS assay did not indicate any interaction between carbon nanomaterials and the fluorescent probe to cause a false signal of ROS (Figure 12.20b).

In a previous study, Jia et al. [109] reported that the degree of cytotoxicty of CNTs on a mass basis was SWCNTs > MWCNTs > quartz > C60. When evaluating the role of dimension in this study, we found that after 24 h, MWCNT and SWCNT showed increased cellular toxicity when compared to CF and

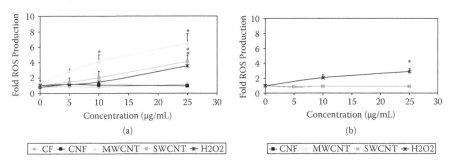

FIGURE 12.20
Reactive oxygen species generation after 24 h exposure to carbon fiber, carbon nanofiber, multiwalled carbon nanotube, single-walled carbon nanotube, and hydrogen peroxide: (a) with cells and (b) without cells.

CNF. However, there was not a difference in the level of toxicity between the SWCNT and the MWCNT. This indicates that dimension may play a role in toxicity to a certain degree, but once the dimension is below a certain point, other factors have a greater impact on cellular viability.

The mechanism involved in the increased viability after exposure to MWCNT and SWCNT between the 48 and 72 h time point is not completely understood at this time. However, it can be assumed that cellular pathways are not completely destroyed in all cells after interaction with CNTs. It is not clear whether the nanotubes are taken up by the cells in this study, but it is clear that there is significant interaction of SWCNT with the cellular membrane [180]. This interaction may initiate an irritation response from the cells, leading to the release of protective proteins from cellular enzymes, causing the membrane to become impermeable to nanotubes [165,181]. Also, proteins that have been affected may be repaired after initial injury [182,183]. It is also possible that the high surface energy of the nanotubes changes the nature of the media, causing decreased proliferation of cells in the first 48 h, but recovered proliferation after an adjustment period [181]. Furthermore, it is hypothesized that resistant cells release growth factors to help other cells proliferate or use energy from the nanotubes to proliferate [184]. If the nanotubes are taken up into the cells, it is possible that enzymes signal cells to exclude materials from major organelles and trap materials within vacuoles [165,182], allowing cells to recover their ability to proliferate after the nanotubes have been removed from the medium. Previous studies have demonstrated uptake of CNTs into cells. In one study, MWCNT accumulated within the free cytoplasm and cytoplasmic vacuoles of human epidermal keratinocytes with time and dose [116], and in a separate study, CNF (diameter = 60 to 100 nm) were found to be taken up into osteoblasts and enclosed in vacuoles [107].

There is an observable change in morphology in cells near the surface of CF, but this does not seem to affect viability according to MTT data. CNF demonstrated very minor toxicity at any of the time points, and the morphology is unchanged from control. At the 24 h time point, MWCNT and SWCNT caused significant toxicity to HEL-30 cells, and the cell morphology was altered dramatically. Therefore, even though alteration of cell morphology does not always indicate toxicity, it may play a role in the toxicity of CNTs. Although MWCNT and SWCNT formed agglomerates with drastically different morphologies in growth media, both materials caused similar reduction in cell viability and disruption of cellular morphology, possible due to their similar surface area values. It was found by Wick et al. [166] that nanotube samples containing similar amounts of impurities, but differing in agglomerate morphology, induced different toxic effects to cells, where CNT bundles were less toxic than CNT clusters. In the current study, MWCNT and SWCNT exhibited different agglomerate morphology but induced similar cytotoxic effects. The equal toxicity between MWCNT and SWCNT could be attributed to equal ability to bind to cell membranes and disrupt their morphology, inhibiting proliferation in those cells.

In a previous study, ROS generation was verified as a valid test to compare nanomaterial toxicity [185]. Nel et al. [52] reported that when they treated keratinocytes and bronchial epithelial cells with high doses of SWCNTs, this resulted in ROS generation and oxidative stress. Similarly, here we show that HEL-30 cells treated with MWCNT and SWCNT exhibit ROS generation, which was not found in the cells treated with CNF and CF. Furthermore, the ROS generation increased as the concentration of MWCNT and SWCNT increased, implying this was a dose-dependent effect. The increase in ROS after exposure to MWCNT versus SWCNT corresponds with Fe mobilization data and iron content calculated using ICP-OES. The iron mobilization assay showed a twofold increase in available iron in MWCNT over SWCNT, and the ROS assay showed a threefold increase in oxidative stress, which suggests the iron is the likely cause of the oxidative stress mechanism.

Overall, the results of this study suggest that we need to evaluate multiple physicochemical properties when evaluating cytotoxicity of carbon nanomaterials. It is not sufficient to focus on size, surface chemistry, or dimension without considering the morphology of nanomaterials in solution and their impurity content. Furthermore, when decreasing the dimension of a material, the surface area per volume and surface energy increase, so these parameters must also be assessed for their potential to induce toxicity.

12.7.2 Effect of Impurities and Surface Area on Toxicity

In this chapter, it was found that the presence of metal impurities plays an important role in the toxicity of CNT. Specifically, the metal impurity content correlated to the level of oxidative stress in the cells, marked by the production of ROS. However, the viability of the cells, measured by mitochondrial function, did not correlate directly with the metal impurity content, so there must be additional factors that affect the toxicity of carbon-based nanomaterials.

Studies have suggested that high surface area per mass may play a role in the toxicity of materials [58,59]. Therefore, further studies were carried out to test the effect of surface area on cell viability and oxidative stress *in vitro* using three materials with low impurities and increasing surface area: CULOT, Soot, and Nanog. CULOT are quasi-pure CNT synthesized via arc discharge method. Nanog and Soot were synthesized via arc discharge process followed by activation. The carbon arc discharge method, initially used for producing C60 fullerenes, is a common and simple method for producing CNTs. However, this technique produces a complex mixture of carbon components (e.g., soots, nanotubes and fullerenes, etc.), which requires separation of the CNTs from the soot and the residual catalytic metals present in the crude product. This method creates CNTs through arc-vaporization of two carbon rods placed end to end, separated by approximately 1 mm, in an enclosure that is usually filled with inert gas at low pressure. A direct current of 50 to 100 A, driven by a potential difference of approximately 20 V, creates a high-temperature discharge between the two electrodes. The discharge vaporizes the surface of one of the carbon electrodes and forms a small rod-shaped deposit on the other electrode labeled "CULOT." Producing CNTs in high yield depends on the uniformity of the plasma arc and the temperature of the deposit forming on the carbon electrode. Some of the vaporized carbon leads to the formation of carbon soots and fullerenes that are deposited on the reactor wall. To create Nanog, the carbon soot was activated using CO_2 at 900°C for 7 h.

Cells were seeded and dosed the following day with nanoparticles at a concentration of 25 µg/mL. Due to interference of carbon nanomaterials with the insoluble MTT reagent in literature [10,136,176,177] and slight interference with MWCNT, a water-soluble form of the MTT reagent was used in this study (WST-1). First, a control study was performed to verify the consistency of data between these two assays using MWCNT. For the remaining experiments, cell viability was assessed at 24 h using the WST-1 assay. In this assay, the tetrazolium salt was added to live cells after exposure, and the cells were incubated for 2 h while the mitochondria of healthy cells cleaved the ring of the WST-1, reducing it to a water-soluble formazan salt. The resulting solutions were transferred from each well into a fresh 96-well plate to avoid disruption in absorbance reading from nanoparticles settled on the bottom of the plate. The plate was read on a spectrophotometer at a peak absorbance of 450 nm with a reference wavelength of 600 nm. Cadmium oxide (2.5 µg/ mL) was used as a positive control, and cells in culture media were used as a negative control in each experiment. As with the MTT assay, the absorbance values were compared to control values and related directly to cell viability.

After arc discharge experiments, carbon samples were characterized by SEM. CULOT appears to contain agglomerated polyhedral carbon particulates (double arrows in Figure 12.21) and very fine MWCNTs either singly or in a rope (Figure 12.21). As shown in the X-ray spectra (Figure 12.22), CULOT contains some amorphous carbon, and Soot consists of a mixture

FIGURE 12.21
Scanning electron micrographs of carbon-based nanomaterials: (a) CULOT, (b) Soot, and (c) Nanog.   *(Continued)*

of turbostratic and amorphous carbons. The 002 peak at 2θ of 25 degrees is indicative of typical turbostratic carbons, which is more intense for Soot than CULOT, but is present for both. However, there is no signal at 112, which indicates that both samples consist of two-dimensional carbon ordering structure (Figure 12.22a). The SEM micrograph showing Nanog does not contain clearly defined particles. After activation, Nanog became very fragile and susceptible to break into fine powder due to formation of nanoporosity (Figure 12.22c). The X-ray spectrum for Nanog does not contain a peak for turbostratic carbons (Figure 12.22b).

FIGURE 12.21
Continued.

Using a Raman technique for further carbon sample characterization (Table 12.6), both the $I(G)/I(D)$ ratio and crystallite size are higher in the CULOT sample than in the Soot and Nanog which correlates with the previous results.

The ICP analysis shows that CULOT and Soot contain about 100 ppm in impurities (silicon, iron, and nickel). The surface area of each sample was measured using the BET method (Table 12.6). The zeta potential at the surface of each carbon material was determined at pH 7. When compared to the rest of the materials, Nanog exhibited the lowest zeta potential. CULOT and Soot are shown to have very similar zeta potentials (Table 12.6).

WST-1 data were used to show cell viability after 24 h exposures to each material at 25 µg/mL. MWCNT reduced cell viability to around 68%, and CULOT reduced viability to around 90%, a reduction that was not statistically significant from the negative control. For the high surface area materials, Nanog induced the greatest reduction in cell viability that was close to the effect caused by MWCNT. The soot did not significantly reduce cell viability compared to control (Figure 12.23).

The ROS assay was used to assess the ability for materials to produce oxidative stress cells after exposure to each material at 25 µg/mL after 24 h (Figure 12.24). As shown in Chapter 3, MWCNT produced a greater than threefold increase in ROS. CULOT did not produce ROS that was significantly different from control. Nanog produced a nearly 14-fold increase in ROS after a 24 h exposure, which highly exceeded the ability for any of the other materials to induce oxidative stress in cells. Soot also produced a significant amount of ROS compared to control, with a fourfold increase from control.

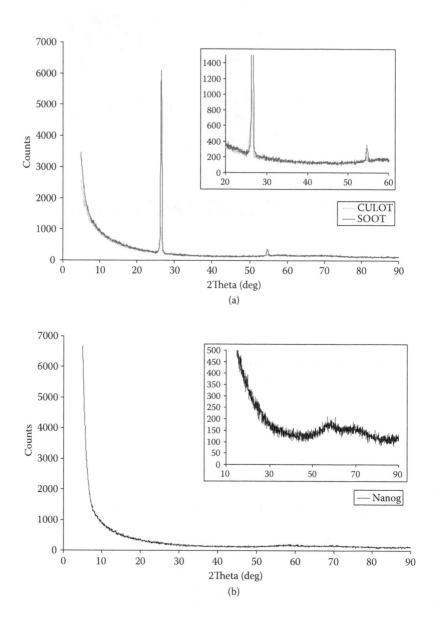

FIGURE 12.22
X-ray diffraction: (a) CULOT and Soot and (b) Nanog.

TABLE 12.6

Crystallite Size; Brunauer, Emmett, and Teller Surface Area; Total
Impurities; and Zeta Potential for CULOT, Soot, and Nanog

Sample	I(G)/I(D)	La (nm)	BET Surface Area (m²/g)	Impurities (ppm)	Zeta Potential (mV)
CULOT	1.134	4.9	25	120	−32.6
Soot	0.479	2.1	125	95	−34.9
Nanog	0.436	1.9	762	115	−8.29

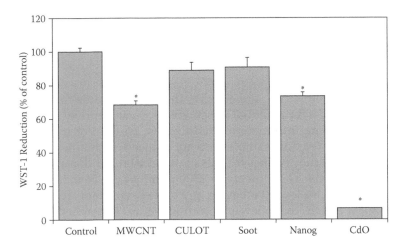

FIGURE 12.23
WST-1 cell viability data for carbon-based samples.

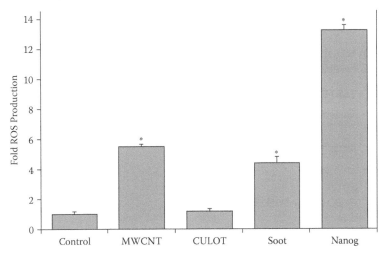

FIGURE 12.24
Reactive oxygen species generation after 24 h exposure to carbon-based nanomaterials.

Nano- and Biocomposites

12.7.3 Toxicity Factor

Based on the results from the previous two studies, the *in vitro* toxicity of carbon-based nanomaterials is a function of impurity content (I), specific surface area (SA), and zeta potential (ZP):

$$TF = f(I, SA, ZP) \tag{12.1}$$

By assuming that TF equals 1 means that the material is not toxic, and it is then possible to develop an equation based on these parameters:

$$TF = \frac{1}{3}\left(\frac{I}{I_{\text{lim}}} + \frac{SA}{SA_{\text{lim}}} + \frac{ZP_{\text{lim}}}{ZP} \right) \tag{12.2}$$

where I_{lim}, SA_{lim}, and ZP_{lim} are equal to the limiting value for impurities, surface area, and zeta potential, respectively. Such limiting values could be calculated using information from toxicity assays. For example, data have shown that the CNF sample does not induce any significant effects in cells, so for this study, values for this material were used as a benchmark for the calculations. Therefore, $I_{\text{lim}} = 2500$ ppm, $SA_{\text{lim}} = 36$ m2/g, and $ZP_{\text{lim}} = -42.7$ mV. Because lower values for any of these constants results in zero toxicological effects, the following conditions must be applied:

$$\text{If } I < I_{\text{lim}}, \text{ then } I = I_{\text{lim}} \tag{12.3}$$

$$\text{If } SA < SA_{\text{lim}}, \text{ then } SA = SA_{\text{lim}} \tag{12.4}$$

$$\text{If } ZP < ZP_{\text{lim}}, \text{ then } ZP = ZP_{\text{lim}} \tag{12.5}$$

Based on this model and data acquired for carbon-based nanomaterials, values for TF were calculated and are displayed in Table 12.7 and Table 12.8.

TABLE 12.7

Toxicity Factor Estimation for Carbon-Based Nanomaterials

Sample	$I \cdot I_{\text{lim}}^{-1}$	$SA \cdot SA_{\text{lim}}^{-1}$	$ZP^{-1} \cdot ZP_{\text{lim}}$	TF
CNF	1.0	1.0	1.0	1.0
MWCNT	15.0	8.3	1.1	8.1
SWCNT[a]	4.5	8.9	1.6	5.0
CULOT (clean MWCNT)	1.0	1.0	1.3	1.1
Soot	1.0	3.5	1.2	1.9
Nanog	1.0	21.2	5.2	9.1

Notes: CNF, carbon nanofiber; MWCNT, multiwalled carbon nanotube; SWCNT, single-walled carbon nanotube.

[a] Mixture of SWCNT and MWCNT.

TABLE 12.8

Comparison of Toxicity Factor (TF) and Reactive Oxygen Species (ROS) Data after a 24 h Exposure to a Concentration of 25 μg/mL of Carbon-Based Nanomaterials

Sample	TF	ROS
CULOT (clean MWCNT)	2.4	1.2
Soot	2.8	4.4
Nanog	10.3	13.2
CNF	1.0	1.1
MWCNT	8.1	6.5
SWCNT[a]	5.0	4.0

Notes: CNF, carbon nanofiber; MWCNT, multiwalled carbon nanotube; SWCNT, single-walled carbon nanotube.
[a] Mixture of SWCNT and MWCNT.

The assumptions associated with these calculations are that the zeta potential is a negative value and that each parameter is of equal importance in toxicity.

The toxicity factor is intended to predict the toxic potential for each material. The factor correlates almost directly with oxidative stress data but does not correlate with cell viability data. Therefore, it is better described as a predictor for the ability to produce oxidative stress, which could play a large role in the chronic affects of exposure to these materials. Therefore, it is possible that such a parameter could minimize the toxicity studies for carbon-based nanomaterials in the future. Previous toxicity experiments have suggested that ROS generation is caused by the amount of metal impurities mobilized in solution, so including this data for I rather than ICP data may generate more accurate values for *TF*.

Based on the similarity between the values for *TF* and ROS, it may be possible to predict values for ROS directly. In this model, additional constants must be added to adjust for the relative effect of each term. The equation can be written as follows:

$$ROS, calculated = \left(a \cdot \frac{I}{I_{\lim}} + b \cdot \frac{SA}{SA_{\lim}} + c \cdot \frac{ZP_{\lim}}{ZP} \right) \qquad (12.6)$$

where a, b, and c are constants determined to be 0.1, 0.5, and 0.4, respectively, by minimizing the error between the calculated and experimental ROS values. The calculated ROS values were plotted versus experimental ROS values along a line with a slope equal to 1 to visualize the accuracy of the model (Figure 12.25).

Based on these results, the ROS value for Soot was predicted to be too low, which could be related to the small size of the particles and the possibility of

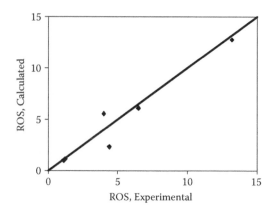

FIGURE 12.25
Plot of reactive oxygen species (ROS), calculated versus ROS, experimental.

being taken up into cells in a greater amount than high-aspect-ratio materials. The ROS values for SWCNT were predicted to be too high, which could be related to the impurity data used, which did not take into account that only some of the iron impurities are available to the cell.

12.8 Summary

Although nanotechnology has promised invaluable progress in science and technology, it is up to the scientific community to predict unknown outcomes of exposure to nanomaterials. The establishment of test procedures to ensure safe manufacture and use of nanomaterials in the marketplace is urgently required and achievable. Because nanomaterials can vary with respect to composition, size, shape, surface chemistry, and crystal structure, it is not appropriate to establish general safety regulations for all nanomaterials. Detailed characterization and risk assessment of each property must be performed through toxicity studies, and methods for prediction must be established in order to implement proper safety practices.

Material characterization techniques and toxicity endpoints currently used in nanotoxicity studies were discussed. New developments in optical microscopy were highlighted, demonstrating the ability to view nanomaterial uptake and interactions with live cells using simple methods. One particularly important technique included the ability to illuminate cellular membranes and nanomaterial aggregates using dual-mode fluorescence in combination with dark-field microscopy. Common toxicity endpoints and challenges associated with implementing common assays for the assessment of nanomaterial toxicity were also described. Approaches to developing more advanced toxicity investigation methods, such as imaging analyses and electrochemical measurements, were highlighted as an important direction for research and implementation due to the potential to develop larger data sets.

Once current nanomaterial characterization and toxicity techniques were reviewed, the effect of dimension on carbon-based nanomaterial toxicity was evaluated using an epithelial cell line. High-aspect-ratio materials of different diameters were studied, including CF, CNF, MWCNT, and SWCNT. CNF, MWCNT, and SWCNT were synthesized via catalytic chemical vapor deposition and contained different quantities of iron residue. CF and CNF did not produce any effects on cell viability or induce oxidative stress. However, both CNT samples exhibited significant effects. Both MWCNT agglomerates and SWCNT bundles interacted with the cell membrane and caused equal reduction in cell viability. Surface structure plays a role here because inhibitive interaction with the cell membrane was not observed for CNFs, which exhibit similar dimensions to SWCNT bundles, which is possibly due to the high negative zeta potential of CNF and the negative charge of the cell membrane. Overproduction of ROS induced by exposure to MWCNT and SWCNT increased with iron impurities. The conclusion of this study was that the toxicity of carbon-based nanomaterials appears to be dependent on a critical dimension, impurity content, and surface characteristics.

Further investigation was performed to develop a better understanding of the parameters involved in carbon-based nanomaterial toxicity using three materials with low impurity content synthesized via arc discharge. Crystallite size did not demonstrate a clear trend relating to particle toxicity; however, the production of ROS increased with increasing surface area and was greater for more positive values of zeta potential. MWCNT with low impurities did not induce significant effects in cells, whereas MWCNT containing iron impurities reduced cell viability and induced production of ROS. Based on the data collected in toxicity studies, a toxicity factor was derived to quantitatively express the relative toxicity of carbon-based nanomaterials based on impurity content, surface area, and zeta potential. It was found that this toxicity factor successfully predicted toxicity trends and correlated closely with the ability for materials to produce ROS in cells.

Also, the relative toxicity of catalysts commonly used in the synthesis of carbon-based nanomaterials was evaluated. It was found that the toxicity of particles functionalized with Al_2O_3 did not induce significant effects in cells. For the metal catalyst particles that did not contain Al_2O_3, toxicity was not a function of particle size but seemed to be related to zeta potential, where cobalt and nickel demonstrated the highest zeta potential and also the highest level of toxicity. Oxidative stress after exposure to cobalt, nickel, and iron increased with age of the dispersion, where iron exhibited the most significant effects. This was thought to be related to the solubility of the particles in a cellular environment. Further experiments are required to verify these results; however, based on the data shown in this study, it is assumed that metal catalyst particles with low solubility and low zeta potential will exhibit minimal toxicity to cells.

Future research directions can be divided into a biology approach or engineering approach. A biology approach could be applied to improve the limiting values used to calculate the toxicity factor. The values used in this study were estimated using a material with the greatest values for I and SA without inducing toxic effects in cells. However, there may be more appropriate values that could be determined with additional systematic studies. Additionally, the source of data for impurities could be improved. For this study, ICP data were used. In this process, the material is dissolved in acid and the total content of artifacts is determined. However, some of these impurities may be encapsulated within the material, so the quantity of impurities exposed to the cell could be much lower. Therefore, in the future, a method evaluating only the impurities on the surface of the material may be more appropriate. In addition to improving the toxicity factor to predict *in vitro* toxicity, it would be useful to test the ability to implement such a factor to predict *in vivo* toxicity.

Another future direction using a biology approach lies in uncovering more information about the surface interaction between the CNTs and the cell membrane. The main research focus here would be to investigate the receptors to which the materials are binding and to determine the mechanism of toxicity caused by surface interaction of CNT. Direct injury to the cell membrane is known to lead to intracellular Ca^{2+} levels, which causes increased calcium uptake into the mitochondria. The increased uptake dissipates the MMP and causes dissociation of actin from proteins that promote anchoring of the filament to the plasma membrane, which results in disruption of cellular morphology. Dissipation of the MMP can also hinder the production of ATP, resulting in reduction of cellular viability [73]. Such hypotheses could be investigated to explain the observed results in this study.

Using an engineering approach to investigate toxicity, molecular dynamics, or a combination of molecular dynamics and computational fluid dynamics could be used to predict the behavior of nanomaterials in the body. Such an approach could be useful in several cases, such as investigating the flow behavior of nanomaterials in the respiratory system, diffusive behavior through epithelial layers, or transport phenomena at the cellular level. In addition to implementing a modeling approach, it would be useful to study the kinetics of material properties in a biological environment. For example, the kinetics of surface interactions between materials and biological constituents could be studied. Additionally, the kinetics metal-based material dissolution could be investigated to determine the mechanism of increased ROS production with increasing dispersion age as seen in Chapter 4. There are many future research areas associated with nanomaterial toxicity, and each is important to develop a solid understanding of the structure/activity relationships necessary for progress in this field.

The goal of this chapter was evaluate the physicochemical parameters that affect carbon-based nanomaterial toxicity *in vitro* and develop insight into the mechanisms behind these effects. Based on the results, it is possible that carbon-based nanomaterial toxicity can be predicted based on impurity content (and the impurity composition), surface area, and zeta potential. As advancements in the field of nanotechnology continue to progress, it is important to continue a commitment to systematic research on toxicity and safety with the goal of understanding and managing health risks for new material developments. In the future, nanomaterial synthesis may be optimized for minimal health impact based on an understanding of nanomaterial toxicity derived from *in vitro* and *in vivo* studies. In the meantime, scientists must protect themselves from inhalation and direct exposure to the nanomaterials with which they work.

References

1. Banfield JF, Navrotsky A, Eds. *Nanoparticles and the Environment.* Washington, DC: Mineralogical Society of America, 2001.
2. Henig RM. Our silver-coated future. *OnEarth* 2007; 29(3): 22–29.
3. West JL, Halas NJ. Engineered nanomaterials for biophotonics applications: Improving sensing, imaging, and therapeutics. *Annu Rev Biomed Eng* 2003; 5: 285–292.
4. Jung H, Kittelson DB, Zachariah MR. The influence of a cerium additive on ultrafine diesel particle emissions and kinetics of oxidation. *Comb Flame* 2005; 142: 276–288.
5. AZoNanotechnology.com. Carbon Nanotubes—Applications of Carbon Nanotubes (Buckytubes). 2007; October 2007. www.azonano.com/details.asp?ArticleID=980.
6. Shi Kam NW, O'Connell M, Wisdom JA, Dai H. Carbon nanotubes as multifunctional biological transporters and near-infrared agents for selective cancer cell destruction. *Proc Natl Acad Sci USA* 2005; 102: 11600–11605.
7. Shi Kam NW, Dai H. Carbon nanotubes as intracellular protein transporters: Generality and biological functionality. *J Am Chem Soc* 2005; 127: 6021–6026.
8. Warheit DB. What is currently known about the health risks related to carbon nanotube exposures? *Carbon* 2006; 44: 1064–1069.
9. Oberdörster G, Maynard A, Donaldson K, Castranova V, Fitzpatrick J, Ausman K, et al. Principles for characterizing the potential human health effects from exposure to nanomaterials: Elements of a screening strategy. *Part Fibre Toxicol* 2005; 2: 8.
10. Merrill RA. Regulatory Toxicology. In "Casarett and Doull's Toxicology: The Basic Science of Poisons" (Klaassen CD, Ed.). Chicago: McGraw Hill, 2001: pp. 1141-1153.
11. Lippman M. Biophysical Factors Affecting Toxicity. In "Fiber Toxicology" (Warheit DB, Ed.). New York: Academic Press, 1993: pp. 259–303.

12. Hu EL, Shaw CT. Synthesis and Assembly. In "Nanostructure Science and Technology: R&D Status and Trends in Nanoparticles, Nanostructured Materials, and Nanodevices." (Siegel RW, Hu E, Roco MC, Eds.) Baltimore, MD: Kluwer Academic, 1999: pp. 15–34.
13. Suslick K, Hyeon T, Fang M. Nanostructured materials generated by high intensity ultrasound: Sonochemical synthesis and catalytic studies. *Chem Mater* 1996; 8: 2172–2179.
14. Little R, Wang X, Goddard R. Nano-diamond synthesis in strong magnetic field. *J Cluster Sci* 2005; 16(1): 53–63.
15. Vul YA, Eidelman ED, Dideikin AT. Thermoelectric Effects in Field Electron Emission from Nanocarbon. In "Synthesis, Properties and Applications of Ultrananocrystalline Diamond" (Gruen DM et al., Eds.). Netherlands: Springer, 2005: pp. 383–394.
16. Kulakova II. Chemical Properties of Nanodiamond. In "Innovative Superhard Materials and Sustainable Coatings for Advanced Manufacturing" (Lee J, Novikov N, Eds). Netherlands: Springer, 2005: pp. 365–379.
17. Shevchenko VY, Madison AE, Mackay AL. Coherent coexistence of nanodiamonds and carbon onions in icosahedral core-shell particles. *Acta Cryst* 2007; A63: 172-176.
18. Sano N, Wang H, Alexandrou I, Chhowalla M, Teo KBK, Amaratunga GA. Properties of carbon onions produced by an arc discharge in water. *J Appl Phys* 2002; 92(5): 2783-2788.
19. Viculis LM, Mack JJ, Mayer OM, Hahn T, Kaner RB. Intercalation and exfoliation routes to graphite nanoplatelets. *J Mater Chem* 2005; 15: 974–978.
20. Lee BJ. Characteristics of exfoliated graphite prepared by intercalation of gaseous SO3 into graphite. *Bull Korean Chem Soc* 2002; 23(12): 1801.
21. Inagaki M. *New Carbons: Control of Structure and Functions.* Oxford: Elsevier Science, 2000.
22. Lueking AD, Deepa LP, Narayanan L, Clifford CB. Exfoliated graphite nanofibers for hydrogen storage. *J Phys Chem B* 2005; 109(26); 12710–12717.
23. Nic M, Jirat J, Kosata B. IUPAC Compendium of Chemical Terminology, Electronic version. 2005; July 2007. http://goldbook.iupac.org/C00824.html.
24. Mantell CL. *Carbon and Graphite Handbook.* New York: Interscience, 1968.
25. Hutcheon JM. Polycrystalline Carbon and Graphite. In "Modern Aspects of Graphite Technology" (Blackman LCF, Ed.) New York: Academic Press, 1970: pp. 37–39.
26. Davidson HW. *Manufactured Carbon.* New York: Pergamon Press, 1968.
27. Department of Defense Director. Defense Nanotechnology Research and Development. 2005; July 2007. www.nano.gov/html/res/DefenseNano2005.pdf.
28. Luchtel DL. Carbon/Graphite Toxicology. In "Fiber Toxicology" (Warheit DB, Ed.). New York: Academic Press, 1993: pp. 493-521.
29. Pierson HO. *Handbook of Carbon, Graphite, Diamonds and Fullerenes.* Norwich, NY: William Andrew, 1994.
30. AZoNanotechnology.com. Buckminsterfullerene—A Review Covering the Discovery, Structure, Production, Properties and Applications of Buckyballs. 2007; November 2007. www.azonano.com/details.asp?ArticleID=1641.
31. Iijima S. Helical microtubules of graphitic carbon. *Nature* 1991; 354: 56–58.
32. Dresselhaus M, Dresselhaus G, Eklund P., Saito R. Carbon nanotubes. *Phys World* 1998; January: 33–44.

33. Peigney A, Laurent CH, Flahaut E, Bacsa R, Rousset A. Specific surface area of carbon nanotubes and bundles of carbon nanotubes. *Carbon* 2001; 39(4): 507–514.

34. Mikó Cs, Milas M, Seo JW, Couteau E, Barišiae N, Gaál R, Forró. Electrical Properties of Single-Walled Carbon Nanotube Fiber under Electron Irradiation. In "Nano-Architectured and Nanostructured Materials" (Champion Y, Fecht H, Eds.). Germany: Wiley-VCH, 2004, pp. 17–20.

35. Ye Y, Ahn CC, Witham C, Fultz B, Liu J, Rinzler AG, Colbert D, Smith KA, Smalley RE. Hydrogen adsorption and cohesive energy of single-walled carbon nanotubes. *Appl Phys Lett* 1999; 74(16): 2307–2309.

36. Cox DM. High Surface Area Materials. In "Nanostructure Science and Technology: R&D Status and Trends in Nanoparticles, Nanostructured Materials, and Nanodevices." (Siegel RW, Hu E, Roco MC, Eds.) Baltimore, MD: Kluwer Academic, 1999: pp. 49–66.

37. Yoong H, Jabbar T. Carbon nanofibers get new respect. *R&D Magazine* 2005; December: 21.

38. Chłopek J, Czajkowska B, Szaraniec B, Frackowiak E, Szostak K, Béguin F. *In vitro* studies of carbon nanotubes biocompatibility. *Carbon* 2006; 44: 1106–1111.

39. Zhao B, Hu H, Mandal SK, Haddon RC. A bone mimic based on the self-assembly of hydroxyapatite on chemically functionalized single-walled carbon nanotubes. *Chem Mater* 2005; 17(12): 3235–3241.

40. Webster TJ, Waid MC, McKenzie JL, Price RL, Ejiofor JU. Nano biotechnology: Carbon nanofibres as improved neural and orthopaedic implants. *Nanotechnology* 2004; 15: 48–54.

41. Nguyen-Vu TDB, Chen H, Cassell AM, Andrews RJ, Meyyappan M, Li J. Vertically aligned carbon nanofiber architecture as a multifunctional 3-D neural electrical interface. *IEEE Trans Biomed Eng* 2007; 54(6): 1121–1128.

42. Hu H, Ni Y, Montana V, Haddon RC, Parpura V. Chemically functionalized carbon nanotubes as substrates for neuronal growth. *Nano Lett* 2004; 4: 507–511.

43. Yao G, Wang L, Wu Y, Smith J, Xu J, Zhao W, Lee E, Tan W. FloDots: Luminescent nanoparticles. *Anal Bioanal Chem* 2006; 385: 518–524.

44. Csőgör Z, Nacken M, Sameti M, Lehr CM, Schmidt H. Modified silica nanoparticles for gene delivery. *Mater Sci Eng C* 2003; 23: 93–97.

45. Hu S, Liu J, Yang T, Liu H, Huang J, Lin Q, Zhu G, Huang X. Determination of human alpha-fetoprotein (AFP) by solid substrate room temperature phosphorescence enzyme-linked immune response using luminescent nanoparticles. *Microchim Acta* 2005; 152(1–2): 53–59.

46. Peng J, He X, Wang K , Tan W , Wang Y, Liu Y. Noninvasive monitoring of intracellular pH change induced by drug stimulation using silica nanoparticle sensors. *Anal Bioanal Chem* 2007; 388(3): 645–654.

47. Alt V, Bechert T, Steinrucke P, Wagener M, Seidel P, Dingeldein E, Domann E, Schnettler R. An *in vitro* assessment of the antibacterial properties and cytotoxicity of nanoparticulate silver bone cement. *Biomater* 2004; 25: 4383–4391.

48. Anshup J, Venkataraman S, Subramaniam C, Rajeev Kumar RR, Priya S, Santhosh Kumar TR, Omkumar RV, John A, Pradeep T. Growth of gold nanoparticles in human cells. *Langmuir* 2005; 21(25): 11562–11567.

49. Miziolek A. Nanoenergetics: An emerging technology area of national importance. *AMPTIAC Q* 2002; 6(1): 43–48.

50. Pathak P, Katiyar VK, Shibashish G. Cancer research—Nanoparticles, nanobiosensors and their use in cancer research. *J Nanotech* Online 2007; doi: 10.2240/ azojono0116.

51. Wallace W, Keane M, Murray D, Chisholm W, Maynard A, Ong T. Phospholipid lung surfactant and nanoparticle surface toxicity: Lessons from diesel soots and silicate dusts. *J Nanoparticle Res* 2007; 9: 23-28.

52. Nel A, Xia T, Mädler L, Li N. Toxic potential of materials at the nanolevel. *Science* 2006; 331(5761): 622–627.

53. Gogotsi Y. How safe are nanotubes and other nanofilaments? *Mat Res Innovat* 2003; 7: 192–194.

54. Hoet P, Briusle-Hohlfeld I, Salata O. Nanoparticles—Known and unknown health risks. *J Nanobiotechnol* 2004; 2(1): 12.

55. Enderle J, Blanchard S, Bronzino J. *Introduction to Biomedical Engineering*. New York: Elsevier Academic Press, 2005.

56. Rozman KK, Klaassen CD. Absorption, Distribution, and Excretion of Toxicants. In "Casarett and Doull's Toxicology: The Basic Science of Poisons" (Klaassen CD, Ed.). Chicago: McGraw-Hill, 2001: pp. 107–132.

57. Tinkle SS, Antonini JM, Rich BA, Roberts JR, Salmen R, DePree K, et al. Skin as a route of exposure and sensitization in chronic beryllium disease. *Environ Health Perspect* 2003; 111: 1202–1208.

58. Gopee NV, Roberts DW, Webb P, Cozart CR, Siitonen PH, Warbritton AR, Yu WW, Colvin VL, Walker NJ, Howard PC. Migration of intradermally injected quantum dots to sentinel organs in mice. *Toxicol Sci* 2007; 98(1): 249–257.

59. Luther W, Ed. Industrial application of nanomaterials—Chances and risks. *Future Technologies* 2004; 54: 1–112.

60. Cook MJ, Nasset ES, Karhausen LR, Howells GP, Tipton IH. *Report of the Task Group on Reference Man*. Oxford, New York: Pergamon Press, 1975.

61. Möller W, Haussinger K, Winkler-Heil R, Stahlhofen W, Meyer T, Hofmann W, Heyder J. Mucociliary and long-term particle clearance in the airways of healthy nonsmoker subjects. *J Appl Physiol* 2004; 97: 2200–2206.

62. Rogers DF, Lethem MI, Eds. *Airway Mucus: Basic Mechanisms and Clinical Perspectives*. Germany: Birkhauser Verlag, 1997.

63. Schürch S, Geiser M, Lee MM, Gehr P. Particles at the airway interfaces of the lung. *Colloids Surf B: Biointerfaces* 1999; 15: 339–353.

64. Warheit DB. Preface. In "Fiber Toxicology" (Warheit DB, Ed.). New York: Academic Press, 1993: pp. xxiii–xxiv.

65. Kotin P. A Historical Perspective of Fiber-Related Disease. In "Fiber Toxicology" (Warheit DB, Ed.). New York: Academic Press, 1993: pp. 1–14.

66. Oberdörster G, Ferin J, Lehnert BE. Correlation between particle size, *in vivo* particle persistance, and lung injury. *Environ Health Perspect* 1994; 102 (Suppl 5): 173–179.

67. Khanvilkar K, Donovan MD, Flanagan DR. Drug transfer through mucus. *Adv Drug Delivery Rev* 2001; 48: 173–193.

68. Yamago S, Tokuyama H, Nakamura E, Kikuchi K, Kananishi S, Sueki K, et al. *In vivo* biological behavior of a water-miscible fullerene: 14C labeling, absorption, distribution, excretion and acute toxicity. *Chem Biol* 1995; 2: 385–389.

69. Jani PU, McCarthy DE, Florence AT. Titanium dioxide (rutile) particle uptake from the rat GI tract and translocation to systemic organs after oral administration. *Int J Pharm* 1994; 105: 157–168 (abstract).

70. Witschi HR, Last HA. Toxic Responses of the Respiratory System. In "Casarett and Doull's Toxicology: The Basic Science of Poisons" (Klaassen CD, Ed.). Chicago: McGraw-Hill, 2001: pp. 515–534.

71. Nemmar A, Vanbilloen H, Hoylaerts MF, Hoet PH, Verbruggen A, Nemery B. Passage of intratracheally instilled ultrafine particles from the lung into the systemic circulation in hamster. *Am J Respir Crit Care Med* 2001; 164: 1665–1668.

72. Nemmar A, Hoet PH, Vanquickenborne B, Dinsdale D, Thomeer M, Hoylaerts MF, Vanbilloen H, Mortelmans L, Nemery B. Passage of inhaled particles into the blood circulation in humans. *Circulation* 2002; 105: 411–414.

73. Gragus Z, Klaassen CD. Mechanisms of Toxicity. In "Casarett and Doull's Toxicology: The Basic Science of Poisons" (Klaassen CD, Ed.). Chicago: McGraw-Hill, 2001: pp. 35–81.

74. Elder A, Gelein R, Silva V, Feikert T, Opanashuk L, Carter J, Potter R, Maynard A, Ito Y, Finkelstein J, Oberdörster G. Translocation of inhaled ultrafine manganese oxide particles to the central nervous system. *Environ Health Perspect* 2006; 114(8): 1172–1178.

75. Eaton DL, Klaassen CD. Principles of Toxicology. In "Casarett and Doull's Toxicology: The Basic Science of Poisons" (Klaassen CD, Ed.). Chicago: McGraw-Hill, 2001: pp. 11–34.

76. Hesterberg TW, Hart GA, Bunn WB. *In Vitro* Toxicology of Fibers: Mechanistic Studies and Possible Use for Screening Assays. In "Fiber Toxicology" (Warheit DB, Ed.). New York: Academic Press, 1993: pp. 139–170.

77. Rausch LJ, Bisinger Jr EC, Sharma A. Carbon black should not be classified as a human carcinogen based on rodent bioassay data. *Regul Toxicol Pharm* 2004; 40: 28–41.

78. Johnson NF. The Limitations of Inhalation, Intratracheal, and Intracoelomic Routes of Administration for Identifying Hazardous Fibrous Materials. In "Fiber Toxicology" (Warheit DB, Ed.). New York: Academic Press, 1993: pp. 43–72.

79. Cohn LA, Akley NJ, Adler KB. Study of Xenobiotics with Airway Eptihelium *In Vitro*. In "Fiber Toxicology" (Warheit DB, Ed.). New York: Academic Press, 1993: pp. 171–194.

80. Freshney RI. *Culture of Animal Cells: A Manual of Basic Technique*. New Jersey: Wiley-Liss, 2005.

81. Ferguson DG. Ultrastructure of Cells. In "Cell Physiology Source Book," 2nd Ed. (Sperelakis N, Ed.) New York: Academic Press, 1995, pp. 75–90.

82. McGrath JL, Dewey CF. Cell Dynamics and the Actin Cytoskeleton. In "Cytoskeletal Mechanics: Models and Measurements" (Mofrad MRK, Kamm RD, Eds.) New York: Cambridge University Press, 2006, pp. 170–203.

83. Carlier MF. Actin: Protein structure and filament dynamics. *J Biol Chem* 1991; 266: 1–4.

84. Bray D. *Cell Movements*. New York: Garland, 1992, pp. 406.

85. Schmidt A, Hall MN. Signaling to the actin cytoskeleton. *Annu Rev Cell Dev Biol* 1998; 14: 305–338.

86. Schoenwaelder SM, Burridge K. Bidirectional signaling between the cytoskeleton and integrins. *Curr Opin Cell Bio* 1999; 11: 274–286.

87. Ekwall B. Screening of Toxic Compounds in Mammalian Cell Cultures. In "Cellular Systems for Toxicity Testing" (Williams GM, Dunkel VC, Ray VA, Eds.). Annals of the New York Academy of Sciences, Vol. 407, 1983, pp. 64–77.

88. Fubini B. The Possible Role of Surface Chemistry in the Toxicity of Inhaled Fibers. In "Fiber Toxicology" (Warheit DB, Ed.). New York: Academic Press, 1993: pp. 229–257.
89. Kennedy GL, Kelly DP. Introduction to Fiber Toxicology. In "Fiber Toxicology" (Warheit DB, Ed.). New York: Academic Press, 1993: pp. 15–42.
90. Oberdörster G, Ferin J, Gelein R, Finkelstein J. Role of alveolar macrophages in lung injury: Studies with ultrafine particles. *Environ Health Perspect* 1992; 97: 193–199.
91. Bridges JW, Benford DJ, Hubbard SA. Mechanisms of Toxic Injury. In "Cellular Systems for Toxicity Testing" (Williams GM, Dunkel VC, Ray VA, Eds.) Annals of the New York Academy of Sciences, Vol. 407, 1983: pp. 42–63.
92. Guerrero A, Arias JM. Apoptosis. In "Cell Physiology Source Book, 2nd ed." (Sperelakis N, Ed.). New York: Academic Press, 1995, pp. 1031–1043.
93. Tsuji JS, Maynard AD, Howard PC, James JT, Lam CW, Warheit DB, Santamaria AB. Research strategies for safety evaluation of nanomaterials, part IV: Risk assessment of nanoparticles. *Toxicol Sci* 2006; 89: 42–50.
94. Gilmour PS, Ziesenis A, Morrison ER, Vickers MA, Drost EM, Ford I, et al. Pulmonary and systemic effects of short-term inhalation exposure to ultrafine carbon black particles. *Toxicol Appl Pharmacol* 2004; 195: 35–44.
95. Stone V, Shaw J, Brown DM, Macnee W, Faux SP, Donaldson K. Role of oxidative stress in the prolonged inhibitory effect of ultrafine carbon black on epithelial cell function. *Toxicol In Vitro* 1998; 12: 649–659.
96. Wilson MR, Lightbody JH, Donaldson K, Sales J, Stone V. Interactions between ultrafine particles and transition metals *in vivo* and *in vitro*. *Toxicol Appl Pharmacol* 2002; 184: 172–179.
97. Schrand AM, Huang H, Carlson C, Schlager JJ, Ohsawa E, Hussain SM, Dai LD. Are diamond nanoparticles cytotoxic? *J Phys Chem B* 2007; 111: 2–7.
98. Ding L, Stilwell J, Zhang T, Elboudwarej O, Jiang H, Selegue JP, Cooke PA, Gray JW, Chen FF. Molecular characterization of the cytotoxic mechanism of multi-wall carbon nanotubes and nano-onions on human skin fibroblast. *Nano Lett* 2005; 5: 2448–2464.
99. Sayes CM, Gobin AM, Ausman KD, Mendez J, West JL, Colvin VL. Nano-C60 cytotoxicity is due to lipid peroxidation. *Biomater* 2005; 26(36): 7587–7595.
100. Isakovic A, Markovic Z, Todorovic-Markovic B, Nikolic N, Vranjes-Djuric S, Mirkovic M, et al. Distinct cytotoxic mechanisms of pristine versus hydroxylated fullerene. *Toxicol Sci* 2006; 91(1): 173–183.
101. Oberdorster E. Manufactured nanomaterials (fullerenes, C60) induce oxidative stress in the brain of juvenile largemouth bass. *Environ Health Persp* 2004; 112: 1058–1062.
102. Usenko CY, Harper SL, Tanguay RL. *In vivo* evaluation of carbon fullerene toxicity using embryonic zebrafish. *Carbon* 2007; 45(9): 1891–1898.
103. Sayes CM, Fortner JD, Guo W, Lyon D, Boyd AM, Ausman KD, Tao YJ, Sitharaman B, Wilson LJ, Hughes JB, West JL, Colvin V. The differential cytotoxicity of water-soluble fullerenes. *Nano Lett* 2004; 4: 1881–1887.
104. Sayes CM, Marchione A, Reed K, Warheit D. Comparative pulmonary toxicity assessments of C60 water suspensions in rats: Few differences in fullerene toxicity *in vivo* in contrast to *in vitro* profiles. *Nano Lett* 2007; 7(8): 2399–2406.

105. Hilaski RJ, Bergmann, JD, Burnett DC, Muse WT, Thomson SA. Acute inhalation of explosively disseminated carbon fibers in rats. Edgewood Research Development and Engineering Center Aberdeen Proving Ground MD. 1994; July 2007. http://handle.dtic.mil/100.2/ADA280556.

106. Vu VT, Dearfield KL. Biological Effects of Fibrous Materials in Experimental Studies and Related Regulatory Aspects. In "Fiber Toxicology" (Warheit DB, Ed.). New York: Academic Press, 1993: pp. 449–492.

107. Price RL, Haberstroh KM, Webster TJ. Improved osteoblast viability in the presence of smaller nanometer dimensioned carbon fibres. *Nanotechnol* 2004; 15: 892–900.

108. Herzog E, Casey A, Lyng FM, Chambers G, Byrne HJ, Davoren H. A new approach to the toxicity testing of carbon-based nanomaterials—The clonogenic assay. *Toxicol Lett* 2007: 174: 49–60.

109. Jia G, Wang H, Yan L, Wang X, Pei R, Yan T, Zhao Y, Guo X. Cytotoxicity of carbon nanomaterials: Single-wall nanotube, multi-wall nanotube, and fullerene. *Environ Sci Technol* 2005; 39: 1378–1383.

110. Soto KF, Carrasco A, Powell TG, Garza KM, Murr LE. Comparative *in vitro* cytotoxicity assessment of some manufactured nanoparticulate materials characterized by transmission electron microscopy. *J Nanopart Res* 2005; 7: 145–169.

111. Magrez A, Kasas S, Salicio V, Pasquier N, Seo JW, Celio M, Catsicas S, Schwaller B, Forro L. Cellular toxicity of carbon-based nanomaterials. *Nano Lett* 2006; 6: 1121–1125.

112. Pulskamp K, Diabate S, Krug HF. Carbon nanotubes show no sign of acute toxicity but induce intracellular reactive oxygen species in dependence on contaminants. *Toxicol Lett* 2007; 168: 58–74.

113. Monteiro-Riviere NA, Nemanich RJ, Inman AO, Wang YY, Riviere JE. Multi-walled carbon nanotube interactions with human epidermal keratinocytes. *Toxicol Lett* 2005; 155: 377–384.

114. Cui D, Tian F, Ozkan CS, Wang M, Gao H. Effect of single wall carbon nanotubes on human HEK293 cells. *Toxicol Lett* 2007; 155: 73–85.

115. Flahaut E, Durrieu MC, Remy-Zolghadri M, Bareille R, Baquey C. Investigation of the cytotoxicity of CCVD carbon nanotubes towards human umbilical vein endothelial cells. *Carbon* 2006; 44: 1093–1099.

116. Huczko A, Lange H. Carbon nanotubes: Experimental evidence for a null risk of skin irritation and allergy. *Fullerenes, Nanotubes, Carbon Nanostruct* 2001; 9: 247–250 (abstract).

117. Lam, CW, James JT, McCluskey R, Hunter RL. Pulmonary toxicity of single-wall carbon nanotubes in mice 7 and 90 days after intratracheal instillation. *Resp Toxicol* 2004; 77: 126–134.

118. Muller J, Huaux F, Moreau N, Misson P, Heilier JF, Delos M, et al. Respiratory toxicity of multi-wall carbon nanotubes. *Toxicol Appl Pharmacol* 2005; 207: 221–231.

119. Warheit DB, Laurence BR, Reed KL, Roach DH, Reynolds GA, Webb TR. Comparative pulmonary toxicity assessment of single-wall carbon nanotubes in rats. *Toxicol Sci* 2004; 77: 117–125.

120. Shvedova AA, Kisin ER, Mercer R, Murray AR, Johnson VJ, Potapovich AI, et al. Unusual inflammatory and fibrogenic pulmonary responses to single walled carbon nanotubes in mice. *Am J Physiol Lung Cell Mol Physiol* 2005; 289(5): L696–L697.

121. Mitchell LA, Gao J, Vander Wal R, Gigliotti A, Burchiel SW, McDonald JD. Pulmonary and systemic immune response to inhaled multi-walled carbon nanotubes. *Toxicol Sci* 2007; 100(1): 203–214.

122. Sayes CM, Liang F, Hudson JL, Mendez J, Guo W, Beach JM, Moore VC, Doyle CD, West JL, Billups WE, Ausman KD, Colvin VL. Functionalization density dependence of single-walled carbon nanotubes cytotoxicity *in vitro*. *Toxicol Lett* 2005; 161: 135–142.

123. Dutta D, Sundaram SK, Teeguarden JG, Riley BJ, Fifield LS, Jacobs JM, Addleman SR, Kaysen GA, Moudgil BM, Weber TJ. Adsorbed proteins influence the biological activity and molecular targeting of nanomaterials. *Toxicol Sci* 2007; 100(1): 303–315.

124. Raja PMV, Connolley J, Ganesan GP, Ci L, Ajayan PM, Nalamasu O, Thompson DM. Impact of carbon nanotube exposure, dosage and aggregation on smooth muscle cells. *Toxicol Lett* 2007; 169(1): 51–63.

125. Tian F, Cui D, Schwarz H, Estrada GG, Kobayashi H. Cytotoxicity of single-wall carbon nanotubes on human fibroblasts. *Toxicol In Vitro* 2006; 20(7): 1202–1212.

126. Zhang Q, Kusaka Y, Sato K, Nakakuki K, Kohyama N, Donaldson K. Differences in the extent of inflammation caused by intratracheal exposure to three ultrafine metals: Role of free radicals. *J Toxicol Environ Health A* 1998; 53: 423–438.

127. Zhang Q, Kusaka Y, Sato K, Nakakuki K, Kohyama N, Donaldson K. Toxicity of ultrafine nickel particles in lungs after intratracheal instillation. *J Occup Health* 1998; 40: 171–176.

128. Zhou YM, Zhong CY, Kennedy IM, Pinkerton KE. Pulmonary responses of acute exposure to ultrafine iron particles in health adult rats. *Environ Toxicol* 2003; 18: 227–235.

129. Wagner AJ, Bleckmann CA, Murdock RC, Schrand AM, Schlager JJ, Hussain SM. Cellular interaction of different forms of aluminum nanoparticles in rat alveolar macrophages. *J Phys Chem B* 2007; 111: 7353–7359.

130. De Boeck M, Lardau S, Buchet JP, Kirsch-Volders M, Lison D. 2000. Absence of significant genotoxicity in lymphocytes and urine from workers exposed to moderate levels of cobalt-containing dust: A cross-sectional study. *Environ Molecular Mutagenesis* 2000; 36: 151–160.

131. Hussain SM, Hess KL, Gearhart JM, Geiss KT, Schlager JJ. *In vitro* toxicity of nanoparticles in BRL 3A rat liver cells. *Toxicol In Vitro* 2005; 19: 975–983.

132. Brown DM, Wilson MR, MacNee W, Stone V, Donaldson K. Size dependent proinflammatory effects of ultrafine polystyrene particles: A role for surface area and oxidative stress in the enhanced activity of ultrafines. *Toxicol Appl Pharmacol* 2001: 175; 191–199.

133. Donaldson K, Stone V, Gilmour PS, Brown DM, MacNee W. Ultrafine particles: Mechanisms of lung injury. *Phil Trans R Soc Lond A* 2000: 358; 2741–2749.

134. Oberdörster G, Ferin J, Lehnert BE. Correlation between particle size, *in vivo* particle persistence, and lung injury. *Environ Health Perspect* 1994; 102(Suppl 5): 173–179.

135. Sayes CM, Wahi R, Kurian PA, Liu Y, West JL, Ausman KD, Warheit DB, Colvin VL. Correlating nanoscale titania structure with toxicity: A cytotoxicity and inflammatory response study with human dermal fibroblasts and human lung epithelial cells. *Toxicol Sci* 2006; 92(1): 174–185.

136. Light WG, Wei ET. Surface charge and asbestos toxicity. *Nature* 1977; 26: 537–539.
137. Jakab GJ. The toxicologic interactions resulting from inhalation of carbon black and acrolein on pulmonary antibacterial and antiviral defenses. *Toxicol Appl Pharmacol* 1993; 121: 167–175.
138. Jakab GJ, Clarke RW, Hemenway DR, Longphre MV, Kleeberger SR, Frank R. Inhalation of acid coated carbon black particles impairs alveolar macrophage phagocytosis. *Toxicol Lett* 1996; 88: 243–248.
139. Oberdörster, G. Toxicology of ultrafine particles: *in vivo* studies. *Philos Trans R Soc, A* 2000; 358: 2719–2740.
140. Renwick LC, Donaldson K, Clouter A. Impairment of alveolar macrophage phagocytosis by ultrafine particles. *Toxicol Appl Pharmacol* 2001; 172(2): 119–127.
141. Lundborg M, Johard U, Lastbom L, Gerde P, Camner P. Human alveolar macrophage phagocytic function is impaired by aggregates of ultrafine carbon particles. *Environ Res* 2001; 86(3): 244–253.
142. Fiorito S, Serafino A, Andreola F, Bernier P. Effects of fullerenes and single-wall carbon nanotubes on murine and human macrophages. *Carbon* 2006; 44: 1100–1105.
143. Wörle-Knirsch JM, Pulskamp K, Krug HF. Oops they did it again! Carbon nanotubes hoax scientists in viability assays. *Nano Lett* 2006; 6: 1261–1268.
144. Stone V, Donaldson K. Nanotoxicology: Signs of stress. *Nature Nanotechnol* 2006; 1: 23–24.
145. Nimmagadda A, Thurston K, Nollert MU, McFetridge PS. Chemical modification of SWNT alters *in vitro* cell-SWNT interactions. *J Biomed Mater Res* 2005; 76A(3): 614–625.
146. Lippmann M. Review. Asbestos exposure indices. *Environ Res* 1988; 46: 86–106.
147. Powers KW, Brown SC, Krishna VB, Wasdo SC, Moudgil BM, Roberts SM. Research strategies for safety evaluation of nanomaterials. Part VI. Characterization of nanoscale particles for toxicological evaluation. *Toxicol Sci* 2006; 90(2): 296–303.
148. Fultz B, Howe JM. *Transmission Electron Microscopy and Diffractometry of Materials.* Germany: Springer, 2001.
149. Ernst F, Sigle W. Microcharacterisation of Materials. In "High-Resolution Imaging and Spectrometry of Materials" (Ernst F, Ruhle, Eds.). Germany: Springer, 2003; pp. 4–5.
150. Smart L, Moore EA. *Solid State Chemistry: An Introduction.* New York: Taylor & Francis, 2005.
151. Papp RB. X-ray Fluorescence Spectrometry. In "Instrumental Methods for Determining Elements" (Taylor LR, Papp RB, Pollard BD, Eds.). New York: VCH, 1994; pp. 211–261.
152. Murthy NS, Reidinger F. X-Ray Analysis. In "A Guide to Materials Characterization and Chemical Analysis" (Sibilia JP, Ed). New York: VCH, 1996; pp. 161–163.
153. Williams RJ, Bause DE. Elemental and Chemical Analysis. In "A Guide to Materials Characterization and Chemical Analysis" (Sibilia JP, Ed.). New York: VCH, 1996; pp. 130–132.
154. Davidson MW, Abramowitz M. Optical Microscopy. Olympus Microscopy Resource Center. December 1999; May 2006.

155. PerkinElmer Life and Analytical Sciences. Guide to Inorganic Analysis. Shelton, CN: 2004.
156. Spectroscopy. 2008. In Encyclopaedia Britannica. Retrieved March 7, 2008, from *Encyclopaedia Britannica* Online: www.britannica.com/eb/article-80630.
157. Ko FK. Nanofiber Technology: Bridging the Gap between Nano and the Macro World. Guceri S, Gogotsi YG, Kutznetsov V. (Eds) Nanoengineered Nanofibrous Materials. NATO Science Series Vol 169. Springer 2004 Turkey. Proceedings of NATO Advanced Study Institute on Nanoengineered Nanofibrous Materials, pp. 1–18.
158. Malvern Instruments Ltd. Zetasizer User Manual. MAN0172 Issue 1.1; Worcestershire, UK: 1996.
159. Skebo JE, Grabinski CM, Schrand AM, Schlager JJ, Hussain SM. Assessment of metal nanoparticle agglomeration, uptake, and interaction using high-illuminating system. *Intl J Toxicol* 2007; 26: 135–141.
160. Vainrub A, Pustovyy O, Vodyanoy V. Resolution of 90 nm (lambda/5) in an optical transmission microscope with an annular condenser. *Opt Lett* 2006; 31: 2855–2857.
161. Steyger P. Assessing confocal microscopy systems for purchase. *Methods* 1999; 18: 435–446.
162. Garside JR, Somekh MG, See CW. Biological imaging using fast laser scanning heterodyne differential phase confocal microscopes. *J Microsc* 1997; 185: 385–395.
163. Stringer B, Imrich A, Kobzik L. Flow cytometric assay of lung macrophage uptake of environmental particulates. *Cytometry* 1995; 20: 23–32.
164. Carmichael J, DeGraff WG, Gazdar AF, Minna JD, Mitchell JB. Evaluation of a tetrazolium-based semiautomated colorimetric assay: Assessment of chemosensitivity testing. *Cancer Res* 1987; 47: 936–942.
165. Monteiro-Riviere NA, Inman AO. Challenges for assessing carbon nanomaterial toxicity to the skin. *Carbon* 2006; 44: 1070–1078.
166. Wick P, Manser P, Limbach LK, Dettlaff-Weglikowskab U, Krumeich F, Roth S, Stark WJ, Bruinink A. The degree and kind of agglomeration affect carbon nanotube cytotoxicity. *Toxicol Lett* 2007; 168: 121–131.
167. Kocbach A, Totlandsdal AI, Låg M, Refsnes M, Schwarze PE. Differential binding of cytokines to environmentally relevant particles: A possible source for misinterpretation of *in vitro* results? *Toxicol Lett* 2008; 176: 131–137.
168. Toescu EC, Verkhratsky A. Assessment of mitochondrial polarization status in living cells based on analysis of the spatial heterogeneity of rhodamine 123 fluorescence staining. *Pflügers Archiv* 2000; 440(6): 941–947.
169. Halliwell B, Gutteridge JMC. Oxygen free radicals and iron in relation to biology and medicine: Some problems and concepts. *Arch Biochem Biophys* 1986; 246(2): 501–514.
170. Wang H, Joseph JA. Quantitating cellular oxidative stress by dichlorofluorescein assay using microplate reader. *Free Rad Biol Med* 1999; 27: 612–616.
171. Poser I, Rahman Q, Lohani M, Yadav S, Becker HH, Weiss DG, Schiffmann D, Dopp E. Modulation of genotoxic effects in asbestos-exposed primary human mesothelial cells by radical scavengers, metal chelators and a glutathione precursor. *Mutation Res/Genet Toxicol Environ Mutagenesis* 2004; 559: 19–27.

172. Lund LG, Aust AE. Iron mobilization from crocidolite asbestos greatly enhances crocidolite-dependent formation of DNA single-strand breaks in φX174 RFI DNA. *Carcinogensis* 1992; 13: 637–642 (abstract).

173. Guo L, Morris DG, Liu X, Vaslet C, Kane AB, Hurt RH. Iron bioavailability and redox activity in diverse carbon nanotube samples. *Chem Mater* 2007; 19(14): 3472–3478.

174. Kino-oka M, Prenosil JE. Development of an on-line monitoring system of human keratinocyte growth by image analysis and its application to bioreactor culture. *Biotechnol Bioeng* 2000; 67(2): 234–239.

175. Ehret R, Baumann W, Brischman M, Schwinde A, Stegbauer K, Wolf B. Monitoring of cellular behavior by impedance measurements on interdigitated electrode structures. *Biosens Bioelectron* 1997; 12(1): 29–41.

176. Sohn LL, Saleh OA, Facer GR, Beavis AJ, Allan RS, Notterman DA. Capacitance cytometry: Measuring biological cells one by one. *PNAS* 2000; 97(20): 10687–10690.

177. Yeon JH, Park JK. Cytotoxicity test based on electrochemical impedance measurement of HepG2 cultured in microfabricated cell chip. *Anal Biochem* 2005; 341: 308–315.

178. Ceriotti L, Ponti J, Colpo P, Sabbioni E, Rossi F. Assessment of cytotoxicity by impedance spectroscopy. *Biosens Bioelectron* 2007; 22: 3057–3063.

179. Atienza Jm, Zhu J, Bowang X, Xu X, Abassi Y. Dynamic monitoring of cell adhesion and spreading on microelectronic sensor arrays. *J Biomol Screening* 2005; 10(8): 795–805.

180. Grabinski C, Saber H, Lafdi K, Effect of dimension on biocompatibility of carbon-based nanomaterials. *Carbon* 2007; 45: 2828–2835.

181. Beyersmann D, Hechtenberg S. Cadmium, gene regulation, and cellular signalling in mammalian cells. *Toxicol Appl Pharmacol* 1997; 144(2): 247–261.

182. Hall JL. Cellular mechanisms for heavy metal detoxification and tolerance. *J Exp Bot* 2002; 53(366): 1–11.

183. Ernest DM, Ristich VL, Ray S, Lober RM, Bollag WB. Regulation of protein kinase d during differentiation and proliferation of primary mouse keratinocytes. *J Invest Dermatol* 2005; 125(2): 294–306.

184. Hashimoto K. Regulation of keratinocyte function by growth factors. *J Dermatol Sci* 2000; 24(1): S46–S50.

185. Xia T, Kovochich M, Brant J, Hotze M, Sempf J, Oberley T, et al. Comparison of the abilities of ambient and manufactured nanoparticles to induce cellular toxicity according to an oxidative stress paradigm. *Nano Lett* 2006; 6(8): 1794–1807.

Index